Spectral Atlas for Amateur Astronomers
A Guide to the Spectra of Astronomical Objects and Terrestrial Light Sources

Featuring detailed comments on spectral profiles of more than 100 astronomical objects, in color, this spectral guide documents most of the important and spectroscopically observable objects accessible with typical amateur equipment. It allows you to read and interpret the recorded spectra of the main stellar classes, as well as most of the steps from protostars through to the final stages of stellar evolution as planetary nebulae, white dwarfs or the different types of supernovae. It also presents integrated spectra of stellar clusters, galaxies and quasars, and the reference spectra of some terrestrial light sources, for calibration purposes. Whether used as the principal reference for comparing with your recorded spectra or for inspiring independent observing projects, this atlas provides a breathtaking view into our Universe's past. The atlas is accompanied and supplemented by *Spectroscopy for Amateur Astronomers*, which explains in detail the methods for recording, processing, analyzing and interpreting your spectra.

Richard Walker has been an avid astronomer since he was aged 12. He spent his career in civil engineering, planning large projects such as power plants, dams and tunnels. Now retired, in the last 10 years he has focused increasingly on stellar astronomy and on the indispensable key to this topic – spectroscopy. He undertook a large observing project to record and document the spectra of the most important astronomical objects, and chose to share this gathered information for the benefit of other amateurs worldwide. The *Spectral Atlas* and *Spectroscopy for Amateur Astronomers* are the fruits of his labor. He lives near Zürich, in Switzerland.

Spectral Atlas for Amateur Astronomers

A Guide to the Spectra of Astronomical Objects and Terrestrial Light Sources

Richard Walker

Amateur astronomer, Switzerland

CAMBRIDGE
UNIVERSITY PRESS

CAMBRIDGE
UNIVERSITY PRESS

University Printing House, Cambridge CB2 8BS, United Kingdom

One Liberty Plaza, 20th Floor, New York, NY 10006, USA

477 Williamstown Road, Port Melbourne, VIC 3207, Australia

314-321, 3rd Floor, Plot 3, Splendor Forum, Jasola District Centre, New Delhi - 110025, India

79 Anson Road, #06-04/06, Singapore 079906

Cambridge University Press is part of the University of Cambridge.

It furthers the University's mission by disseminating knowledge in the pursuit of education, learning and research at the highest international levels of excellence.

www.cambridge.org
Information on this title: www.cambridge.org/9781107165908

First published 2017

A catalogue record for this publication is available from the British Library

Library of Congress Cataloging in Publication data
Names: Walker, Richard, 1951-
Title: Spectral atlas for amateur astronomers : a guide to the spectra of astronomical objects and terrestrial light sources / Richard Walker, amateur astronomer, Switzerland.
Description: Cambridge : Cambridge University Press, 2017. | Includes bibliographical references and indexes.
Identifiers: LCCN 2016036435 | ISBN 9781107165908 (Hardback : alk. paper)
Subjects: LCSH: Stars–Spectra–Atlases. | Stars–Atlases. | Astronomy–Charts, diagrams, etc. | Stars–Observers' manuals. | Amateur astronomy–Observers' manuals. | Spectrum analysis. | Large type books.
Classification: LCC QB883 .W277 2017 | DDC 522/.67–dc23 LC record available at https://lccn.loc.gov/2016036435

ISBN 978-1-107-16590-8 Hardback
9781316642566 – 2 volume Hardback set
9781107165908 – Volume 1
9781107166189 – Volume 2

CONTENTS

PREFACE

Photons are generated in stars, and carry valuable information over immense periods of time and unimaginable distances. Some of them are finally ending on the CCD chip of our cameras. By a selective saturation of individual pixels they deposit their valuable information. By switching a spectrograph between the telescope and camera the photons will provide a wealth of information which surpasses by far the simple photographic image of an object. Therefore today – not surprisingly – most of the larger professional observatories use a significant part of the telescope time for the recording of spectra. Spectroscopy is not just the main key to astrophysics – without this tool, our current understanding of the atomic structure and quantum mechanics would be unthinkable.

It was not until a few years after the turn of the millennium quite affordable spectrographs triggered a substantial, albeit relatively late boom of this highly important discipline, also within the amateur field. Nowadays, CCD cameras, combined with well-documented software packages, allow an easy recording, processing and analyzing of the spectra. Furthermore, compared with astrophotography, most spectroscopic disciplines are pretty robust, both with regard to light pollution and rather poor seeing.

With such relatively modest equipment not only most of the significant discoveries of modern astronomy can be reproduced but also valuable contributions, even to professional research programs, are enabled. Further, the night sky provides an overwhelming number of objects, today allowing even for amateurs to follow spectroscopically nearly all significant stages of stellar evolution from birth to death, a highly rewarding experience for a deeper understanding of stellar astronomy. But even on the extragalactic and cosmological level many fascinating fields of activities are waiting, for example the measurement and analysis of larger redshifts.

Broad comments on spectra can actually just be found in *Stars and their Spectra*, a somewhat older but excellent publication by J. Kaler [1]. The modern standard work *Stellar Spectral Classification* by O. Gray and C. J. Corbally [2], however, is primarily aimed at professional astronomy and rather advanced students. Furthermore, both of these publications comment on larger scale spectra in the UV and IR domain, which are not accessible for amateurs. This atlas, however, is primarily intended as a spectral guide focused on important astronomical objects and on wave bands, which are recordable by average amateur equipment. Further, spectra of some terrestrial light sources and interesting telluric phenomena are presented.

All objects are documented here at least by broadband and low-resolution spectra, identifying most of the recognizable absorptions and emissions. Interesting details are additionally presented by slightly higher resolved profile sections. Today a fast growing number of amateurs have access to powerful telescopes and high-resolution spectrographs. Thus, the comments to some suitable objects are complemented with highly resolved spectral details and for "cosmological excursions," Appendix H provides a list of distant active galactic nuclei (AGN) and quasars, brighter than $m_v \approx 16$. However, the atlas is not limited to a collection of labeled plates with spectral profiles, but provides also object-related astrophysical backgrounds, appropriate classification systems and in some cases even historically interesting details. The aim is, typically after the first successfully calibrated Sirius profile, to excite the interest for further rewarding objects and to support the recording and analyzing of the profiles with appropriate information. The extensive bibliography is provided here not only as a source of references, but is also intended as proposal for a "further reading" to specific topics. Many of the professional publications are written in a surprisingly comprehensible way, at least for rather advanced amateurs.

This atlas is supported by the book *Spectroscopy for Amateur Astronomers: Recording, Processing, Analysis and Interpretation* [3]. It provides basic information, definitions and formulas which are presupposed in this atlas. At an undergraduate level it contains the theoretical basics for a deeper understanding of the involved processes, related to the specific astronomical objects. Further, it provides some important astrophysical applications of spectroscopy and practical hints for the recording, processing and analyzing of the spectra.

Finally, we hope that this *Spectral Atlas*, combined with the supplementary book, will be helpful to many amateurs worldwide to discover the fascinating world of astronomy by means of spectroscopy.

ACKNOWLEDGEMENTS

The *Spectral Atlas* was launched in 2011 as a nonprofit project. The aim was first to provide an Internet document with fully commented, low resolution spectra for amateurs, covering the seven basic spectral classes. At that time, this way also a significant publication gap should be reduced. Anyway, Urs Flükiger, member of the Swiss Spectroscopic Group, deserves first thanks, for convincing the author to publish the very first version of the atlas on his "Ursus Maior" homepage. Otherwise it could even be possible that the manuscript would still remain as "personal notes" in any folder on his laptop.

The following worldwide responses to the very first online version have been surprisingly numerous and so positive that the project was soon extended to further special classes. Thus, further thanks are here addressed to all amateurs and even professionals, who have contributed with their valuable feedback to the benefit of this atlas.

In 2011, Thomas Eversberg from the German Aerospace Center (Deutsches Zentrum für Luft und Raumfahrt, DLR) wrote a review in the *VDS Journal für Astronomie*, No. 38 and he was also the very first to propose that this atlas should be published as a book. Anyway, a further five years have been required to record the spectra for all presently documented special classes and to collect the necessary background information for the description of the numerous objects. At this stage I also got valuable support from Martin Huwiler who assisted me not only with his immense optical knowledge, but also by discussing relevant spectroscopic and astronomical topics and by reviewing the text. In this respect I would also like to include Helen Wider. Further, Erik Wischnewski helped me with the chapter about novae and contributed excellent, self-recorded spectra of nova Delphini V339, which annoyingly erupted exactly during a time when the author was prevented from observing.

In 2014, during the NEAF (Northeast Astronomy Forum) convention near New York, Olivier Thizy of Shelyak Instruments met Vince Higgs from Cambridge University Press and called his attention to the atlas project. It was, however, not yet finished at that time and required a further year to be completed. So Olivier made in this way a small but nevertheless decisive contribution to the creation of this book. Further, the excellent programs Visual Spec by Valerie Désnoux and IRIS by Christian Buil have been of great help, not only for analyzing the recorded profiles but also for the illustration of the numerous colored plates in this atlas.

While transforming the Internet document into a book it became clear quite soon that a separate publication would be necessary, supporting the atlas with practical and theoretical aspects. This inevitably required a co-author with complementary knowledge, for which fortunately Marc Trypsteen could be won. He also contributed valuable reviews and inputs to the atlas, not least a stunning reflectance spectrum of the lunar eclipse from September 2015.

Finally, I generally want to express here again my warmest thanks to everyone who contributed in any way to this atlas.

Richard Walker

1 Directory of Plates

Plate	*Page*	Topic	Objects	Wavelength domain/λ	Grating
1	21	Overview of the spectral classes	1D-spectra of all basic spectral classes	3950–6690	200 L
2	22	Intensity profiles of all basic spectral classes		3950–6690	200 L
3	27	Spectral features of the late O class	Alnitak, ζ Ori Mintaka, δ Ori	3920–6710	200 L
4	28	Detailed spectrum of a late O-class star	Alnitak, ζ Ori	3950–4750 5740–6700	900 L
5	29	Spectral features of the early to middle O class, luminosity effect	Θ^1 Ori C 68 Cyg	3800–6700	200 L
6	33	Development of spectral features within the B class	Alnilam, ε Ori Gienah Corvi, γ Crv	3900–6700	200 L
7	34	The effect of luminosity on spectra of the late B class	Regulus, α Leo Rigel, β Ori φ Sgr	3920–4750	900 L
8	35	Detailed spectrum of an early B-class star	Spica, α Vir	3800–6750 3900–4750 4800–5100 5700–6050 6450–6600	200 L 900 L 900 L 900 L 900 L
9	39	Development of spectral features within the A class	Castor, α Gem Altair, α Aql	3900–6800	200 L
10	40	Detailed spectrum of an early A-class star	Sirius A, α CMa	3900–6700 3900–4700 4780–5400	200 L 900 L 900 L
11	41	Effects of the luminosity on spectra of the early A class	Vega, α Lyr Ruchbah, δ Cas Deneb, α Cyg	3900–4700	900 L

(*cont.*)

Plate	*Page*	Topic	Objects	Wavelength domain/λ	Grating
12	44	Development of spectral features within the F class	Adhafera, ζ Leo Procyon, α CMi	3830–6700	200 L
13	45	Effect of the luminosity on spectra of the early F class	Porrima, γ Vir Caph, β Cas Mirfak, α Per	3920–4750	900 L
14	49	Development of spectral features within the G class	Muphrid, η Boo Vindemiatrix, ε Vir	3800–6600	200 L
15	50	Detailed spectrum of an early G-class star	Sun	3800–7200 3900–4800	200 L 900 L
16	51			4700–5700 5650–6700	900 L
17	52	Highly resolved Fraunhofer lines in the solar spectrum	Sun	5884–5899 5265–5274 5161–5186 4849–4871 4300–4345 3915–3983	SQUES echelle
18	57	Development of spectral features within the K class	Arcturus, α Boo Alterf, λ Leo	3900–6800	200 L
19	58	Detailed spectrum of an early K-class star	Pollux, β Gem	3900–6800 3800–4800	200 L 900 L
20	59	Detailed spectrum of a late K-class star	Aldebaran, α Tau	5150–5900 5850–6700	900 L
21	60	Luminosity effects on spectra of the late K class	Alsciaukat, α Lyn 61 Cyg B	4000–4900	900 L
22	64	Development of spectral features within the M class	Antares, α Sco Ras Algethi, α Her	3900–7200	200 L
23	65	Red dwarfs and flare stars	Gliese388 Gliese411	4200–7200	200 L
24	66	Effects of the luminosity on spectra of the late M class	Ras Algethi, α Her Gliese699	3900–7200	200 L
25	71	Mira variable M(e), comparison to a late-classified M star	Mira, o Cet Ras Algethi, α Her	3900–7200	200 L
26	75	Extreme S-class star, comparison to a Mira variable M(e)	R Cygni Mira, o Cet	4100–7300	200 L
27	76	Development of spectral features within the S class	Omicron1 Ori Chi Cyg R Cyg	4100–7300	200 L
28	77	Comparison of an "intrinsic" and "extrinsic" S-class star	BD Cam HR Peg	4300–7200	200 L
29	82	Comparison of differently classified carbon stars	WZ Cas Z Psc W Ori	4600–7300	200 L

(cont.)

Plate	Page	Topic	Objects	Wavelength domain/λ	Grating
30	83	Merrill–Sanford bands, details at a higher resolved spectrum	W Ori	4730–5400	900 L
31	84	Carbon star with Hα emission line	R Lep, Hind's Crimson Star	5100–7300	200 L
32	88	White dwarfs with different spectral classifications	WD 0644 +375 40 Eri B Van Maanen's Star	3900–6600	200 L
33	94	Wolf–Rayet stars, types WN and WC	WR133 WR140	3850–7250	200 L
34	95	Wolf–Rayet stars, types WN and WO	WR136 WR142	3860–7200 3750–7200	200 L
35	98	LBV star, early B class, P Cygni profiles	P Cyg, 34 Cyg	3900–6950 6000–6800	200 L 900 L
36	99	Detailed spectrum of LBV star, early B class, P Cygni profiles	P Cyg, 34 Cyg	3850–4650 4700–6050	900 L
37	103	Be star, early B class	Dschubba, δ Sco	3650–7000 4820–4940 6500–6700 6670–6690	200 L 900 L 900 L 900 L
38	104	Be star, early B class	Tsih, γ Cas	3970–6750	200 L
39	107	Be shell star, comparison with an "ordinary" Be star	ζ Tau Dschubba, δ Sco	3800–6800	200 L
40	112	Herbig Ae/Be protostar	R Mon NGC 2261	3900–7200	200 L
41	113	Prototype of the T Tauri stars	T Tau	3900–7000	200 L
42	114	FU Orionis protostar, comparison with K0 giant	FU Ori Algieba, γ Leo	3900–6800	200 L
43	120	Comparison of metallicity	Vega, α Lyr Sirius A, α CMa	3920–4700	900 L
44	121	Comparison of spectral classes Bp–Ap Hg–Mn and the reference profile of Vega	Alpheratz, α And Vega, α Lyr Cor Caroli, α² CVn	4100–4220	SQUES echelle
45	122	Estimation flux density of mean magnetic field modulus	Babcock's Star, HD 215441	small sections 4555–6380	SQUES echelle
46	129	Spectroscopic binaries: Detached SB2 systems	Spica, α Vir Mizar A, ζ UMa	6540–6580 6660–6690	SQUES echelle
47	130	Semi-detached system and eclipsing binary	Sheliak, β Lyr	3800–7200	200 L
48	131	Semi-detached system and eclipsing binary, spectral details	Sheliak, β Lyr	Hα, Hβ, He I 6678, 5876, 5016, 4921	SQUES echelle

(*cont.*)

Plate	*Page*	Topic	Objects	Wavelength domain/λ	Grating
49	138	Classical nova: development of the spectrum during outburst	Nova V339 Del	4200–6800	Trans. 100 L
50	139	Recurrent nova in quiescence	T CrB	4100–7300	200 L
51	140	Dwarf nova	SS Cyg	3800–6800	200 L
52	141	Symbiotic star	EG And	3900–7200	200 L
53	148	Supernova type Ia, characteristic spectrum and chronology	SN2014 J Host galaxy M82	3800–7200	200 L
54	158	Absorption line galaxy: comparison with stellar G class	Andromeda M31 Vindemiatrix ε Vir	3900–6700	200 L
55	159	LINER galaxy	M94	3850–6900	200 L
56	160	Starburst galaxy	M82	4300–7000	200 L
57	161	Seyfert galaxy, AGN	M77	3800–6800	200 L
58	162	Quasar: emission lines	3C273	3800–6600	200 L
59	163	Quasar: redshift	3C273	4400–7600	200 L
60	164	Blazar, BL Lacertae object	Mrk 421	3950–7000	200 L
61	170	Stars of the open Pleiades cluster with emission line classification	17-, 23-, 25- and 28 Tau	3800–6700	200 L
62	171	Stars of the open Pleiades cluster with absorption lines	16-, 18-, 19-, 20-, 21- and 27 Tau	3800–6700	200 L
63	172	Integrated spectra of globular clusters	M3, M5, M13	3900–6700	200 L
64	185	Emission nebula: H II region	M42	3800–7300	200 L
65	186	Intensity profiles of Hβ and [O III] (λ5007) in the central area of M42	M42	n.a.	200 L
66	187	Post-AGB star with protoplanetary nebula	Red Rectangle HD 44179	3800–7000	200 L
67	188	Planetary nebula: excitation class E1	IC418 Spirograph Nebula	4100–7100	200 L
68	189	Planetary nebula: excitation class E3	NGC 6210 Turtle Nebula	3850–6600	200 L
69	190	Planetary nebula: excitation class E8	NGC 7009 Saturn Nebula	3800–6700	200 L
70	191	Planetary nebula: excitation class E10	M57 Ring Nebula	4600–6800	200 L
71	192	Intensity profiles of [O III] and [N II] in the longitudinal axis of M57	M57 Ring Nebula	n.a.	200 L
72	193	Estimation of the mean expansion velocity of a PN	NGC 7662 Blue Snowball Nebula	4950–5010	SQUES echelle
73	194	Emission nebula: SNR Excitation class E>5	Crab Nebula M1 NGC 1952	4600–6800	200 L

(cont.)

Plate	Page	Topic	Objects	Wavelength domain/λ	Grating
74	195	Emission nebula: Wolf–Rayet Excitation Class E1	Crescent Nebula NGC 6888	4700–6800	200 L
75	196	Emission Nebula: Wolf–Rayet Star as its Ionizing Source Excitation Class E3	NGC 2359 Thor's Helmet, WR7, HD 56925	4000–7200 4500–6800	200 L 200 L
76	200	Reflectance Spectra of Mars and Venus	Mars, Venus	4300–7800	200 L
77	201	Reflectance Spectra of Jupiter and Saturn	Jupiter, Saturn	4400–7800	200 L
78	202	Reflectance Spectra of Uranus, Neptune and Titan	Uranus, Neptune, Titan	4500–7900	200 L
79	203	Spectra of Comet	C/2009 P1 Garradd	3800–6400	200 L
80	206	Telluric Absorptions in the Solar Spectrum	Earth's atmosphere	6800–7800	900 L
81	207	Telluric H_2O Absorptions Around H_α in the Spectra of the Sun and δ Sco	Sun δ Sco	6490–6610	SQUES echelle
82	208	Telluric Fraunhofer A and B Absorptions in Spectra of the Sun and δ Sco	Sun δ Sco	7590–7700 6865–6940	SQUES echelle
83	210	The Night Sky Spectrum	Light pollution airglow	4000–7400	200 L
84	217	Gas Discharge Lamp (Ne)	Neon glow lamp	5800–8100	900 L
85	218	Gas Discharge Lamp (ESL)	ESL Osram Sunlux	3900–6400	200 L
86	219	Gas Discharge Lamp (Na)	High pressure sodium vapor lamp	4700–7250	200 L
87	220	Gas Discharge Lamp (Xe)	Xenon high power lamp (Top of Mt. Rigi)	4900–6900	200 L
88	221	Calibration lamp with Ne, Xe	Modified glow starter Philips S10	3900–7200	200 L
89	222	Calibration Lamp with Ar	Modified glow starter OSRAM ST 111	3900–4800 4000–7700	900 L 200 L
90	223	Calibration Lamp with Ar, Ne, He Overview spectrum R~900	Modified glow starter RELCO 480	3800–7200	200 L
91	224	Calibration Lamp with Ar, Ne, He Highly Resolved Spectrum R~4000	Modified glow starter RELCO 480	3850–6300	900 L
92	225	Calibration Lamp with Ar, Ne, He Highly Resolved Spectrum R~4000	Modified glow starter RELCO 480	5900–8000	900 L
93	226	Calibration Lamp with Ar, Ne, He Highly Resolved Spectrum R~20,000	Modified glow starter RELCO 480	Order 28–32 6900–8160	SQUES echelle

(cont.)

Plate	*Page*	Topic	Objects	Wavelength domain/λ	Grating
94	227	Calibration Lamp with Ar, Ne, He Highly Resolved Spectrum R~20,000	Modified glow starter RELCO 480	Order 33–37 5960–6930	SQUES echelle
95	228	Calibration lamp with Ar, Ne, He Highly resolved spectrum R~20,000	Modified glow starter RELCO 480	Order 38–42 5260–6000	SQUES echelle
96	229	Calibration lamp with Ar, Ne, He Highly resolved spectrum R~20,000	Modified glow starter RELCO 480	Order 43–47 4700–5300	SQUES echelle
97	230	Calibration lamp with Ar, Ne, He Highly resolved spectrum R~20,000	Modified glow starter RELCO 480	Order 48–52 4250–4770	SQUES echelle
98	231	Calibration lamp with Ar, Ne, He Highly resolved spectrum R~20,000	Modified glow starter RELCO 480	Order 53–57 3880–4315	SQUES echelle
99	232	Calibration lamp with Ar, Ne, He Image with SQUES echelle orders	Modified glow starter RELCO 480	Order 28–37 5970–8160	SQUES echelle
100	233	Calibration lamp with Ar, Ne, He Image with SQUES echelle orders	Modified glow starter RELCO 480	Order 38–47 4700–6020	SQUES echelle
101	234	Calibration lamp with Ar, Ne, He Image with SQUES echelle orders	Modified glow starter RELCO 480	Order 48–57 3750–4770	SQUES echelle
102	241	Swan bands/hydrocarbon gas flames: Comparison with a montage of spectra	Butane gas torch Comet Hyakutake WZ Cassiopeiae	3800–6400	200 L
103	242	Spectra of terrestrial lightning discharges	Integrated light of several lightning discharges	3750–7200	200 L

CHAPTER

2

Selection, Processing and Presentation of the Spectra

2.1 Selection of Spectra

The main criteria for the selection of the spectra have been the documentation of the spectral characteristics and the demonstration of certain effects, for example due to the different luminosity classes. The consideration of bright "common knowledge stars" was of secondary importance. The ordinary spectral classes are presented here at least with an early and a rather late subtype to show the development of characteristic features in the profile. Further spectra are commented on in separate chapters such as extraordinary star types, emission nebulae, composite spectra of extragalactic objects, star clusters, reflectance spectra of solar system bodies, absorption bands generated by the Earth's atmosphere and finally some profiles of terrestrial and calibration light sources.

2.2 Recording and Resolution of the Spectra

Except of a few clearly declared exceptions all spectra have been recorded by the author with the DADOS spectrograph, equipped with reflection gratings of 200 or 900 lines mm^{-1} [4]. Unless otherwise noted, the recording was made by an 8 inch Schmidt–Cassegrain Celestron C8, applying the 25 μm slit of the DADOS spectrograph. Similar results could surely be achieved by applying other spectrographs, for example from Shelyak Instruments or other suppliers. To display some highly resolved spectral details, the SQUES echelle spectrograph was used, applying slit widths of 25–85 μm [5]. Finally, most of the spectra have been recorded by the cooled monochrome camera ATIK 314L+, equipped with the Sony chip ICX285AL. A few spectra have been

recorded, in cooperation with Martin Huwiler, by the CEDES 36 inch telescope of the Mirasteilas Observatory in Falera. However, applying longer exposure times, these objects are also within the reach of average amateur equipment!

The processing of the profiles with Vspec software yields about the following dispersion values [Å/pixel]: DADOS 200 L mm^{-1}: 2.55, DADOS 900 L mm^{-1}: 0.65 and SQUES echelle: 0.18. Data of the Sony chip ICX285AL: 1.4 megapixel, 2/3" monochrome CCD, pixel size 6.45 μm × 6.45 μm.

2.3 Processing of the Spectra

The monochrome fits-images have been processed with the standard procedure of IRIS [6]. In most cases, about 2–5 spectral profiles have been stacked, to achieve a reasonable noise reduction. The generating and analyzing of the final profile was subsequently performed with Vspec [7]. For longer exposure times a dark frame was subtracted, if necessary also the separately recorded light pollution (by IRIS or fitswork). The processing of "flat fields" was omitted.

For profiles recorded by the echelle spectrograph SQUES, each order was processed and calibrated separately, analogously to the DADOS spectra. Further in rare cases with a significantly increased noise level (recording of very faint objects), the profile was sometimes smoothed, using filters such as "Spline" provided by the Vspec software. The goal of this process here was exclusively to improve the recognizability of the documented lines. A reduction of the telluric H_2O/O_2 absorptions in the yellow/red range of the spectral profile was omitted. Correspondingly, the documentation of lines of lower resolved profiles has been cautious and rather restrained.

2.4 Calibration of the Wavelength

Most of the spectra have been calibrated, based on rest wavelengths of known lines and not absolutely with the calibration lamp. This prevents the profile, as a result of possibly higher radial velocities, appearing to be shifted on the wavelength axis. The focus here is on the presentation of the spectral class and not on the documentation of the individual star. The calibration with the light source was restricted to higher resolved spectra of very late spectral classes, extraordinary stars and of course applied to extragalactic objects. Generally the applied unit for the wavelength is ångstrom [Å]. According to convention such values are indicated with the prefix λ [2]. For instance 5000 Å corresponds to λ5000. The values are mostly taken from the Vspec database "lineident," according to the catalog of lines in stellar objects, ILLSS Catalog (Coluzzi 1993–1999, VI/71A). As usual in the optical spectral domain, the wavelength λ is based on "standard air."

2.5 Display of the Intensity Scale and Normalization of the Profiles

With minor exceptions in all spectra the pseudo-continuum was removed. The according process is outlined in [3]. The profiles have been rectified, divided by the course of their own continuum. Thus, the intensity of the spectral lines becomes visually comparable over the entire range. Otherwise in a raw profile of a pseudo-continuum strong lines at the blue or red end of the spectrum appear to be optically as too weak and vice versa, weak lines in the middle part as too high. But just this reasonable correction may confuse beginners, if they try to find spectral lines of their uncorrected pseudo-continuum in the according flat and rectified atlas profile. This effect is demonstrated in [3].

An absolute flux calibration of the intensity profile would be extremely time consuming and demanding and is not necessary for the purpose of this atlas. However, with minor exceptions, the intensity of the rectified profiles was normalized to unity, so the medium continuum level yields about $I_c = 1$. Hence, for space saving reasons, the displayed wavelength axis is normally shifted upwards, recognizable on the according level of the intensity axis ($I > 0$). In some cases only the intensity value of the shifted wavelength axis is labeled (Figure 2.1). In case of montages, showing several

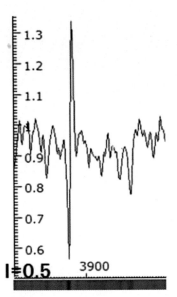

Figure 2.1 Labeling of the intensity scale, located here on the level $I = 0.5$

spectra in one chart, the individual profiles have been normalized to unity, based on the same continuum section, to enable a rough comparison of the line intensities.

2.6 Identification and Labeling of the Spectral Lines

For line identification spectra commented on from the literature were used. In many cases, the sources are referenced accordingly. The aim was to specify two decimal places. If required some positions have been supplemented by the according Vspec databases, based on the ILLSS Catalog, Coluzzi 1993. In some cases NIST has also been consulted [8]. This way, a few intensive but undocumented lines have also been identified. However, this procedure was applied very restrictively, i.e. in few cases with a clear profile and a high line-intensity as well as missing plausible alternative elements in the immediate neighborhood. The labeling of such items is declared with a red "**V**."

If appropriate in highly resolved SQUES echelle profiles, the lines are labeled with an accuracy of three decimal places – mostly based on SpectroWeb [9], or other referenced sources.

Blends of several spectral lines, appearing especially frequently in the mid to late stellar spectral classes, are usually labeled without decimal places. The most intense of the metal lines involved were mostly determined by SpectroWeb based on a comparable spectral class and in a few special cases even counter-checked by the *Bonner Spektralatlas* (BSA) [10].

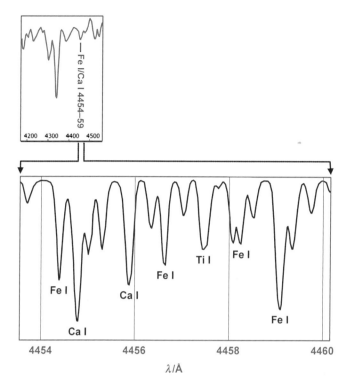

Figure 2.2 Procyon, α CMi, DADOS profile $R \sim 900$ (above), compared to a corresponding resolution of $R \sim 80,000$ (cut-out, schematically after SpectroWeb [9])

If the intensity of a blended line is clearly dominated by a certain element or ion, it is also used to label the whole blend.

Figure 2.2 shows using the example of Procyon how complexly composed even an inconspicuous small absorption can be, if displayed in a high-resolution profile. What in Plate 12 is roughly summarized as "Fe I/Ca l 4454–59," turns out in SpectroWeb to be a highly complex blend of numerous metal lines.

2.7 Presentation

All spectra are at least documented by a broadband profile ($200\,\mathrm{L\,mm^{-1}}$). In the presence of interesting lines or according information, higher resolved spectra are attached, recorded with the $900\,\mathrm{L\ mm^{-1}}$ or in some special cases even with the SQUES echelle spectrograph. The line profiles are supplemented on the wavelength axis by synthetically generated 1D-spectra (by Vspec). Their color gradient is chiefly intended as a rough visual reference for the wavelength domain. Molecular absorption bands are marked with this icon, indicating the direction of fading intensity: ▉ .

2.8 Object Coordinates

With exception of the quasar list in Appendix H, this atlas contains no coordinates for the individual objects. Today the corresponding data for different epochs can easily be found online, either in the CDS [11] or NED database [12], and for brighter objects also in the numerous existing planetarium programs. Many such objects are further implemented even in the control software of today's standard amateur telescopes and can be directly approached this way (Goto).

2.9 Distances

Distance information is given in light years [ly] or parsec [pc], where $1\,\mathrm{pc} \approx 3.26\,\mathrm{ly}$. For extragalactic objects often the z-value is applied, directly depending on the measured red- or blueshift in the spectrum (Section 26.10).

CHAPTER

3 Terms, Definitions and Abbreviations

3.1 Designation and Parameters of the Stars

The information of the spectral and luminosity classes, as well as of the distances are mostly taken from CDS [11] or professional publications, also in a few cases from the Bright Stars Catalog [13] or J. Kaler [14]. The spectral classes of certain objects, provided by different sources, may be considerably different. The indications to the "effective temperature" T_{eff} of the stars originate from different reputable sources. For most of the stars it corresponds very roughly to the temperature of the photosphere, which is mainly responsible for the formation of the spectral lines as well as for the spectral distribution of the stellar radiation intensity. For a definition of T_{eff} refer to [3].

The rotation velocity of a star refers here to the share of the surface velocity, $v \sin i$, which is projected in direction to the Earth and can be spectrographically determined applying the Doppler principle [3]. These values are mainly taken from CDS [11].

Brighter stars are referred to in the atlas with the proper name and in the Bayer system with small Greek letters, combined with the abbreviated Latin constellation name, for example Sirius, α CMa. Fainter stars, lacking proper names, are identified with the Bayer system, or if necessary, with the Flamsteed or HD number (Henry Draper Catalog), for example φ Sagittarii, 61 Cygni, HD 22649. Constellations are referred with the Latin name (Appendix B).

3.2 Galactic Nebulae and Star Clusters

Such parameters originate mainly from NED, the NASA/ IPAC Extragalactic Database [12], or professional publications.

3.3 Extragalactic Objects

Parameters of galaxies and quasars, for example the z-values of the redshift, are taken from NED. From understandable reasons, information about the masses of such objects is missing there. Current estimates are still very uncertain and accordingly the subject of debate – not least about the existence of dark matter. Therefore such values originate from recent publications.

3.4 Labeling of the H-Balmer Series

Absorptions or emissions of the H-Balmer series are labeled with Hα, Hβ, Hγ, Hδ, Hε, H8, H9, H10, etc. Further lines after Hε are labeled with the upper shell number involved in the electron transition.

3.5 Labeling of Elements and Ions

As usual in astrophysics, all the elements, except hydrogen and helium, are called "metals" and identified in the astrophysical notation. The so called "forbidden lines" are labeled within square brackets []. For further details and definitions refer to [3].

3.6 Abbreviations, Symbols and Common Units

Å	unit of wavelength: ångstrom (Appendix G)
AAVSO	American Association of Variable Star Observers
AGB	asymptotic giant branch (HRD)

AGN	active galactic nuclei		IR	infrared range of the electromagnetic spectrum
ALMA	Atacama Large Millimeter/Submillimeter Array in Chile		IRIS	astronomical images processing software by Christian Buil
AU	astronomical unit, 149.6×10^6 km		ISL	integrated starlight (contribution to the brightening of the night sky)
B	magnetic flux density			
BSA	Bonner Spektralatlas		ISM	interstellar matter
c	speed of light 300,000 km s^{-1}		J	joule, SI unit for energy
CDS	Centre de Données astronomiques de Strasbourg		JD	Julian date: number of days after January 1, 4713 BCE
cgs	system of units "centimeter, gram, second" (Appendix G)		K	kelvin, temperature unit K $\approx$ °Celsius + 273
Chandra	X-Ray Telescope, NASA		L	luminosity
CTTS	classical T Tauri star		LBV	luminous blue variable
d	days		LD	line depth (of an absorption line) [3]
Dec	declination		LDR	line depth ratio (LD in relation to the continuum intensity) [3]
DGL	diffuse galactic light (contribution to the brightening of the night sky)		LMC	Large Magellanic Cloud
erg	cgs unit for energy		LPV	long period variable of Mira-type
ESL	energy saving lamp (gas discharge lamp)		ly	light year [ly] 1 ly $\approx 9.46 \times 10^{12}$ km
ESO	European Southern Observatory		M	mass
EW	equivalent width (of a spectral line) [Å] (Appendix G)		M	million
			M	absolute magnitude, apparent brightness at a distance of 10 parsec [mag]
FWHM	full width (of a spectral line) at half maximum height [Å] (Appendix G)		M_B	absolute magnitude in the blue range of the spectrum [mag]
FWZI	full width (of a spectral line) at zero intensity [Å] (Appendix G)		m_b	apparent magnitude in the blue range [mag]
GALEX	Galaxy Evolution Explorer, UV Space Telescope, NASA		M_i	initial stellar mass
			m_v	apparent magnitude in the visual (green) range [mag]
GCL	globular star cluster		M_V	absolute magnitude in the visual (green) range [mag]
Glx	galaxy			
GN	galactic nebula		MK	Morgan, Keenan, Kellman spectral classification system and spectral atlas
GOSSS	Galactic O Star Spectroscopic Survey [15], [16]			
GTR	general theory of relativity		MS	main sequence, position of stars in the HRD with luminosity class V
h	hours			
H$_{(0)}$	Hubble constant, present value of the variable Hubble parameter H$_{(t)}$		NED	NASA Extragalactic Database
HB	horizontal branch (HRD)		nm	nanometer
HD	Henry Draper Catalog		OCL	open star cluster
HRD	Hertzsprung–Russell diagram		pc	parsec [pc] 1 pc $\approx$ 3.26 ly, Mpc = megaparsec
HST	Hubble Space Telescope		pe	"post explosion," indicates for SN the time span since the explosion
HWHD	half width (of a spectral line) at half depth (equivalent to HWHM)		PMS	pre-main sequence star, young protostar, not yet established on main sequence
HWHM	half width (of a spectral line) at half maximum height (equivalent to HWHD)		PN	planetary nebula
HWZI	half width (of a spectral line) at zero intensity [Å] (Appendix G)		R_s	Strömgren radius
			RA	right ascension
IAU	International Astronomical Union		RGB	red giant branch (HRD)

RSG	red supergiant
s	seconds
SB1	SB1 system. Spectroscopic binary stars with a strong brightness difference. Just the spectrum of the brighter component can be observed.
SB2	SB2 system. Spectroscopic binary stars with a small brightness difference. A composite spectrum of both components is observable.
SN	supernova explosion
SNR	signal-to-noise ratio
SNR	supernova remnant
STR	special theory of relativity
UV	ultraviolet range of the electromagnetic spectrum
V	spectral line, identified with help of Vspec or other tools
v_∞	terminal velocity of a stellar wind or an expanding envelope
v_r	spectroscopically measured radial velocity
var	variable
VLT	Very Large Telescope, ESO telescope group on Cerro Paranal, Chile

Vspec	spectroscopic analysis software "Visual Spec" by Valérie Desnoux
WD	white dwarf
WR	Wolf–Rayet stars
WRPN	spectra generated by extremely hot central stars of planetary nebulae, exhibiting similar features like profiles of real WR stars
WTTS	weak T Tauri stars
T_{eff}	effective temperature of a stellar photosphere [3]
y	year
YSO	young stellar objects (PMS stars)
Z	metal abundance Z (metallicity) [3]
z	z-value, a measure for the redshift, distance, and the past [3]
ZAMS	zero age at main sequence, young star just joining the main sequence
ZL	zodiacal light (contribution to the sky brightness)
⊙	comparison to the Sun: $M_\odot$: solar mass, $L_\odot$: luminosity of the Sun
⊗	telluric absorption in the Earth's atmosphere
200 L, 900 L	reflection gratings with a constant of 200 or 900 lines mm^{-1}

4

Overview and Characteristics of Stellar Spectral Classes

4.1 Preliminary Remarks

Basic knowledge of the stellar spectral classes is an indispensable prerequisite for reasonable spectroscopic activity. Combined with appropriate knowledge and tools, this classification system contains a wealth of qualitative and quantitative information about the classified objects. The spectral classes from the brighter stars can nowadays be obtained from Internet sources, planetarium programs, etc. For a deeper understanding of the classification system, a rough knowledge of the historical development is very useful, since from each development step something important has remained until now!

4.2 The Fraunhofer Lines

At the beginning of the nineteenth century, the physicist and optician Joseph von Fraunhofer (1787–1826) investigated, based on the discovery of William H. Wollaston, the sunlight with his home-built prism spectroscope. He discovered over 500 absorption lines in this very complex spectrum. At that time the physical background for this phenomenon was still unknown. Therefore he was just able to denote the more prominent of the huge number of solar absorption lines with the letters A–K. Later on, additional lines were supplemented with small letters. These designations are still in use today, even in current professional papers. In this atlas some of the Fraunhofer lines are labeled in the solar spectrum and neighboring spectral classes. For higher resolved profiles refer to Plates 15, 16 and 17. Table 4.1 shows the listed Fraunhofer lines, rounded to 1 Å. Figure 4.1 shows their position in a lower resolved spectrum.

The famous helium D3 or d-line at λ5876 is just detectable in the spectra of the solar chromosphere. Further, "d-line" is

Table 4.1 Fraunhofer lines

Line Ident.	Element	Wavelength [Å]
A band	O_2	7594–7621
B band	O_2	6867–6884
C	H (α)	6563
a band	O_2	6276–6287
D 1, 2	Na	5896 and 5890
D 3 or d	He	5876
E	Fe	5270
b 4, 2, 1	Mg	5167, −73 and −84
F	H (β)	4861
d	Fe	4668
e	Fe	4384
G (f)	H (γ)	4340
G band	Fe, CH	4307–4313
g	Ca	4227
h	H (δ)	4102
H	Ca II	3968
K	Ca II	3934

here sometimes ambiguously assigned to Fe λ4668. The same applies to the letter "e" which sometimes designates also the non-solar Hg line at λ5461.

Figure 4.2 shows an original drawing of the solar spectrum by Joseph von Fraunhofer. Nowadays the convention requires

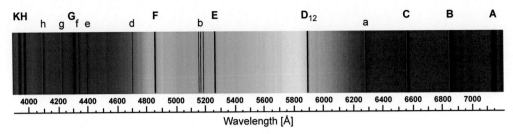

Figure 4.1 Position of the Fraunhofer lines in the solar spectrum (DADOS 200 L mm^{-1})

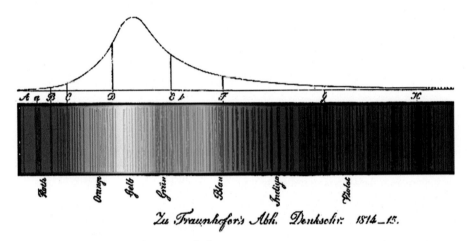

Figure 4.2 Original drawing of the solar spectrum by J. Fraunhofer

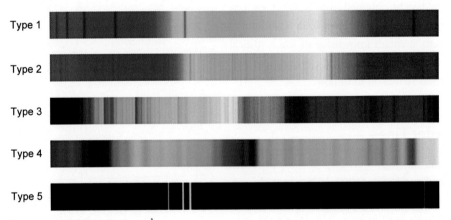

Figure 4.3 The five Secchi classes, DADOS 200 L mm^{-1}

the blue region of the spectrum to be to the left and vice versa the red to be to the right. The sketched estimated course shows the stunning spectral sensitivity of the human eye!

4.3 Further Development Steps: The Five Secchi Classes

In the further course of the nineteenth to the early twentieth century, astronomy, and particularly spectroscopy, benefited from impressive advances in chemistry and physics. So it became increasingly possible to assign the spectral lines to chemical elements – primarily to the merit of Robert Bunsen

(1811–1899) and Gustav Kirchhoff (1824–1887). Father Angelo Secchi (1818–1878) from the Vatican Observatory decisively influenced the future path of stellar spectral classification. By many sources he is therefore referred as the father of modern astrophysics. Figure 4.3 shows his classification system, subdivided into five groups according to specific spectral characteristics. For a view on Secchi's original drawings see Appendix J.

Type 1: Comprises bluish-white stars with relatively simple spectra, which seem to be dominated just by a few, but very bold, lines. These are distributed like thick rungs over the spectral stripe and turned out later to be the hydrogen lines of the famous Balmer series. This simple characteristic allows even beginners to roughly classify such stars into the currently used A or late B class (e.g. Sirius, Vega and Castor).

Type 2: Contains yellow shining stars with complex spectra, dominated by numerous metal lines, like those exhibited by the Sun, Capella, Arcturus and Pollux (today's classes $\approx$ F, G, K).

Type 3: Comprises reddish-orange stars with complex vibrational band spectra and just a few discrete lines. These broad, asymmetric absorptions reach the greatest intensity on the left, short-wave band end and are slowly fading towards the long-wave (red) side. The wavelength of absorption bands always refers to the point of greatest intensity the so called "band heads," also defined as "most distinct edge." Such features show, for example Betelgeuse, Antares and Mira. Not until 1904, did it become clear that these absorption bands are mainly caused by the titanium monoxide molecule TiO (today's class M).

Type 4: Contains very rare, reddish stars with vibrational absorption bands, but their intensity fades towards the short-wave (blue) side (e.g. carbon star Z Piscium). Angelo Secchi already recognized that such features are generated by carbon (see Section 15.2)!

Type 5: Comprises stars or nebulae with "bright lines" – emission lines that we know today.

4.4 "Early" and "Late" Spectral Types

At the beginning of the twentieth century the state of knowledge regarding nuclear physics was still very limited. Therefore a hypothesis, since disproved, was postulated that the stars generate their energy exclusively by contraction and the spectral sequence represents the chronological development stages, starting extremely hot and ending cold. This misleading thesis has subsequently influenced the terminology, which is still applied today. Therefore the classes O, B and A are still called "early," and K, M designated as "late" types. In this atlas the classes F and G are referred as "middle." This terminology is logically also applied within the decimal subclasses. So for example M0 is called an "early" and M8 a "late" M type. In consequence for example A2 is "earlier" than A7.

4.5 Temperature Sequence of the Harvard Classification System

Quite soon it turned out that Secchi's classification system was too rudimentary. Based on a large number of spectra and preliminary work by Henry Draper, Edward Pickering (1846–1919) refined Secchi's system by capital letters from A to Q. The letter "A" corresponded to the Secchi class type 1 for stars with the strongest hydrogen lines. The further letters,

represented spectra with increasingly weakening H lines. Finally, just the A classification has survived up to the present time! As a director of the Harvard Observatory, Pickering employed many women, for its time a truly avant-garde attitude, which proved to be highly successful! Three female employees of his staff took care of the classification problem. Despite many deviations, the system of Annie J. Cannon (1863–1941) became widely accepted around the end of the World War I. Its basic structure has survived until today and is essentially based on the simple sequence of seven letters O, B, A, F, G, K, M. With this system, even today, still over 99% of the stars can be classified.

Cecilia Payne-Gaposchkin discovered that this sequence of letters follows the decreasing atmospheric temperature of the classified stars, starting from the very hot O types with several 10,000 K down to the cool M types with just about 2400–3500 K. This so called temperature sequence reflects the absolutely ground-breaking recognition that the observed spectral features mainly depend on the photospheric temperature T_{eff} of the star and other parameters, such as chemical composition, density of the atmosphere, rotation velocity etc., are just secondary.

This may not be really surprising from today's perspective. We know that, even about 13.7 billion years after the big bang, the share of hydrogen and helium still comprise about 99% of the elements in the Universe, with 75% and 24% respectively. This temperature sequence also forms the horizontal axis of the almost simultaneously developed Hertzsprung–Russell diagram [3]. The temperature sequence was later complemented by the following classes:

– R for cyan (CN) and carbon monoxide (CO)
– N for carbon
– S for very rare stars, whose absorption bands, instead of TiO, are generated, by zirconium monoxide (ZrO), yttrium monoxide (YO) or lanthanum monoxide (LaO).

Further, the individual classes have been subdivided into decimal subclasses mostly of 0–9, for example the Sun G2, Pollux K0, Vega A0, Sirius A1 and Procyon F5.

4.6 Rough Determination of the One-Dimensional Spectral Class

The rough determination of the one-dimensional spectral classes O, B, A, F, G, K, M is easy and even feasible for slightly advanced amateurs. In this system, the Sun with 5800 K is classified as G2. Spectral distinction criteria are prominent features such as lines in absorption or emission, which appear prominently in certain classes, and otherwise are weaker or even entirely absent.

Plate 1 Overview of the Spectral Classes

Plate 2 Intensity Profiles of All Basic Spectral Classes

Plates 1 and **2** show a montage with sample spectra ($200\,L\,mm^{-1}$) of the entire class sequence O–M, which are also used here to present the individual spectral classes. This spectral resolution already enables the identification of the intense and therefore most important spectral lines. Also, the influence of some elements, ions or molecules to the different spectral types, becomes roughly visible here. Displayed in Plate 1 are the synthetically obtained 1D spectra by Vspec, which result from the intensity profiles, normalized to the same sector and shown in Plate 2.

The following become obvious here:

In the upper third of the plate (B2–A5), the strong lines of the H-Balmer series, i.e. Hα, Hβ, Hγ, etc. They appear most pronounced in the class ~A2 and are weakening from here towards earlier and later spectral classes.

In the lower quarter of the plate (K5–M5) the eye-catching, shaded bands of molecular absorption spectra, generated mainly by titanium monoxide (TiO).

In the upper part of the lower half some spectra (F5–K0) show just a few prominent features, but are charged with a large number of fine metal lines. Striking absorptions are only the Na I double line (Fraunhofer D1, 2) and in the "blue" part the impressive Ca II (H+K) absorption, gaining strength towards later spectral classes. Fraunhofer H at λ3968 starts around the early F class to overprint the weakening Hε hydrogen line at λ3970. In addition, the H-Balmer series is further weakening towards later classes.

Finally, at the top of the plate is the extremely hot O class with very few, fine lines, mostly of ionized helium (He II) and multiply ionized metals. The H-Balmer series appears here quite weak, as a result of the extremely high temperatures. The telluric H_2O and O_2 absorption bands are reaching high intensities here because the strongest radiation of the star takes place in the ultraviolet whereas the telluric absorption bands are located in the undisturbed domain near the infrared part of the spectrum.

By contrast the maximum radiation of the late spectral classes takes place in the infrared part, enabling the stellar TiO absorption bands to overprint the telluric lines here.

In the spectra of hot stars (approximately classes from early A to O) the double line of neutral sodium Na I (Fraunhofer D1, 2) must inevitably be of absorptions by interstellar matter. This element has a very low ionization energy of just 5.1 eV (see Appendix F). Neutral sodium can therefore just exist in the atmospheres of relatively cool stars. The wavelengths of the ionized Na II lie already in the ultraviolet range and are therefore not detectable by amateur equipment.

4.7 Flowcharts for Estimation of the Spectral Class

Numerous flowcharts exist to determine the spectral class, mostly based on the intensity comparison between certain lines. In Figure 4.4 are just two examples for a rough determination of the one-dimensional spectral class.

4.8 Further Criteria for Estimation of the Spectral Class

Here follow some additional indications with characteristic spectral features:

Course of the Continuum Already a comparison of non-normalized raw spectra between an early and a late classified star, exhibits a significant shift of the maximum intensity due to Wien's displacement law. Extremely hot O and early B stars radiate mainly in the ultraviolet domain (UV) and cold M-type stars chiefly in the infrared range (IR).

Spectral Class O Singly ionized helium He II – appearing also as emission lines, further neutral He I, doubly ionized C III, N III, O III, triply ionized Si IV. The H-Balmer series only very weak. Maximum intensity of the continuum is here in the UV range. Examples include: Alnitak (ζ Ori): O9.7 Ib, Mintaka (δ Ori): O9V.

Spectral Class B Neutral He I in absorption – strongest at B2, the Fraunhofer K line of Ca II becomes faintly visible, further singly ionized O II, Si II, Mg II. The H-Balmer series becomes stronger. Examples include: Spica: B1 III–IV, Regulus: B8 IVn, Alpheratz (α And): B8 IVp Hg Mn, Algol: B8V, as well as all bright stars within the Pleiades cluster.

Spectral Class A H-Balmer lines are strongest at ~A2, Fraunhofer H+K lines (Ca II) become stronger, neutral metal lines become visible, helium lines (He I) disappear. Examples include: Vega: A0 V, Sirius: A1 Vm, Castor: A1V/A2Vm, Deneb: A2 Ia, Denebola: A3V, Altair: A7Vn.

Spectral Class F H-Balmer lines become weaker, H+K lines (Ca II) as well as neutral and singly ionized metal lines become stronger (Fe I, Fe II, Cr II, Ti II). The striking "line double" of G band (CH molecular) and Hγ line can only be seen here and forms the unmistakable "brand" of the middle F class (see Section 8.3). Examples include: Caph (β Cas): F2III, Mirfak (α Per): F5 Ib, Polaris: F8 Ib, Sadr (γ Cyg): F8 Iab, Procyon: F5 IV–V.

Version 1

– Fraunhofer K: Ca II λ3934
– Fraunhofer H: Ca II λ3968
– G-Band CH: λ4300 – 4310
– Hγ: λ4340
– Mn: λλ 4031/4036
– TiO Bandhead: λ5168

The intensity ratio of the compared lines (e.g. Ca II K/Hγ) is calculated by the peak intensities P= I/I_c

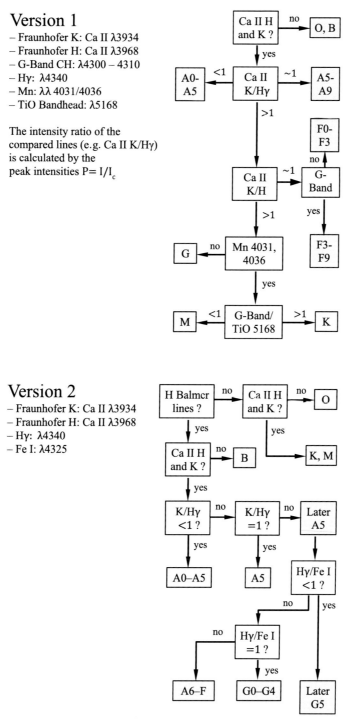

Version 2

– Fraunhofer K: Ca II λ3934
– Fraunhofer H: Ca II λ3968
– Hγ: λ4340
– Fe I: λ4325

Figure 4.4 Flowcharts for estimation of the spectral class (Sources: Lectures from University of Freiburg i. B. [17] and University of Jena [18])

Spectral Class G Fraunhofer H+K lines (Ca II) very strong, H-Balmer lines further weakening, G band becomes stronger as well as many neutral metal lines, for example Fe I and D line (Na I). Examples include: Sun: G2V, the brighter component of Alpha Centauri: G2V, Muphrid (η Boo): G0 IV, Capella: G5IIIe + G0 III (binary star composite spectrum) [13].

Spectral Class K This is dominated by metal lines, H-Balmer lines get very weak, Fraunhofer H+K (Ca II) are still strong, Ca I becomes very strong now as well as the molecular lines CH, CN. At the late K types first appearance of TiO bands. Examples include: Pollux: K0 III, Arcturus: K1.5 IIIpe, Hamal (α Ari): K1 IIIb, Aldebaran: K5 III.

Spectral Class M Molecular TiO bands get increasingly dominant, many strong, neutral metal lines, for example Ca I. Maximum intensity of the continuum is in the IR range. Examples include: Mirach (β And): M0 IIIa, Betelgeuse:

M1–2 Ia–Iab, Antares: M0.5 Ia–b, Menkar (α Cet): M 1.5 IIIa, Scheat (β Peg): M2.5 II–IIIe, Tejat Posterior (μ Gem): M3 IIIab, Ras Algheti: (α Her): M5Ib–II.

4.9 The Two-Dimensional MK (Morgan–Keenan) or Yerkes Classification System

Later on, impressive progress in nuclear physics and the increasing knowledge about the stellar evolution required a further adaptation and extension of the classification system. So it was recognized that within the same spectral class stars can exhibit totally different absolute luminosities, mainly caused by different stages of stellar development.

In 1943, as another milestone, the classification system was extended with an additional Roman numeral by W. Morgan, P. Keenan and S. Kellmann from Mt. Wilson Observatory (see Table 4.2). This second dimension of the classification specifies the eight luminosity classes. As a dwarf star the Sun is classified as G2V since it is still located on the main sequence of the HRD shining with the luminosity class V [3].

4.10 Effect of the Luminosity Class on the Line Width

Figure 4.5 shows that within the same spectral class the FWHM of the H-Balmer series increases with decreasing luminosity. This happens primarily due to the "pressure broadening," of spectral lines due to the higher gas pressure

in the denser stellar atmospheres. In the thin atmospheres of giants noticeably slim absorptions are produced, here clearly visible at the intense fine metal lines. Main sequence stars of the luminosity class V are much smaller, denser and less luminous compared to giants of the classes I–III. Most dense are the atmospheres of the white dwarfs with class VII, least dense among the supergiants of class I. In the spectra this effect is strongest visible at the H-Balmer series of class A (Figure 4.5 left). Already with the F class (same wavelength domain), this effect is visually barely

Table 4.2 Luminosity classes

Luminosity Class	Designation
0	Hypergiants
I	Luminous supergiants
Ia–0, Ia, Iab, Ib	Subdivision of the supergiants according to decreasing luminosity
II	Bright giants
III	Normal giants
IV	Subgiants
V	Dwarfs or main sequence stars
VI	Subdwarfs (hardly ever used, as already specified by prefix)
VII	White dwarfs (hardly ever used, as already specified by prefix)

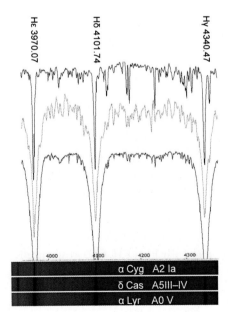
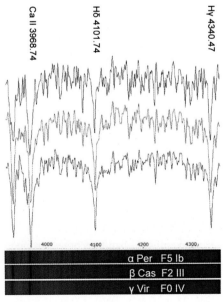

Figure 4.5 Effect of the luminosity class on the line width of H-Balmer absorptions and the intensity of the metal lines. Left: spectral class A. Right: spectral class F

noticeable. This trend continues towards the later spectral classes and thus hampers the determination of the luminosity class in this way. In such cases with high resolution spectroscopy, for example the detection of emission cores in the Ca II H+K lines [3] enables us to roughly distinguish between giants and main sequence stars (see Figure 10.3).

4.11 Suffixes, Prefixes and Special Classes

With additional small letters, placed as a prefix or suffix, extraordinary phenomena, such as a relative overabundance of a metal or the appearance of emission lines in the spectrum, are specified (see Table 4.3). Some additives, however, are over determining, as for example giants, unlike the subdwarfs and white dwarfs, are already specified by the luminosity class. Such labels are therefore barely in use. Further, with additional capital letters some special classes are specified, for example:

- Sirius A: A1 Vm, metal-rich main sequence star (dwarf) spectral class A1
- Sirius B: DA2, white dwarf or "degenerate" of spectral class A2
- o Andromedae: B6 IIIpe
- Mira (o Ceti): M5–9 IIIe
- Kapteyn's star: sd M1, subdwarf of spectral class M1 V
- P Cygni: B1–2 Ia–0 pe

Nowadays the special classes P (planetary nebulae) and Q (novae) are barely in use! The suffixes are not always applied consistently. Often one can see other versions. In the case of shell stars, for example "pe" or "shell" is applied.

The spectral classes L, T and Y which were introduced in the 1990s as a low temperature extension of the M class, including the so-called "brown dwarfs" are not covered by this atlas. These cool objects are extremely faint and radiate primarily in the infrared range. Therefore, these classes remain unreachable even by today's amateur means.

Table 4.3 Suffixes, prefixes and special classes

Suffixes		Prefixes	
a	Normal lines	d	Dwarf
b	Broad lines	sd	Subdwarf
c	Extraordinary sharp ("crisp") lines	g	Giant
comp	Composite spectrum		
e	H emission lines at B and O stars		
em	Metallic emission lines		
f	He and N emission lines in O stars		
k	Interstellar absorption lines	Special Classes	
m	Strong metal lines	Q	Novae (Chapter 24)
n / nn	Diffuse ("nebulous") lines/strongly diffuse lines e.g. due to high rotational velocity	P	Planetary nebulae (Section 28.3)
p, pec	Peculiar spectrum	D	"Degenerate" (white dwarf) + additional letter for spectral classes O, B, A (see Section 16.4)
s	Sharp lines	W	Wolf–Rayet star + additional letter for C, N or O lines (Chapter 17)
sh	Shell	S	Stars with zirconium monoxide absorption bands (Chapter 14)
v	Variation in spectrum	C	Carbon stars (Chapter 15)
wk	Weak lines	L, T	Brown dwarfs
Fe, Mg	Relatively high or low (−) abundance of the specified element	Y	Theoretical class for brown dwarfs < 600 K

Table 4.4 Statistical distribution of the spectral classes in the solar neighborhood

Type	Number of stars	Distribution to MS stars within a radius of ~13 pc	
O	0.00025	0.00003%	1 in ~3.3 Million
B	1	0.12%	1 in ~800
A	5	0.6%	1 in ~160
F	25	3.0%	1 in ~33
G	63	7.7%	1 in ~13
K	100	12.5%	1 in ~8
M	630	76%	

4.12 Statistical Distribution of Spectral Types to the Main Sequence Stars

Table 4.4 shows for the neighborhood of the Sun the percentage distribution of the main sequence (MS) stars to the individual spectral classes [19]. A star spends more than 90% of its total lifetime on the MS. The analyzed sample of 824 MS stars refers to a volume of 10,000 pc^3, i.e. a virtual sphere with a radius of about 13 pc (~43 ly).

By far the largest part consists of main sequence stars of the M class. These are so faint that even the closest one can just be observed telescopically – Proxima Centauri, M6 Ve, $m_v = 11.1$.

PLATE 1

One Dimensional Spectra

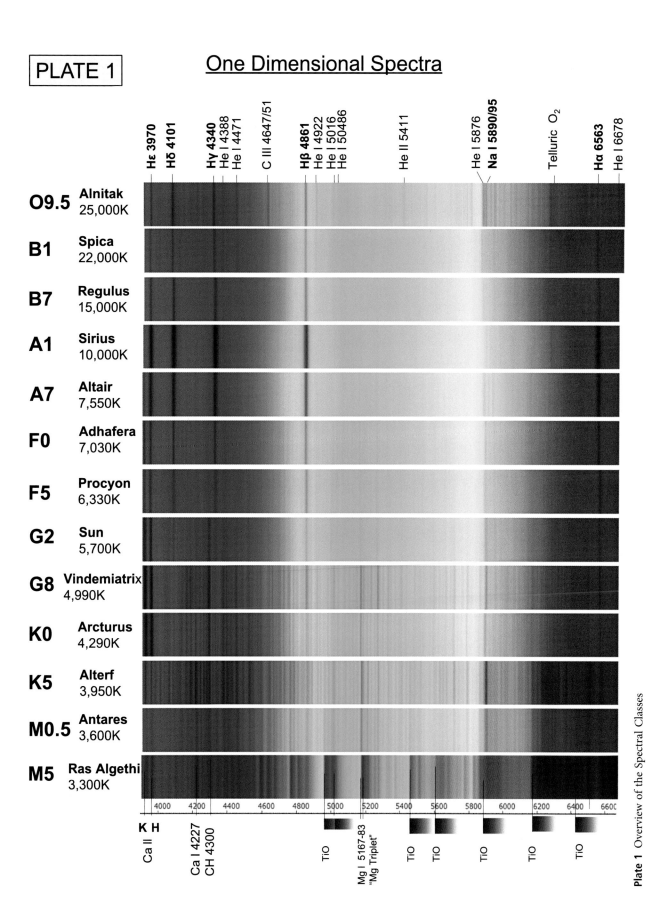

Plate 1 Overview of the Spectral Classes

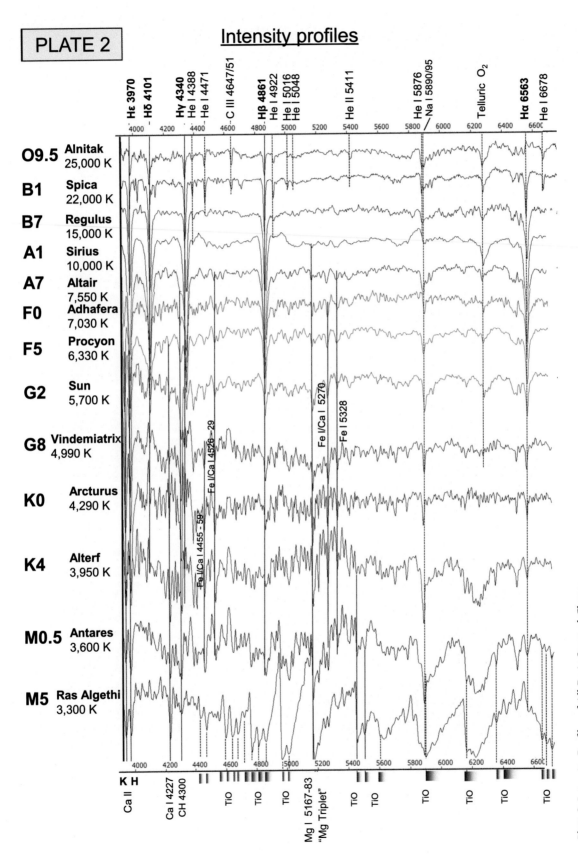

Plate 2 Intensity Profiles of All Basic Spectral Classes

CHAPTER

5 Spectral Class O

5.1 Overview

The O class comprises the hottest, most massive and shortest-living stars of the Universe. Due to their huge masses, at the end of their short lives they will all end in a supernova explosion, most of types Ib or Ic and probably after passing first a short stage as a LBV- and/or Wolf–Rayet (WR) star (see Chapters 17, 18). Subsequently the only remains will consist of a very small, extremely compact neutron star or even a black hole. These blue shining, extreme types of stars are very rare. For the Milky Way only about 20,000 representatives of the O type are estimated. As a result of their tremendous luminosity two bright representatives of this class are visible in a distance of some 1000 ly in the constellation Orion: Alnitak (ζ Ori) $m_v \approx 1.8$ and Mintaka (δ Ori) $m_v \approx 2.4$. All these stars are late O types. Bright representatives of earlier O types are only found in the southern sky, such as Naos, $m_v = 2.3$ (ζ Puppis). Table 5.1 shows O stars with an apparent magnitude brighter than $m_v \approx 5$, which are spectroscopically accessible even for amateurs with average equipment.

Two other Orion stars ε- (Alnilam), and κ- (Saiph), are classified as B0, just scarcely missing the O class. Significantly fainter, but accordingly much further distant, is the multiple star Θ^1 Ori in the Trapezium cluster. Its C-component Θ^1 Ori C is a spectral type O7 V and plays a key role for the ionization of central parts in the Orion Nebula M42. Another famous example, located in the constellation Cassiopeia, is HD 5005 ($m_v \approx 7.8$), a multiple star system containing two O-class stars, which form the ionizing source of NGC 281 ("Pac Man" Nebula).

These striking accumulations of extremely massive stars – known as OB associations – are not yet fully understood. Other slightly smaller clusters are located in the constellations Scorpius, Perseus and Cygnus. Together with other groups, they form the "Gould Belt," (discovered 1879 by Benjamin Gould) which is inclined some $20°$ to the galactic plane and has a diameter of about 2000 ly (Figure 5.1). The Sun is located somewhat off-center but still roughly within the ring plane [20] [21].

Table 5.1 Bright O type stars [11]

HD No.	Name	Spectral Class	m_v
24912	Menkhib, ξ Per	O7.5 III e	4.1
30614	α Cam	O9.5 Ia e	4.3
36486	Mintaka, δ Ori	O9 V + B0 III	2.4
36861	Meissa, λ Ori	O8 III ((f))	3.4
37022	Θ^1 Ori C	O7 V	5.1
37043	Nair al Saif, ι Ori	O9 III	2.8
37468	σ Ori A	O9.5 V	3.8
37742	Alnitak, ζ Ori	O9.7 Ib + B0 III	1.8
47839	15 Mon	O7 V	4.6
57060	29 CMa	O7 Ia	5.0
57061	30 CMa	O9 II	4.4
66811	Naos, ζ Pup	O4 If(n) p	2.3
149757	ζ Oph	O9.2 IV nn	2.6
203064	68 Cyg	O7.5 IIIn ((f))	5.0

Table 5.2 Data for the late to the early subclasses, O class stars

Mass $M/M_\odot$	Lifetime on Main Sequence [My]	Effective Temperature [K]	Radius $R/R_\odot$	Luminosity $L/L_\odot$
20–60	~1–10	$\gtrsim$25,000–50,000	9–15	90,000–800,000

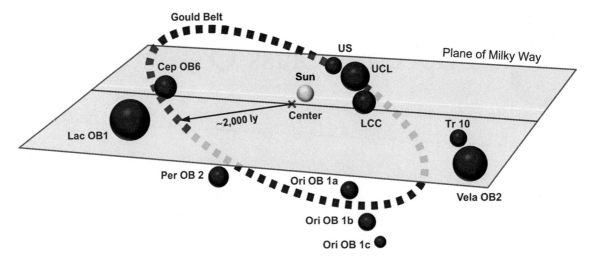

Figure 5.1 Gould Belt, schematically after [20], [21]

5.2 Parameters of the Late to Early O-Class Stars

Table 5.2 shows the data for the late to the early subclasses, exclusively for the main sequence stars, compared to the Sun ($_\odot$). Particularly for the lower temperature limit, there are significant differences within the published values.

The O class is open-ended. Currently, the top ranking is O2 with an effective temperature of about 50,000 K [1]. The late O9 class has been subdivided into further decimal subclasses.

5.3 Spectral Characteristics of the O Class

Spectra of the O class are dominated by relatively low intense absorptions of singly or even multiply ionized elements. The extremely high temperatures cause lines of ionized helium (He II) in addition to neutral helium He I. In the early O classes He II may also appear as an emission line. In earlier times the appearance of He II in the spectrum was used as the main criterion for the definition of the O class [2]. Today, in higher-resolved spectra, it can be detected already in the B0 class. Further, multiply ionized metals also appear, as C III, N III, O III and Si IV. Due to the extreme temperatures, the degree of ionization is here too high for the H-Balmer series and their line intensity is therefore only weak. If H lines appear in emission, the suffix "e" is added to the classification letter (Oe). If He and/or N are observed in

emission, the suffix "f" is added. "Of" stars seem to form the link between the O class and the Wolf–Rayet stars, probably also to the LBV stars (see Chapter 18). The graph in Figure 5.2 shows the theoretical continuum for a synthetic O9V standard star (Vspec Tools/Library). The maximum intensity of the real continuum is in the UV range.

5.4 General Remarks on the Classification of O Stars

In no other spectral class, even among reputable databases, can such different classifications be found. Perhaps this can also be explained by the typically highly variable spectral features. This particularly affects the suffixes. But even by the decimal subclasses we see significant differences. Further, in the early O classes an indication for the luminosity class as a Roman numeral is often missing and may just be recognized here by the suffix f, according to Table 5.3.

5.5 Comments on Observed Spectra

Plate 3 Spectral Features of the Late O Class

Mintaka (δ Ori) and Alnitak (ζ Ori)

This plate shows two broadband spectra (200 L mm^{-1}) providing an overview on the spectral features of the late O class. Both stars are dominant components in multiple

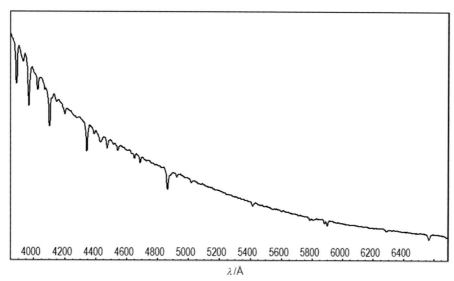

Figure 5.2 Continuum of a synthetic O9V standard star

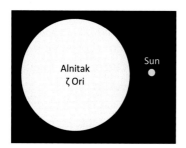

Figure 5.3 Size comparison: Alnitak and the Sun

star systems and their effective temperature is about 25,000 K. In these low-resolution profiles, some differences can be observed.

Mintaka (1200 ly), classified as O9V (+B0 III), is still established on the main sequence.

Alnitak (1200 ly) belongs to the spectral class O9.7 Ib (+B0 III), representing the lower level of the supergiants. Figure 5.3 shows approximately the size of the star compared to the Sun.

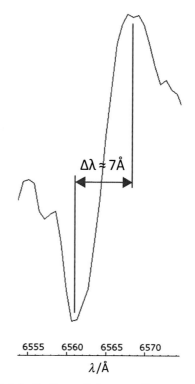

Figure 5.4 Alnitak: Hα line as P Cygni profile

Plate 4 Detailed Spectrum of a Late O-Class Star

Alnitak (ζ Ori)

Plate 4 shows two higher resolved spectra in the blue and red wavelength domains (900 L mm^{-1}) for Alnitak (1200 ly), a late O type star. Here, the main distinguishing feature between the two spectra of Plate 3 is clearly visible (Figure 5.4). The Hα line at λ6563 forms here a textbook example of a P Cygni profile with a redshifted emission and a blueshifted absorption line (see Section 18.2). This is an indication for radially ejected matter by a stellar wind – a common process for some members of this extreme stellar class. In such low resolved spectra, instead of measuring the width of the absorption trough [3], the wavelength difference can simplistically be measured as peak–peak distance, yielding here ~7 Å. According to the spectroscopic Doppler equation (see Section 18.2 and Equation {18.1}), this results in an expansion velocity of $v_r \approx 300$ km s^{-1}. The line identification is based here mainly on [2], [10], [22], [15], [16], [23].

Plate 5 Spectral Features of the Early to Middle O Class, Luminosity Effect

68 Cygni (HD 203064) and Θ^1 Ori C (HD 37022)

Plate 5 shows two apparently rather faint representatives of the early to middle O class, which are easily recordable by amateurs with average equipment.

68 Cygni (~2300 ly) is surrounded by the weakly developed H II region Sharpless 119. Here, it is the brightest of a total of six ionizing stars. CDS [11] classifies this star as O7.5 IIIn ((f)), with ~30 $M_\odot$ and an effective temperature of about 35,000 K. The suffix f is here an indicator for the luminosity class, mainly for the earlier to middle O classes (Table 5.2; [16]). As classification lines for the Of type the blends N III at $\lambda\lambda\lambda4634$–40–42 and He II at $\lambda4686$ are applied [2]. Since the emission intensity of N III decreases with increasing luminosity class, this criterion is called "negative luminosity effect."

According to [13] 68 Cygni passed in about 1959–70 a phase with Hα in emission. Because of the high proper motion, it is regarded as a "runaway star" and originates probably from the OB2 region of Cepheus. As a possible scenario, the acceleration by the supernova explosion of a companion star is discussed. It is possible that the remaining black hole with ~3 $M_\odot$ still orbits the star very closely. Observed fluctuations of equivalent widths at certain spectral lines indicate a possible orbital period of about five days [24].

Θ^1 Ori C (~1350 ly) is the brightest component of the famous Trapezium in M42. This stellar giant with its ever-changing spectral characteristics is being intensively investigated. The interferometric study [25] provides a good overview on these efforts. The indicated data for this spectral type shows a wide variation range, for example O7V [11], O5–O7 [25] or O6 [14]. Depending on the source, the stellar mass is estimated to be about 31–34 $M_\odot$. Like nearly all other stars of the Trapezium, Θ^1 Ori C also has at least one companion of spectral type O9.5, maybe also B0, with an orbital period of about 11 years. Recently, still another very closely orbiting companion, with about 1 $M_\odot$ and a period of ~50 days is

presumed. The data for the effective temperature vary approximately in the range of the proposed spectral types between 39,000 K and 45,000 K. The C component generates some 80% of the total amount of ionizing photons [26], exciting the H II region of the Orion Nebula (see Plates 64 and 65)!

Most of the absorption lines appear here similarly to those of Alnitak and Mintaka. The Hβ appears striking, because the emission line, generated in the surrounding nebula, grows out of a broad, photospheric absorption dip of the star. The two [O III] emissions must be generated by the surrounding nebula, because the hot, dense and turbulent stellar atmospheres can impossibly generate shock sensitive "forbidden" lines [3]. The Hα emission is produced by the recombination of the encircling H II region, totally veiling here the stellar line. These H II emission lines have intentionally not been subtracted here, to show the original spectrum obtained. It is strongly recommended to record Θ^1 Ori C with autoguiding to ensure the tracking on the accurate Trapezium star. The orientation of the slit axis should be optimized accordingly. Figure 5.5 is also intended to facilitate the orientation within the Trapezium cluster.

"Of" stars are considered as a transitional stage to Wolf–Rayet (WR) stars and show, for example He and N lines in emission – a clear sign that matter is being repelled (Table 5.3). Since only N III appears in emission, 68 Cygni is considered as a "mild" Of star, which is manifested by the double bracket in the suffix ((f)) [11].

In contrast to the late O9-types of Plate 3 the C IV lines at $\lambda\lambda5801/5812$ appear quite intensive here. This roughly indicates that 68 Cygni, as well as Θ^1 Ori C, are stars of the rather early O class. With 47.9 eV, C IV requires almost twice the ionization energy as He II (Appendix F).

Table 5.3 Luminosity classes according to the f phenomenon, depending on the spectral type [16]

Spectral Type	((f))	(f)	f
O2–O5.5	V	III	I
O6–O6.5	V–IV	III–Ib	Iab–Ia
O7–O7.5	V –III	II–Ib	Iab–Ia
O8	V–II	Ib	Iab–Ia
O8.5	V–II	Ib–Iab	Ia

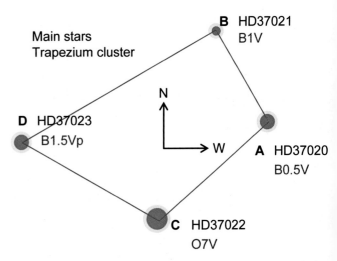

Figure 5.5 Main stars of the Trapezium cluster in the center of M42

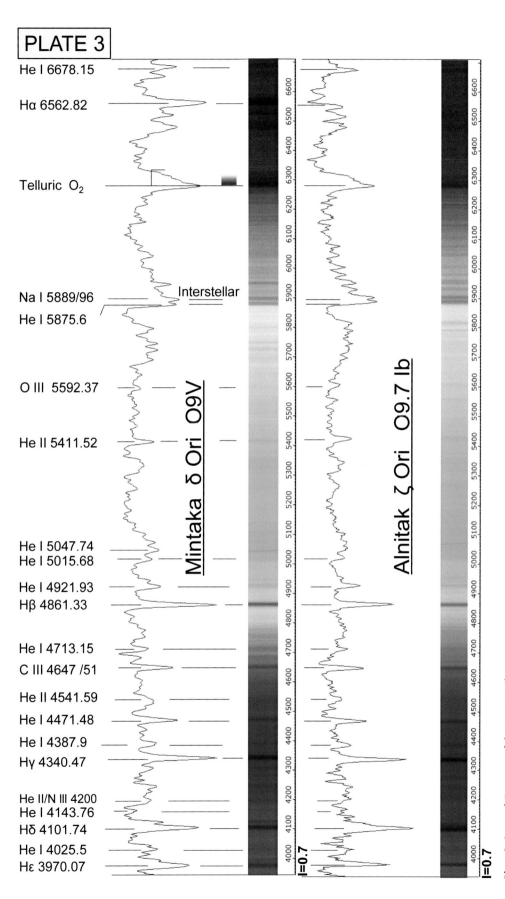

PLATE 3

He I 6678.15

Hα 6562.82

Telluric O₂

Na I 5889/96 — Interstellar
He I 5875.6

O III 5592.37

He II 5411.52

Mintaka δ Ori O9V

He I 5047.74
He I 5015.68

He I 4921.93
Hβ 4861.33

He I 4713.15

C III 4647 /51

He II 4541.59

He I 4471.48

He I 4387.9

Hγ 4340.47

He II/N III 4200
He I 4143.76

Hδ 4101.74

He I 4025.5
Hε 3970.07

I=0.7

Alnitak ζ Ori O9.7 Ib

I=0.7

Plate 3 Spectral Features of the Late O Class

PLATE 4

He I 4713.15

He II 4685.68

C III 4647−51

N II/NIII 4630−34

OII 4609/10 **V**
O II 4596.17
O II 4590.97
Si III 4574.78
Si III 4567.87
Si III 4552.65
He II 4541.59

N III 4511/15

Mg II 4481
He I 4471.48

O II 4416.98
O II 4414.91

He I 4387.9
N III 4379.09
O II 4366.9 **V**

Hγ 4340.47

O II 4315/17

O II 4302.81 **V**

O II 4276−95

O II 4254

He II/N III 4200
O II 4185.46

Fe II 4151.6 **V**
He I 4143.76

Si IV 4116.1   He I 4120.81

Hδ 4101.74

Si IV 4088.86

O II 4070   O II 4075.87

He I 4025.5

He I 4009.27
N II 3995
O II 3983
Hε 3970.07
He I/O III 3962/65

Alnitak ζ Ori   O9.7 Ib

4700
4650
4600
4550
4500
4450
4400
4350
4300
4250
4200
4150
4100
4050
4000
3950

I=0.6

He I 6678.15

Hα 6562.82

Telluric O₂

N II 5932/42
Na I 5895.92
Na I 5889.95   Interstellar
He I 5876.0

C IV 5812.14
C IV 5801.51

6650
6600
6550
6500
6450
6400
6350
6300
6250
6200
6150
6100
6050
6000
5950
5900
5850
5800
5750

I=0.6

Plate 4 Detailed Spectrum of a Late O-Class Star

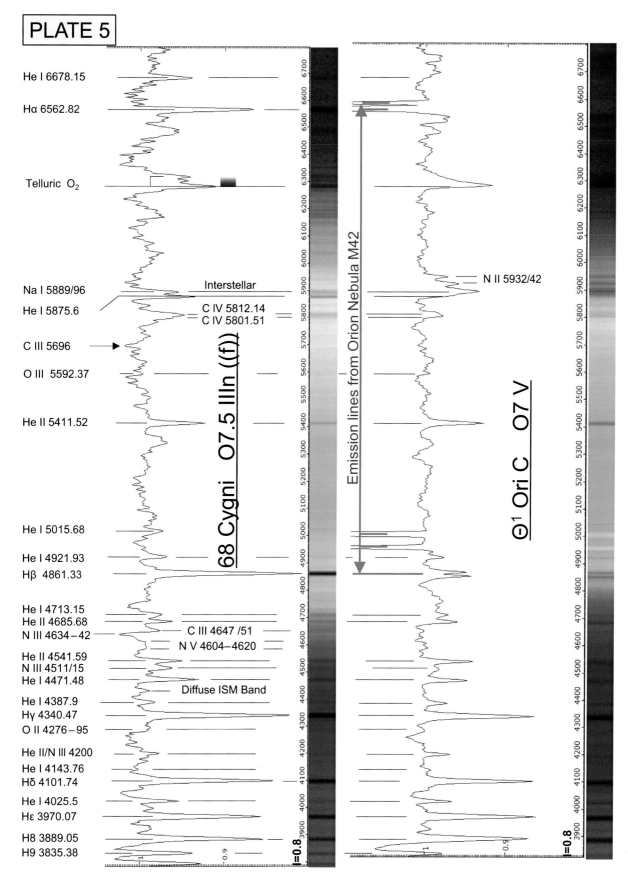

PLATE 5

He I 6678.15

Hα 6562.82

Telluric O₂

Na I 5889/96 — Interstellar

He I 5875.6

C IV 5812.14
C IV 5801.51

C III 5696

O III 5592.37

He II 5411.52

He I 5015.68

He I 4921.93

Hβ 4861.33

He I 4713.15
He II 4685.68
N III 4634 − 42 — C III 4647 /51
N V 4604 − 4620

He II 4541.59
N III 4511/15
He I 4471.48 — Diffuse ISM Band

He I 4387.9
Hγ 4340.47
O II 4276 − 95

He II/N III 4200
He I 4143.76
Hδ 4101.74

He I 4025.5
Hε 3970.07

H8 3889.05
H9 3835.38

68 Cygni   O7.5 IIIn ((f))

Emission lines from Orion Nebula M42

N II 5932/42

Θ¹ Ori C   O7 V

Plate 5 Spectral Features of the Early to Middle O Class, Luminosity Effect

CHAPTER

6

Spectral Class B

6.1 Overview

The B-classified stars seamlessly follow on below the extreme O types. Several of the bright Orion stars are early B0 types and differ therefore, from the previously presented late O9.5 stars, just by small nuances within the spectral profiles. Generally these blue-white luminous B stars are less massive and hot and therefore live much longer. Further, they are much more numerous, dominating a considerable part of the brighter constellations. Some examples are all the bright members of the Pleiades (see Section 27.6), all the bright Orion stars except Alnitak, Mintaka and Betelgeuse, all bright stars in the head of the Scorpion, except for the reddish Antares. Also worthy of mention are Regulus, Spica, Alpheratz and the weaker blue component of the well-known double star Albireo B (B8Ve). Even the famous, unstable giant P Cygni, as well as most of the Be and Be shell stars are special members of the B class.

6.2 Parameters of the Late to Early B-Class Stars

Table 6.1 shows the data for the late to early subclasses, exclusively for the main sequence stars, compared to the Sun ($\odot$), according to different sources. Considering the parameters of the B0–B9 types, the enormous spread becomes evident, covered by this class. Particularly impressive is the vast luminosity difference. The according ratio yields L_{B0}/L_{B9} ~550! Within this class, in the range of about 8–10 $M_{\odot}$, it is decided whether single stars explode as a SN of type II or end up as white dwarfs.

6.3 Spectral Characteristics of the B Class

This class is characterized by the absorption lines of neutral helium, He I, reaching their maximum intensity at about class B2 and weakening down the subclasses to B9. Further dominating are spectral lines of singly ionized metals O II, Si II, Mg II. Towards later subclasses, the Fraunhofer K line of Ca II becomes faintly visible and the H-Balmer series gets significantly stronger. Due to lower temperatures and thereby decreasing degree of ionization the simply ionized He II is only visible in the top B0 subclass, but limited here to the main sequence stars of the luminosity class V. Absorption lines of higher ionized silicon Si III and Si IV appear until down to type B2. Highly abundant in the B class are "fast rotators." Such high rotation velocities of about 150–400 km s^{-1} influence significantly the appearance of the spectrum.

The maximum intensity of the real continuum is still in the UV range, but with a significantly higher share in the visible spectrum. Figure 6.1 shows the theoretical continuum for a B6 IV standard star (Vspec Tools/Library).

Table 6.1 Data for the late to early subclasses, B-class stars

Mass $M/M_{\odot}$	Lifetime on Main Sequence [My]	Effective Temperature [K]	Radius $R/R_{\odot}$	Luminosity $L/L_{\odot}$
3–18	~10–300	~10,500–25,000	3.0–8.4	95–52,000

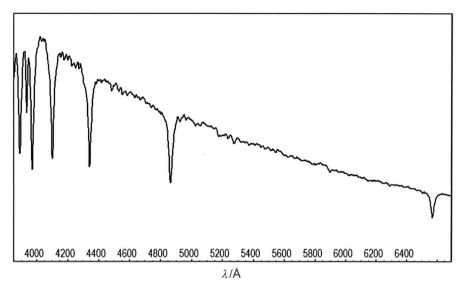

4000 4200 4400 4600 4800 5000 5200 5400 5600 5800 6000 6200 6400

$\lambda\,/\mathrm{\mathring{A}}$

Figure 6.1: Continuum of a synthetic B6 IV standard star

6.4 Comments on Observed Spectra

Plate 6 Development of Spectral Features within the B Class

Alnilam (ε Ori) and Gienah Corvi (γ Crv)

Plate 6 demonstrates the development within the B subclasses by a montage of two broadband overview spectra ($200\,\mathrm{L\,mm^{-1}}$) representing an early and late subtype.

Alnilam (1200 ly) belongs to the spectral class B0 Ia, and is a so-called supergiant. The effective temperature is about 25,000 K. The star is permanently ejecting matter, visible here by the somewhat stunted P Cygni profile of the Hα line (see Section 18.2).

Gienah Corvi (165 ly) is classified with B8IIIp Hg Mn [13]. Thus the star is in the giant stage and shows overabundances of mercury (Hg) and manganese (Mn), which, due to the low resolution, are not recognizable in the profile (Hg Mn class refer to Section 22.6). This giant of the luminosity class III shows an apparent rotation velocity of $\sim$30 km s^{-1} [11].

The comparison of these spectra clearly demonstrates the remarkable change in the spectra within the enormous span from the early to the late B class. The He I lines of Gienah Corvi appear just very faint. Moreover, they are now blended by numerous, but still weak, singly ionized or even neutral metal absorptions, such as Fe I/II, Mg II and Ca I/II. These lines appear to be very diffuse and that barely can be attributed to the rather moderate rotation velocity of the giant. Clearly visible, however, is the striking increase in the intensity of the H-Balmer series towards the late subclasses.

Here appears, for the first time and tentatively only, the Fraunhofer K line, Ca II at λ3933. The remarkable small kink in the continuum peak between Hε and H8, will grow dramatically within the following A class and become the most dominant spectral feature for the middle to late spectral types! Their "twin-sister," the Fraunhofer H line, Ca II at λ3968, has no chance to compete against the strong Balmer absorption line Hε at λ3970.

Plate 7 The Effect of Luminosity on Spectra of the B Class

Rigel (β Ori), φ Sagittarii and Regulus (α Leo)

Plate 7 demonstrates the effect of the luminosity on spectra of the B-class

Rigel (800 ly) is classified as B8 Iae, belonging to the supergiants. It has a small companion, consisting of a spectroscopic binary. Rigel B is even visible in amateur telescopes but barely affects the spectrum of the much brighter A component. The effective temperature is about 11,500 K. The apparent rotation speed is indicated in the range of about 40–60 km s^{-1}, depending on the info source. Rigel is in a transitional phase to a blue super giant. Its expanding and complexly oscillating gas envelope temporarily affects the spectrum with emission lines, sometimes appearing as ordinary or even inverse P Cygni profiles – for amateurs a rewarding object for long-term monitoring.

φ Sagittarii (230 ly) is classified with B8III and therefore a so-called normal giant. Its effective temperature is about 12,300 K, the apparent rotation velocity 35 km s^{-1} [11]. The latter has hardly any noticeable influence on the line width of this lowly resolved spectrum.

Regulus (77 ly) is classified as B8 IVn and, like our Sun, as a so-called dwarf, still established on the main sequence of

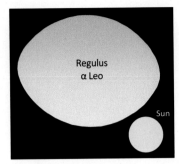

Figure 6.2 Size comparison, Regulus and the Sun, schematically after W. Huang

the HRD. It's the dominant component of a quadruple star system with an effective temperature at the poles of about 15,000 K. Its apparent rotation velocity at the equator is about 300 km s^{-1}, some 5–6 times higher, compared with Rigel and φ Sagittarii. This way the shape of Regulus looks rather like a rugby football than a sphere. Figure 6.2 shows Regulus in size compared to the Sun.

Anyway this very fast rotation is not exceptional for stars of the late B class – Albireo B (B8 Ve) for example has an apparent rotational velocity of some 215 km s^{-1} [11]. This has also an impact on the spectrum due to the so-called rotational broadening. Thereby, particularly the fine metal lines of the Regulus spectrum become additionally broadened, flattened and their intensity considerably reduced. This effect is visible well at the Mg II absorption, λ4481 (marked with red ellipse). In [3] methods are presented to estimate the rotation velocity, based on the EW and FWHM values of certain metal lines.

The comparison of these three equally normalized spectra shows a clear increase of the width and intensity of the

H-Balmer lines with decreasing luminosity. This phenomenon is called the negative luminosity effect [2]. Conversely to this, the metal lines in the giants become more intense and narrower, caused by the less dense stellar atmosphere and thus a reduced pressure and collision broadening. In this late B class the Hα line appears already well developed.

Plate 8 Detailed Spectrum of an Early B-Class Star

Spica (α Vir)

Plate 8 shows a broadband overview spectrum (200 L mm^{-1}) and higher resolved profiles (900 L mm^{-1}).

Spica (260 ly) is a spectroscopic binary with an orbital period of about four days (Section 23.5). The dominant A component is in the giant stage with an effective temperature of about 22,000 K, corresponding to the spectral type B1 III–IV [13]. The smaller B component with the spectral type B2V is still on the main sequence and with a temperature of about 18,500 K significantly less hot and luminous. Due to the Doppler effect this very close orbit of the two stars causes a periodic splitting of spectral lines (SB2 system), which is not recognizable in these profiles. For a higher resolved documentation of a full orbit by the SQUES echelle spectrograph refer to Section 23.4. Ionized helium He II is here no longer recognizable, however, there are still several absorption lines of multiply ionized metals. The line identification is based amongst others on [2], [22], [23], [27].

For comments on spectra of all B-classified stars in the Pleiades cluster (~390–460 ly) see Section 27.6.

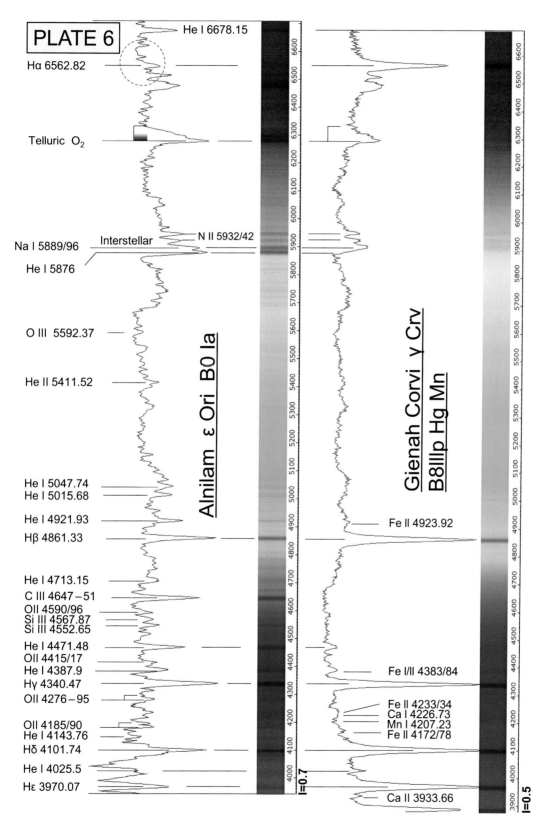

PLATE 6

He I 6678.15

Hα 6562.82

Telluric O₂

Na I 5889/96   Interstellar   N II 5932/42
He I 5876

O III 5592.37

He II 5411.52

Alnilam   ε Ori   B0 Ia

He I 5047.74
He I 5015.68

He I 4921.93
Hβ 4861.33

He I 4713.15
C III 4647−51
OII 4590/96
Si III 4567.87
Si III 4552.65

He I 4471.48
OII 4415/17
He I 4387.9
Hγ 4340.47
OII 4276−95

OII 4185/90
He I 4143.76
Hδ 4101.74

He I 4025.5
Hε 3970.07

I=0.7

Gienah Corvi γ Crv
B8IIIp Hg Mn

Fe II 4923.92

Fe I/II 4383/84

Fe II 4233/34
Ca I 4226.73
Mn I 4207.23
Fe II 4172/78

Ca II 3933.66

I=0.5

Plate 6 Development of Spectral Features within the B Class

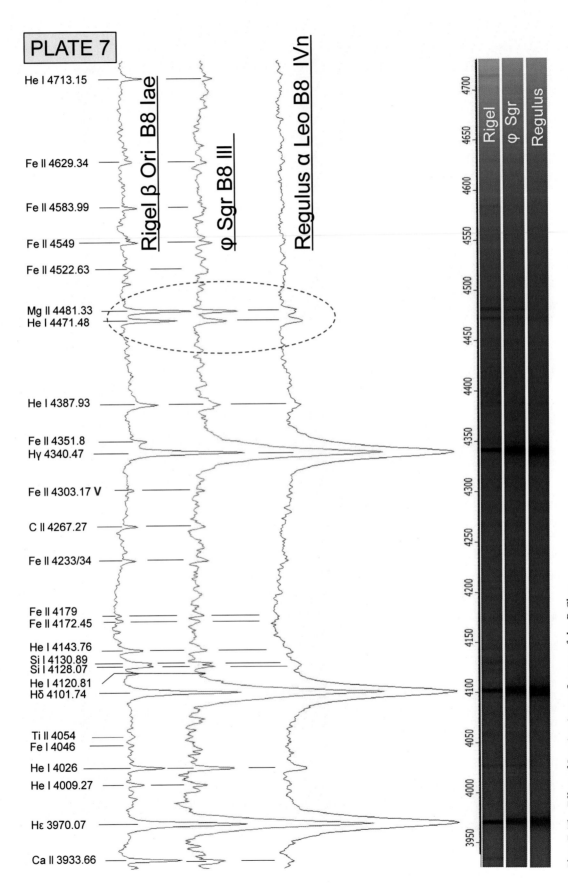

PLATE 7

He I 4713.15

Fe II 4629.34

Fe II 4583.99

Fe II 4549

Fe II 4522.63

Mg II 4481.33
He I 4471.48

He I 4387.93

Fe II 4351.8
Hγ 4340.47

Fe II 4303.17 **V**

C II 4267.27

Fe II 4233/34

Fe II 4179
Fe II 4172.45

He I 4143.76
Si I 4130.89
Si I 4128.07
He I 4120.81
Hδ 4101.74

Ti II 4054
Fe I 4046

He I 4026

He I 4009.27

Hε 3970.07

Ca II 3933.66

Rigel β Ori B8 Iae

φ Sgr B8 III

Regulus α Leo B8 IVn

Rigel

φ Sgr

Regulus

Plate 7 The Effect of Luminosity on Spectra of the B Class

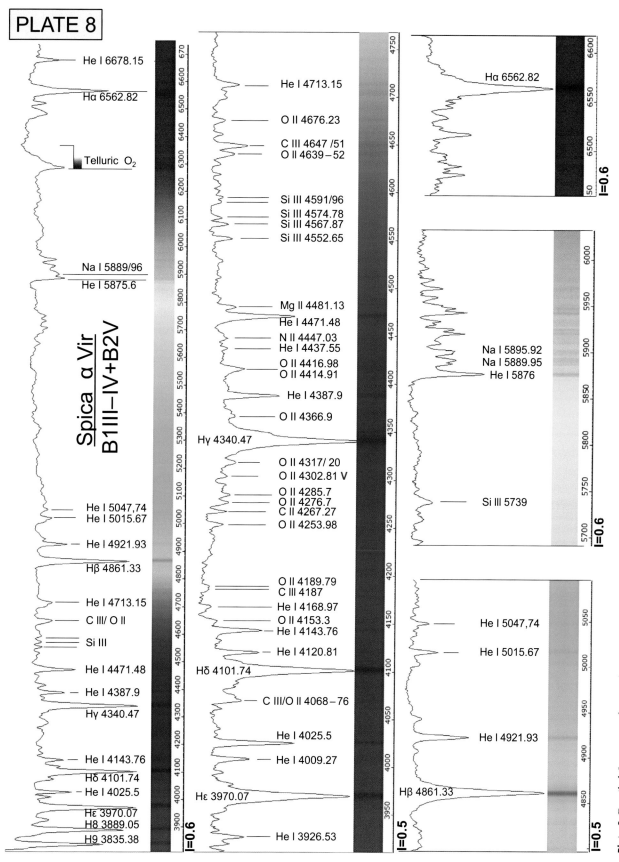

PLATE 8

He I 6678.15

Hα 6562.82

Telluric O₂

Na I 5889/96
He I 5875.6

Spica α Vir
B1III–IV+B2V

He I 5047,74
He I 5015.67

He I 4921.93

Hβ 4861.33

He I 4713.15

C III/ O II

Si III

He I 4471.48

He I 4387.9

Hγ 4340.47

He I 4143.76

Hδ 4101.74
He I 4025.5

Hε 3970.07
H8 3889.05
H9 3835.38

I=0.6

He I 4713.15

O II 4676.23

C III 4647 /51
O II 4639−52

Si III 4591/96
Si III 4574.78
Si III 4567.87
Si III 4552.65

Mg II 4481.13
He I 4471.48
N II 4447.03
He I 4437.55
O II 4416.98
O II 4414.91
He I 4387.9

O II 4366.9

Hγ 4340.47

O II 4317/ 20
O II 4302.81 V
O II 4285.7
O II 4276.7
C II 4267.27
O II 4253.98

O II 4189.79
C III 4187
He I 4168.97
O II 4153.3
He I 4143.76
He I 4120.81

Hδ 4101.74

C III/O II 4068−76

He I 4025.5

He I 4009.27

Hε 3970.07

He I 3926.53

I=0.6

Hα 6562.82

I=0.6

Na I 5895.92
Na I 5889.95
He I 5876

Si III 5739

I=0.6

He I 5047,74

He I 5015.67

He I 4921.93

Hβ 4861.33

I=0.5

I=0.5

Plate 8 Detailed Spectrum of an Early B-Class Star

CHAPTER

7 Spectral Class A

7.1 Overview

Several of the best known and most striking, bright white stars, like Sirius, Vega, Castor, Deneb, Denebola, Altair and most of the stars in the constellation Ursa Major are classified as A types. From this, and all subsequent later classes, these main sequence stars will pass at the end of their life a giant phase, combined with an impressive "farewell tour" through almost the entire HRD. They will end their lives as extremely dense white dwarfs. During this final process at least some of them will repel a photogenic planetary nebula.

7.2 Parameters of the Late to Early A-Class Stars

Table 7.1 shows the data for the late to early subclasses, exclusively for the main sequence stars, compared to the Sun ($\odot$). However, compared to the enormously broad B class, the relatively small mass range (factor ~1.5) is striking here. Nevertheless, the huge influence on the luminosity and life expectancy is impressive.

7.3 Spectral Characteristics of the A Class

Since the beginning of spectroscopy in the nineteenth century this class has fascinated by their impressive hydrogen lines, but otherwise very "tidy" and esthetically looking spectra. This was at least one reason for numerous wrong hypotheses. Father Angelo Secchi already classified these spectra in the mid nineteenth century as "Type I" (see Section 4.3). Edward Pickering designated these in his later refined system as "A-Type." Today, this class still has Pickering's "A-label," but inconspicuously within the upper middle of the MK stellar classes.

These impressive and clear spectra are very well suited as a didactic introduction to practical spectroscopy. Moreover, the pattern of the strong H-Balmer lines is an excellent aid for first calibration attempts. This feature gains intensity since the late B class and reaches its maximum at type ~A1. Quantum mechanically, this can be explained with the effective temperature of about 9800 K [3]. In addition, the H-Balmer lines are highly sensitive to the linear Stark effect. Due to the splitting of the degenerate energy levels by electric fields of neighboring ions, a massive extra broadening of these absorptions takes place [28].

From here on, the later A classes are characterized by a gradual, but still moderate fading of the H-Balmer series. Conversely, the two Fraunhofer H + K lines of ionized calcium (Ca II) get significantly stronger. The Fraunhofer K line penetrates deeper now into the continuum peak between Hε and H8 and exceeds between A7 and F0 the intensity of the hydrogen absorption. Within the same range also the growing Fraunhofer H line ($\lambda3968$) now overprints the weakening Hε absorption. In the late A classes the "blue wing" of the Hγ line shows a small kink. At the latest in the F0 class, it is revealed as the fast growing molecular CH-absorption of the Fraunhofer G band (in the graph in Figure 7.1 this and the K line are shown marked with red arrows).

Table 7.1 Data for the late to early subclasses, A-class stars

Mass $M/M_\odot$	Lifetime on Main Sequence [My]	Effective Temperature [K]	Radius $R/R_\odot$	Luminosity $L/L_\odot$
2–3	~300–2000	7500–10,500	1.7–2.7	8–55

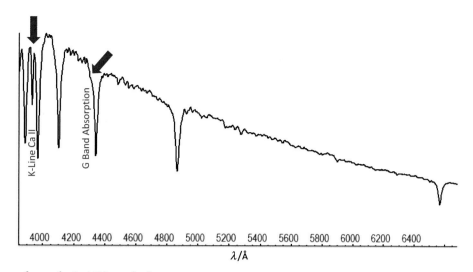

Figure 7.1 Continuum of a synthetic A5V standard star

Higher resolved spectra show immediately that the first impression of the "simple spectrum" is deceptive. The continuum between the hydrogen lines is interrupted by numerous metal absorptions. The absorptions of neutral atoms now become more intensive at the expense of the singly ionized ones. In the early subclasses, a few He I lines may still appear but just very faintly. The intensity maximum of the real continuum moves here to the "blue" short-wave edge of the visible spectrum. Figure 7.1 shows the theoretical continuum for a synthetic A5V standard star (Vspec Tools/Library). Marked with red arrows are the above-mentioned details.

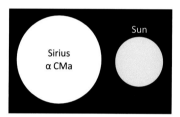

Figure 7.2 Size comparison: Sirius and the Sun

diameter is therefore enhanced by about 22% (interferometric survey in 2007 by J. Monnier *et al.* [30]).

7.4 Comments on Observed Spectra

Plate 9 Development of Spectral Features within the A Class

Castor (α Gem) and Altair (α Aql)

Plate 9 demonstrates The development within the A subclasses by a montage of two broadband overview spectra (200 L mm^{-1}), representing an early and late subtype. The line identification is generally based on [1], [2], [10], [22], [29] and SpectroWeb [9] (Omi Peg, A1 IVm and Canopus F0 II).

Castor (52 ly) is a very rare six star system. The two brightest components can be resolved even in smaller amateur telescopes. They dominate the spectrum and their early A classifications, A1V and A2Vm, are very similar [11]. The effective temperature is about 9800 K.

Altair (17 ly) is an A7Vn main sequence star of the late A class. The effective temperature is therefore correspondingly lower at 7550 K. For this late A class it shows a very high apparent rotation velocity of 210 km s^{-1} [14]. The equatorial

Plate 10 Detailed Spectrum of an Early A-Class Star

Sirius A (α CMa)

Plate 10 shows higher resolved profiles of Sirius A (8.6 ly), spectral type A1 Vm [13] with an effective temperature of about 9880 K, is similar to Vega and Regulus, a dwarf star on the main sequence of the HRD. It forms the main component of a binary system with Sirius B (white dwarf). The plate shows a broadband overview spectrum (200 L mm^{-1}) and two higher-resolved profiles in the blue/green and red wavelength domain (900 L mm^{-1}). Within the blue/green part most of the interesting metal absorptions are concentrated. They look very slim, compared to the huge H-Balmer lines. Here, as well as in the spectrum of Castor, the so-called magnesium triplet appears, still faintly, for the first time at λλ5168–5183 (Fraunhofer b). In fact Sirius belongs to the above average metal-rich stars of the Am–Fm class, which are labeled with the suffix "m" (see Section 22.4). Figure 7.2 shows Sirius compared in size to the Sun.

Plate 11 Effects of the Luminosity on Spectra of the Early A Class

Deneb (α Cyg), Ruchbah (δ Cas) and Vega (α Lyr)

Plate 11 shows the luminosity effect on spectra of the early A class (900 L mm^{-1}).

Deneb (~2000 ly) is classified with A2 Ia (sometimes also A2 Iae) and thus belongs to the supergiants. The effective temperature is about 8500 K. The apparent rotation velocity is indicated with about 39 km s^{-1} [11]. During its former time on the main sequence this supergiant was an early B or even a late O star [14].

Ruchbah (100 ly) is classified with A5 III-IV. Thus it is moving in the HRD on the way from the main sequence to the giant branch. Its effective temperature is about 8400 K and the rotation velocity 123 km s^{-1} [11], an inconspicuous value for this class.

Vega (25 ly) is classified with A0V and, like our Sun, a so-called dwarf on the main sequence of the HRD. Its effective temperature is about 9500 K its apparent rotational speed at the equator with some 15 km s^{-1} is very low. However, recent interferometrical studies show that is why we see Vega almost "pole on" and the effective rotational speed is >200 km s^{-1}.

The comparison of these three equally normalized spectra shows clearly a decrease of the intensity and width of the H lines by increasing luminosity. Conversely, the metal lines in the giant become more intense, which is expected due to the less dense stellar atmosphere and thus a lower pressure and reduced collision broadening. Due to the relatively moderate $v \sin i$ values the rotational broadening here has no dominant influence on the appearance of the profiles.

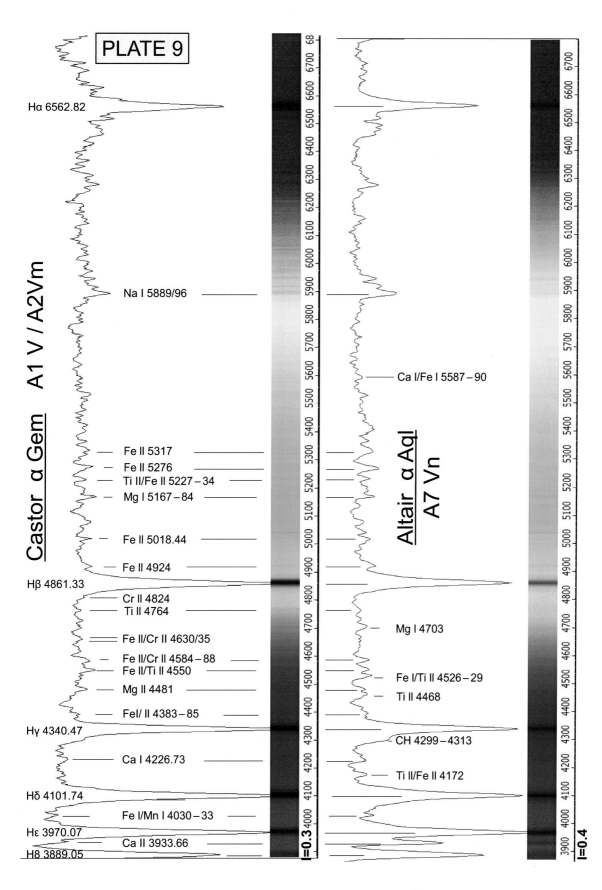

PLATE 9

Castor α Gem   A1 V / A2Vm

Hα 6562.82

Na I 5889/96

Fe II 5317
Fe II 5276
Ti II/Fe II 5227−34
Mg I 5167−84

Fe II 5018.44

Fe II 4924

Hβ 4861.33
Cr II 4824
Ti II 4764

Fe II/Cr II 4630/35

Fe II/Cr II 4584−88
Fe II/Ti II 4550

Mg II 4481

FeI/ II 4383−85

Hγ 4340.47

Ca I 4226.73

Hδ 4101.74

Fe I/Mn I 4030−33

Hε 3970.07
Ca II 3933.66
H8 3889.05

I=0.3

Altair α Aql
A7 Vn

Ca I/Fe I 5587−90

Mg I 4703

Fe I/Ti II 4526−29

Ti II 4468

CH 4299−4313

Ti II/Fe II 4172

I=0.4

Plate 9 Development of Spectral Features within the A Class

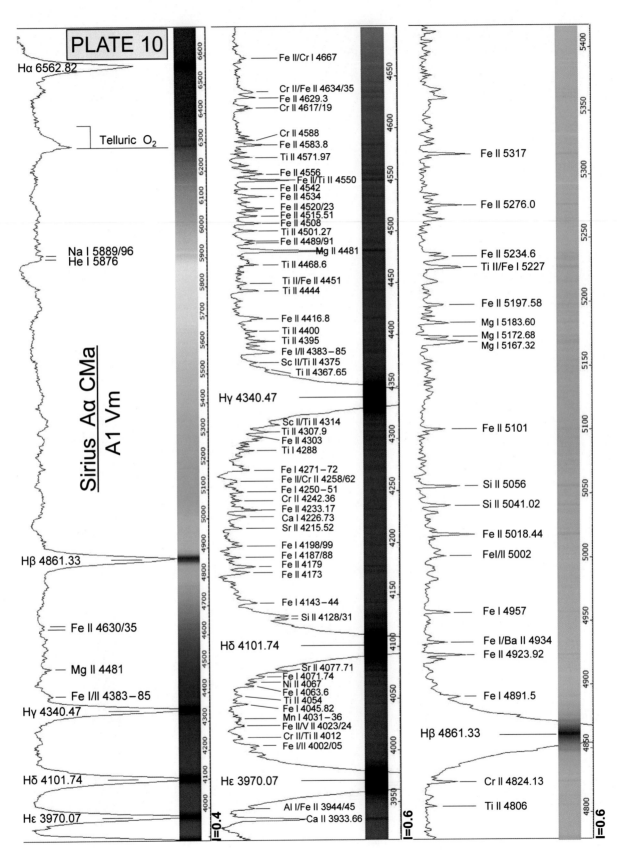

PLATE 10

Hα 6562.82

Telluric O₂

Na I 5889/96
He I 5876

Sirius Aα CMa
A1 Vm

Hβ 4861.33

Fe II 4630/35

Mg II 4481

Fe I/II 4383−85

Hγ 4340.47

Hδ 4101.74

Hε 3970.07

I=0.4

Fe II/Cr I 4667

Cr II/Fe II 4634/35
Fe II 4629.3
Cr II 4617/19

Cr II 4588
Fe II 4583.8
Ti II 4571.97

Fe II 4556
Fe II/Ti II 4550
Fe II 4542
Fe II 4534
Fe II 4520/23
Fe II 4515.51
Fe II 4508
Ti II 4501.27
Fe II 4489/91
Mg II 4481
Ti II 4468.6

Ti II/Fe II 4451
Ti II 4444

Fe II 4416.8
Ti II 4400
Ti II 4395
Fe I/II 4383−85
Sc II/Ti II 4375
Ti II 4367.65

Hγ 4340.47

Sc II/Ti II 4314
Ti II 4307.9
Fe II 4303
Ti I 4288

Fe I 4271−72
Fe II/Cr I 4258/62
Fe I 4250−51
Cr II 4242.36
Fe II 4233.17
Ca I 4226.73
Sr II 4215.52

Fe I 4198/99
Fe I 4187/88
Fe II 4179
Fe II 4173

Fe I 4143−44
Si II 4128/31

Hδ 4101.74

Sr II 4077.71
Fe I 4071.74
Ni II 4067
Fe I 4063.6
Ti II 4054
Fe I 4045.82
Mn I 4031−36
Fe II/V II 4023/24
Cr II/Ti II 4012
Fe I/II 4002/05

Hε 3970.07

Al I/Fe II 3944/45
Ca II 3933.66

I=0.6

Fe II 5317

Fe II 5276.0

Fe II 5234.6
Ti II/Fe I 5227

Fe II 5197.58
Mg I 5183.60
Mg I 5172.68
Mg I 5167.32

Fe II 5101

Si II 5056

Si II 5041.02

Fe II 5018.44

FeI/II 5002

Fe I 4957

Fe I/Ba II 4934
Fe II 4923.92

Fe I 4891.5

Hβ 4861.33

Cr II 4824.13

Ti II 4806

I=0.6

Plate 10 Detailed Spectrum of an Early A-Class Star

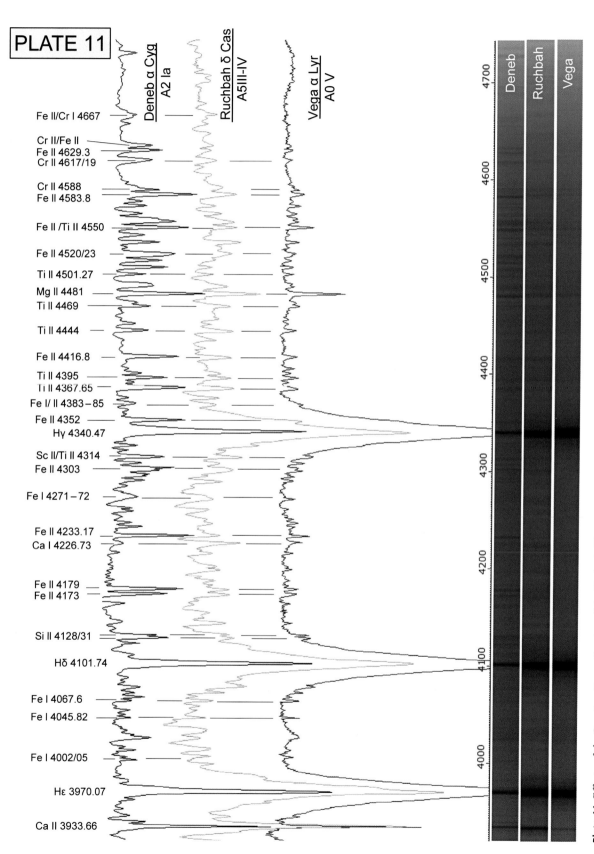

PLATE 11

Deneb α Cyg
A2 Ia

Ruchbah δ Cas
A5III-IV

Vega α Lyr
A0 V

Fe II/Cr I 4667
Cr II/Fe II
Fe II 4629.3
Cr II 4617/19
Cr II 4588
Fe II 4583.8
Fe II /Ti II 4550
Fe II 4520/23
Ti II 4501.27
Mg II 4481
Ti II 4469
Ti II 4444
Fe II 4416.8
Ti II 4395
Ti II 4367.65
Fe I/ II 4383−85
Fe II 4352
Hγ 4340.47
Sc II/Ti II 4314
Fe II 4303
Fe I 4271−72
Fe II 4233.17
Ca I 4226.73
Fe II 4179
Fe II 4173
Si II 4128/31
Hδ 4101.74
Fe I 4067.6
Fe I 4045.82
Fe I 4002/05
Hε 3970.07
Ca II 3933.66

Deneb
Ruchbah
Vega

4700
4600
4500
4400
4300
4200
4100
4000

Plate 11 Effects of the Luminosity on Spectra of the Early A Class

CHAPTER

8

Spectral Class F

8.1 Overview

The F class is located directly above the G category, where our Sun is classified. It includes several well-known, bright yellow shining stars like Procyon, Caph (β Cas), Porrima (γ Vir), Mirfak (α Per), and the Pole Star. On the giant branch of the HRD we find here several pulsation variables, belonging to the categories of δ Cephei and RR Lyrae.

8.2 Parameters of the Late to Early F-Class Stars

Table 8.1 shows the data for the late to early subclasses, exclusively for the main sequence stars, compared to the Sun (⊙).

8.3 Spectral Characteristics of the F Class

The H-Balmer lines are now much weaker and the Fraunhofer H + K lines (Ca II) become the dominant features, so that the Fraunhofer H absorption now clearly displaces the Hε line. Towards the late subclasses the neutral elements, for example Fe I, Cr I, now increasingly replace the absorption of the ionized states. The Ca I line at λ4227 clearly intensifies, as well as the G band (CH molecular) which surpasses within the F class the intensity of the neighboring Hγ line. This

striking "line double" can therefore exclusively be seen here, so it forms essentially the unmistakable "brand" of the F class! Compared to the A class the magnesium triplet (λλ5168–83) becomes stronger here.

The intensity maximum of the real continuum is now clearly located within the visible range of the spectrum. Figure 8.1 shows the theoretical continuum for a synthetic F5 V standard star (Vspec/Tools/Library [7]). Highlighted with a red circle is here the "line-double" of the G band and Hγ, the striking "brand" of the F class.

8.4 Comments on Observed Spectra

Plate 12 Development of Spectral Features within the F Class

Adhafera (ζ Leo) and Procyon A (α CMi)

Plate 12 demonstrates The development of the F subclasses by a montage of two broadband overview spectra (200 L mm^{-1}) representing an early and middle subtype. The line identification is here generally based on [1], [2], [10], [22], and Spectro-Web [9] (Canopus F0II, Procyon F5 IV–V).

Adhafera (260 ly), classified as F0 III, is located at the top of the F class and has already reached the giant stage (III). The effective temperature is about 7030 K and its apparent

Table 8.1 Data for the late to early subclasses, F class stars

Mass $M/M_\odot$	Lifetime on Main Sequence [Gy]	Effective Temperature [K]	Radius $R/R_\odot$	Luminosity $L/L_\odot$
1.1–1.6	~2–7	6200–7500	1.2–1.6	2.0–6.5

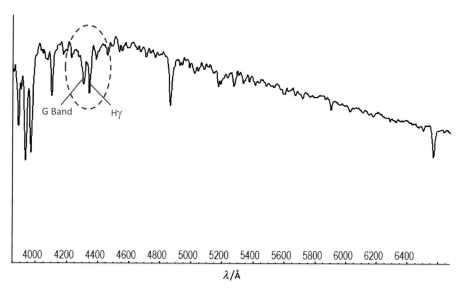

Figure 8.1 Continuum of a synthetic F5V standard star

rotation speed some 72 km s^{-1} [11]. Even within this early F class, the Fraunhofer H and K lines have already surpassed the intensity of the H-Balmer series, where the K line at λ3968 has overprinted the Hε absorption.

Procyon A (11 ly) is classified as F5 IV–V and thus representing approximately the "center" of the F class. Like Sirius, it has a white dwarf companion. The luminosity class IV–V reveals that it has begun to leave the main sequence in the HRD and is moving towards the giant branch. The apparent rotation speed is only ~5 km s^{-1}. The effective temperature of about 6630 K is as expected lower than for the earlier classified Adhafera. At the first glance this has little effect on the spectral profile.

The most noticeable and important difference relates to the "brand" of the F class, the increasing intensity of the CH molecular absorption band (at λ4300) and conversely the shrinking of the neighboring Hγ line. Further, the growing intensity of the Ca I line (λ4227), at this resolution clearly visible only by the somewhat colder Procyon. Some further differences may be caused by the different luminosity classes.

Plate 13 Effect of the Luminosity on Spectra of the F Class

Mirfak (α Per), Caph (β Cas) and Porrima (γ Vir)

Mirfak (600 ly) is classified with F5 Ib and therefore one of the supergiants. The effective temperature is about 6180 K. The apparent rotation velocity is reported to be ~18 km s^{-1}. Mirfak is a former main sequence star of the B class [14].

Caph (55 ly) is a giant, classified as F2 III. Its effective temperature is about 6700 K and its apparent rotation speed 71 km s^{-1}.

Porrima (39 ly) is a detached binary system, consisting of two identical classified components F0 IV [11]. The effective temperature is about 7100 K.

The comparison of these three equally normalized spectra shows no striking differences between the luminosity classes. A difference regarding the intensity and width of the H lines is here not recognizable purely visually. The metal lines of the supergiant Mirfak are somewhat more intensive than in the profile of the giant Caph. Between Caph and the main sequence star Porrima differences are hardly discernible.

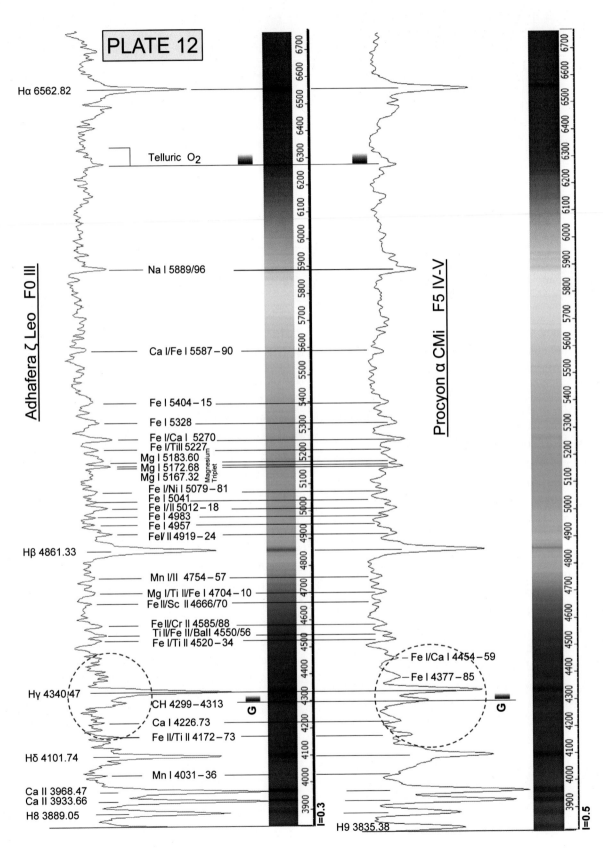

PLATE 12

Adhafera ζ Leo   F0 III

Procyon α CMi   F5 IV-V

Hα 6562.82

Telluric O₂

Na I 5889/96

Ca I/Fe I 5587−90

Fe I 5404−15
Fe I 5328
Fe I/Ca I 5270
Fe I/TiII 5227
Mg I 5183.60
Mg I 5172.68   Magnesium Triplet
Mg I 5167.32
Fe I/Ni I 5079−81
Fe I 5041
Fe I/II 5012−18
Fe I 4983
Fe I 4957
FeI/ II 4919−24

Hβ 4861.33

Mn I/II 4754−57
Mg I/Ti II/Fe I 4704−10
Fe II/Sc II 4666/70

Fe II/Cr II 4585/88
Ti II/Fe II/BaII 4550/56
Fe I/Ti II 4520−34

Fe I/Ca I 4454−59
Fe I 4377−85

Hγ 4340.47
CH 4299−4313   G
Ca I 4226.73
Fe II/Ti II 4172−73

Hδ 4101.74
Mn I 4031−36

Ca II 3968.47
Ca II 3933.66
H8 3889.05

H9 3835.38

G

I=0.3

I=0.5

Plate 12 Development of Spectral Features within the F Class

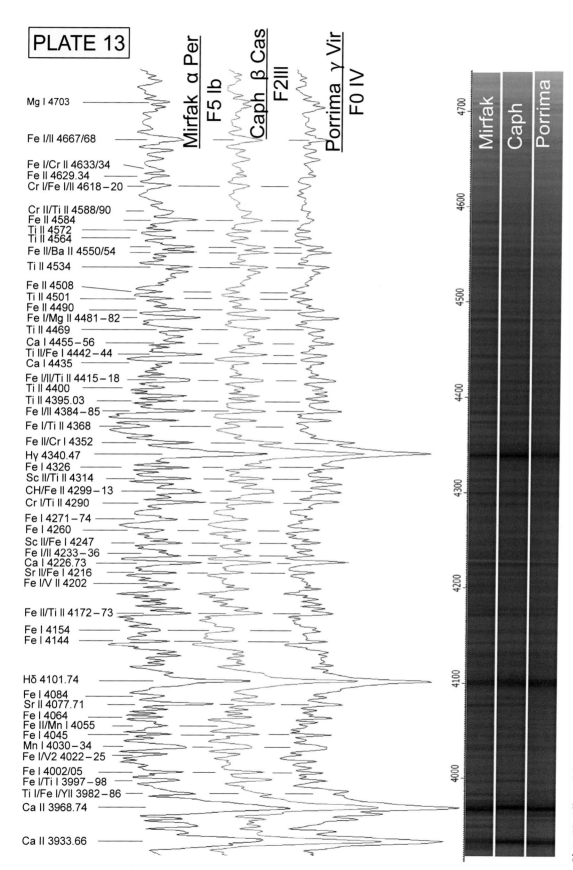

Plate 13 Effect of the Luminosity on Spectra of the F Class

CHAPTER

9 Spectral Class G

9.1 Overview

From a spectroscopic view, the yellow shining stars of the G class have a special status, because their spectra are more or less similar to that of our Sun, which is probably one of the best explored and documented stars. α Centauri in the southern sky as well as 18 Scorpii ($m_v = 5.5$), also called the "solar twin," are with G2V equally classified as the Sun. They show therefore nearly the same effective temperature as our central star. Further, in the northern sky also Muphrid (η Boo) with G0 IV is classified relatively closely to the Sun. Otherwise, among the bright stars no further similarly classified can be found here. Capella (α Aur) with its two binary components of G8III and G0III is already settled on the giant branch of the HRD. The same applies to Sadalsuud (β Aqr) with G0Ib. Other well-known G stars can only be found in the classes G7 and later, such as Kornephoros (β Her), γ Leo, γ Per, δ Boo, Vindemiatrix (ε Vir).

9.2 Parameters of the Late to Early G-Class Stars

Table 9.1 shows the data for the late to early subclasses, exclusively for the main sequence stars, compared to the Sun (⊙).

The very low mass range covered by the G class is striking. Nevertheless, all other parameters of the star respond amazingly sensitively to this difference. Our Sun with an effective temperature of about 5800 K (G2V), belongs to the early G class, and will spend some 10 billion years on the main sequence.

9.3 Spectral Characteristics of the G Class

The Fraunhofer H + K lines of ionized Ca II become here impressively strong achieving theoretically the maximum intensity within the late G classes. This also becomes evident on Plates 1 and 2. In the solar spectrum (G2V) they are by far the strongest lines generated by the star itself. For main sequence stars of the G class, the K line is always slightly more intense than the H line.

The H-Balmer series becomes significantly weaker, so these lines are now surpassed even by various metal absorptions. Thus from here on they lose their function as welcome orientation marks, for example for the calibration and line identification. The intensity of the so-called magnesium triplet ($\lambda\lambda5167, -73, -84$) has increased during the F class and achieves a considerable strength here, so with "b" it's even labeled with its own Fraunhofer letter. The Ca I line at $\lambda4227$ has also impressively gained intensity since the early F classes and becomes here a striking spectral feature also denoted with its own Fraunhofer letter "g."

In general, the trend here continues by growing intensity of neutral metals for example Fe I and the Fraunhofer D lines (Na I). Towards the later subclasses they increasingly replace the absorption of the ionized elements. Due to the dominance of

Table 9.1 Data for the late to early subclasses, G-class stars

Mass $M/M_\odot$	Lifetime on Main Sequence [Gy]	Effective Temperature [K]	Radius $R/R_\odot$	Luminosity $L/L_\odot$
0.9–1.05	7–20	5000–6200	0.85–1.1	0.66–1.5

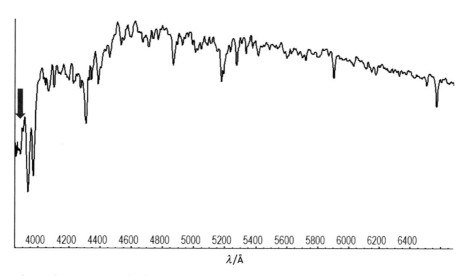

Figure 9.1 Continuum of a synthetic G5V standard star

fine metal lines, the spectra become now more and more complex. Therefore, our Sun is not really a suitable object for complete beginners.

The temperature has dropped here to such a low level that simple and robust diatomic molecules can survive in these stellar atmospheres. Most prominent of such features is the Fraunhofer G band of the CH molecule, which already surpassed in the late F class the intensity of the Hγ line. Further mentionable are also the strong CN and CH absorption bands in the violet spectral range, overprinting here the H8 and H9 Balmer lines beyond recognition. In highly resolved spectra even finer absorption bands of carbon monoxide CO may appear.

The intensity maximum of the real continuum shifts now more towards the green part of the visible spectrum, hence evolution has optimized the sensitivity of the human eyes to this wavelength domain (Sun G2V). Figure 9.1 shows the theoretical continuum for a synthetic G5V standard star (Vspec/Tools/Library [7]). Highlighted with a red arrow is the area on the violet side of the H+K Fraunhofer lines, where strong molecular CN and CH absorption bands overprint the former H8- and H9-Balmer lines.

Muphrid (37 ly) with G0 IV has already moved away from the main sequence towards the giant branch of the HRD. Of all brighter stars in the northern sky, it is the closest "class neighbor" to the Sun (G2V). Its effective temperature is about 6100 K, for instance slightly hotter than our central star – hence the earlier classification. It is probably a part of a spectroscopic binary (not yet confirmed). Muphrid is classified by several sources as above average metal rich. Nevertheless, the suffix "m" is missing here in the classification [11].

Vindemiatrix (103 ly) with G8 III is a late representative of the G class and has already developed to the giant stage. Its apparent rotation speed is ~0 km s^{-1} [11]. According to [14] Vindemiatrix was formerly a B-class main sequence star.

The comparison of these spectra demonstrates mainly the fading of the H-Balmer lines, well visible at the Hβ line. The profile of Vindemiatrix shows also a massively shrunken Hγ line, which is now difficult to identify, particularly in the immediate neighborhood of the dominant G band (CH). Otherwise, no spectacular changes can be seen. However, in highly resolved spectra, significant intensity differences of individual lines would be detectable, particularly within the blends.

9.4 Comments on Observed Spectra

Plate 14 Development of Spectral Features within the G Class

Muphrid (η Boo) and Vindemiatrix (ε Vir)
Plate 14 demonstrates the development of the G subclasses by a montage of two superposed broadband overview spectra (200 L mm^{-1}), representing an early and late subtype.

Plates 15 and 16 Detailed Spectra of an Early G-Class Star

The Sun, Spectra Recorded from Reflected Daylight
Our central star (G2V), with an effective temperature of ~5800 K, is a normal dwarf on the main sequence. On **Plates 15 and 16** the Sun is documented with a broadband overview spectrum (200 L mm^{-1}) and three higher-resolved

Figure 9.2 The limb of the Sun in Hα light
(Credit: Urs Fluekiger)

profiles in the blue, green and red domain (900 L mm^{-1}). The Sun has a very low rotational velocity of slightly below 2 km s^{-1}, a value which is usual for middle and late spectral types. The increased thickness of the convective shell generates strong magnetic fields with a considerable breaking effect on the stellar rotation. Figure 9.2 shows the limb of the Sun with prominences recorded in the Hα light by Urs Fluekiger (using a refractor 150 mm/f15, Daystar Quantum). The line identification is based here mainly on SpectroWeb [9] and [31], [32].

Plate 17 Highly Resolved Fraunhofer Lines in the Solar Spectrum

As a supplement to Plates 15 and 16, **Plate 17** presents details of the Fraunhofer lines D, E, b, F, f, G, H, K with high-resolution spectra, recorded by the SQUES echelle spectrograph. The corresponding profiles of the A, B, and C (Hα) lines are commented on in Plates 81 and 82. The D1 and D2 lines are surrounded by many telluric H_2O water vapor absorptions. The E line appears here highly resolved as two separate Fe absorptions. At this resolution, the magnesium triplet turns out to consist of four sublines b_1–b_4. The G line is mainly generated by a blend of the CH absorption in the range ~ λλ4308–4313 and the Fe I lines at λ4308 and λ4315. The line identification and all wavelengths are based here on SpectroWeb [9].

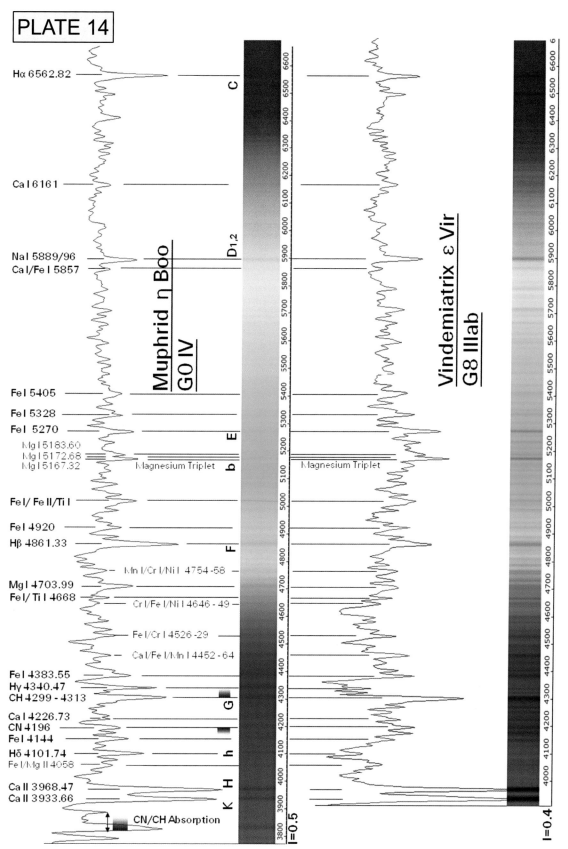

PLATE 14

Hα 6562.82

Ca I 6161

Na I 5889/96
Ca I/Fe I 5857

Muphrid η Boo
G0 IV

Fe I 5405
Fe I 5328
Fe I 5270
Mg I 5183.60
Mg I 5172.68
Mg I 5167.32
Magnesium Triplet

Fe I/ Fe II/Ti I

Fe I 4920
Hβ 4861.33

Mn I/Cr I/Ni I 4754-58

Mg I 4703.99
Fe I/ Ti I 4668
Cr I/Fe I/Ni I 4646 - 49

Fe I/Cr I 4526 -29

Ca I/Fe I/Mn I 4452 - 64

Fe I 4383.55
Hγ 4340.47
CH 4299 - 4313

Ca I 4226.73
CN 4196
Fe I 4144
Hδ 4101.74
Fe I/Mg II 4058

Ca II 3968.47
Ca II 3933.66

CN/CH Absorption

I=0.5

C

D₁,₂

E
b

F

G

h

H

K

Vindemiatrix ε Vir
G8 IIIab

Magnesium Triplet

I=0.4

Plate 14 Development of Spectral Features within the G Class

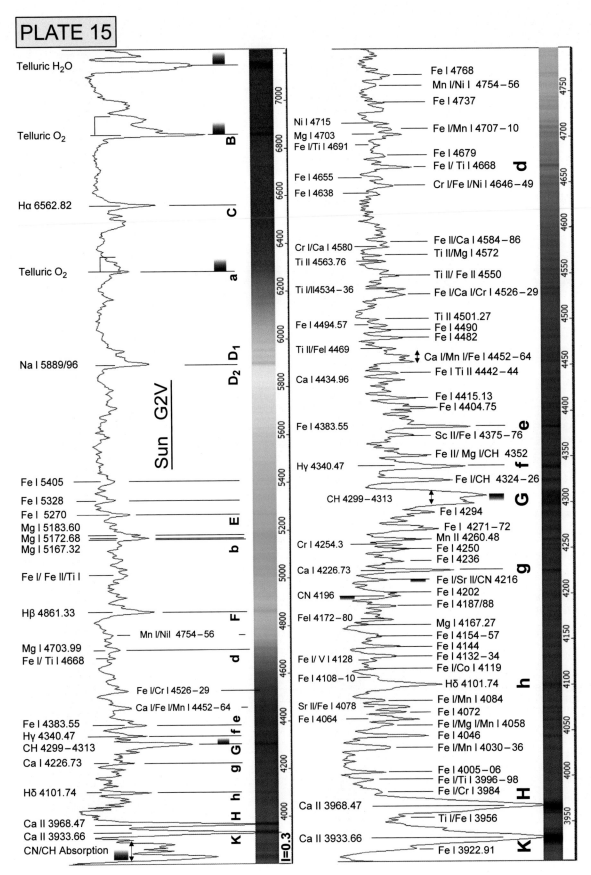

PLATE 15

Telluric H₂O

Telluric O₂

Hα 6562.82

Telluric O₂

Na I 5889/96

Sun G2V

Fe I 5405
Fe I 5328
Fe I 5270
Mg I 5183.60
Mg I 5172.68
Mg I 5167.32

Fe I/ Fe II/Ti I

Hβ 4861.33

Mn I/NiI 4754–56

Mg I 4703.99
Fe I/ Ti I 4668

Fe I/Cr I 4526–29

Ca I/Fe I/Mn I 4452–64

Fe I 4383.55
Hγ 4340.47
CH 4299–4313
Ca I 4226.73

Hδ 4101.74

Ca II 3968.47
Ca II 3933.66
CN/CH Absorption

I=0.3

Ni I 4715
Mg I 4703
Fe I/Ti I 4691

Fe I 4655

Fe I 4638

Cr I/Ca I 4580
Ti II 4563.76

Ti I/II 4534–36

Fe I 4494.57

Ti II/FeI 4469

Ca I 4434.96

Fe I 4383.55

Hγ 4340.47

CH 4299–4313

Cr I 4254.3

Ca I 4226.73

CN 4196

FeI 4172–80

Fe I/ V I 4128

Fe I 4108–10

Sr II/Fe I 4078
Fe I 4064

Ca II 3968.47

Ca II 3933.66

Fe I 4768
Mn I/Ni I 4754–56
Fe I 4737
Fe I/Mn I 4707–10
Fe I 4679
Fe I/ Ti I 4668
Cr I/Fe I/Ni I 4646–49

Fe II/Ca I 4584–86
Ti II/Mg I 4572
Ti II/ Fe II 4550
Fe I/Ca I/Cr I 4526–29
Ti II 4501.27
Fe I 4490
Fe I 4482
Ca I/Mn I/Fe I 4452–64
Fe I Ti II 4442–44
Fe I 4415.13
Fe I 4404.75
Sc II/Fe I 4375–76
Fe II/ Mg I/CH 4352
Fe I/CH 4324–26
Fe I 4294
Fe I 4271–72
Mn II 4260.48
Fe I 4250
Fe I 4236
Fe I/Sr II/CN 4216
Fe I 4202
Fe I 4187/88
Mg I 4167.27
Fe I 4154–57
Fe I 4144
Fe I 4132–34
Fe I/Co I 4119
Hδ 4101.74
Fe I/Mn I 4084
Fe I 4072
Fe I/Mg I/Mn I 4058
Fe I 4046
Fe I/Mn I 4030–36
Fe I 4005–06
Fe I/Ti I 3996–98
Fe I/Cr I 3984
Ti I/Fe I 3956
Fe I 3922.91

Plates 15 and 16 Detailed Spectra of an Early G-Class Star

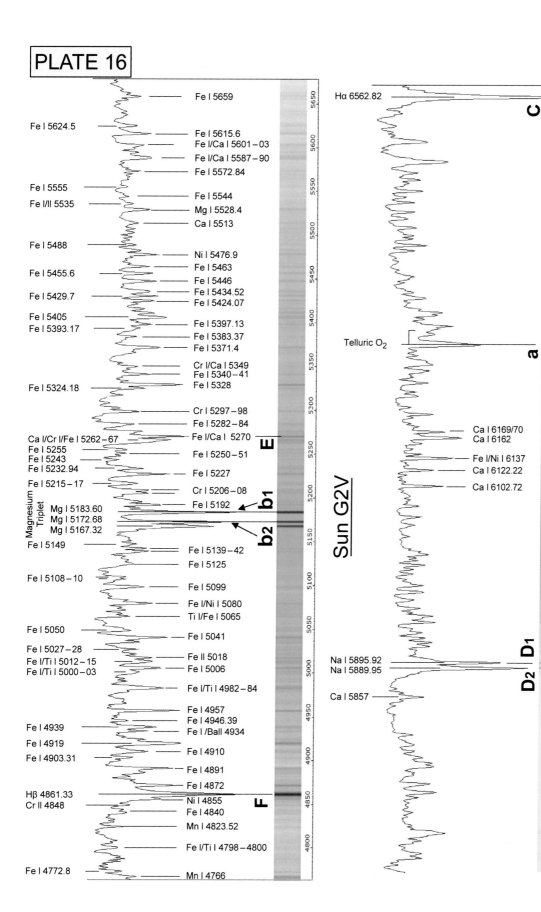

PLATE 16

Fe I 5659
Fe I 5624.5
Fe I 5615.6
Fe I/Ca I 5601–03
Fe I/Ca I 5587–90
Fe I 5572.84
Fe I 5555
Fe I 5544
Fe I/II 5535
Mg I 5528.4
Ca I 5513
Fe I 5488
Ni I 5476.9
Fe I 5463
Fe I 5455.6
Fe I 5446
Fe I 5429.7
Fe I 5434.52
Fe I 5424.07
Fe I 5405
Fe I 5397.13
Fe I 5393.17
Fe I 5383.37
Fe I 5371.4
Cr I/Ca I 5349
Fe I 5340–41
Fe I 5324.18
Fe I 5328
Cr I 5297–98
Fe I 5282–84
Ca I/Cr I/Fe I 5262–67
Fe I/Ca I 5270   **E**
Fe I 5255
Fe I 5250–51
Fe I 5243
Fe I 5232.94
Fe I 5227
Fe I 5215–17
Cr I 5206–08
Magnesium
Triplet   Mg I 5183.60   Fe I 5192   **b₁**
           Mg I 5172.68
           Mg I 5167.32   **b₂**
Fe I 5149
Fe I 5139–42
Fe I 5125
Fe I 5108–10
Fe I 5099
Fe I/Ni I 5080
Ti I/Fe I 5065
Fe I 5050
Fe I 5041
Fe I 5027–28
Fe I/Ti 5012–15
Fe II 5018
Fe I/Ti 5000–03
Fe I 5006
Fe I/Ti I 4982–84
Fe I 4957
Fe I 4946.39
Fe I 4939
Fe I /Ba II 4934
Fe I 4919
Fe I 4910
Fe I 4903.31
Fe I 4891
Fe I 4872
Hβ 4861.33
Ni I 4855   **F**
Cr II 4848
Fe I 4840
Mn I 4823.52
Fe I/Ti I 4798–4800
Fe I 4772.8
Mn I 4766

5650
5600
5550
5500
5450
5400
5350
5300
5250
5200
5150
5100
5050
5000
4950
4900
4850
4800

Hα 6562.82   **C**

Telluric O₂   **a**

Sun G2V

Ca I 6169/70
Ca I 6162
Fe I/Ni I 6137
Ca I 6122.22
Ca I 6102.72

Na I 5895.92   **D₁**
Na I 5889.95   **D₂**
Ca I 5857

6550
6500
6450
6400
6350
6300
6250
6200
6150
6100
6050
6000
5950
5900
5850
5800
5750
5700
5650

Plates 15 and 16 (*cont.*)

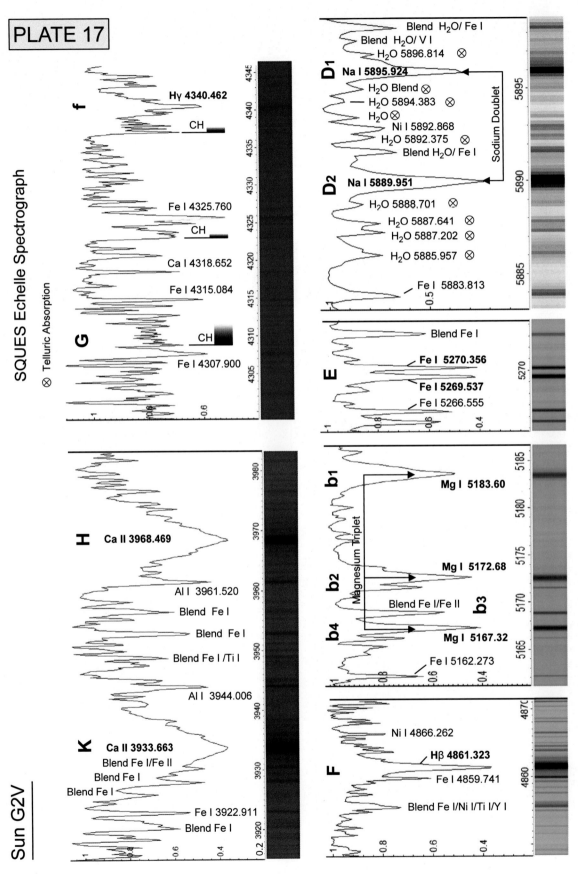

Plate 17 Highly Resolved Fraunhofer Lines in the Solar Spectrum

CHAPTER

10 Spectral Class K

10.1 Overview

The stars of the orange-yellow shining K class include familiar names such as Pollux, Aldebaran, Arcturus, Hamal (α Ari), Alphard (α Hya), as well as the brighter, yellow-orange component of the famous double star Albireo A.

10.2 Parameters of the Late to Early K-Class Stars

Table 10.1 shows the data for the late to early subclasses, exclusively for the main sequence stars, compared to the Sun ($\odot$).

The low mass range which is covered here by this class is also striking. The expected duration for all K-class dwarfs on the main sequence is longer than the estimated age of the Universe which is some 13.7 billion years. Therefore, in the entire Universe not one K star has yet migrated from the main sequence to the giant branch! Further, the luminosity of such K dwarfs is so low that they are only visible at relatively short distances. Only a few real main sequence stars of this class are visible with the naked eye. In the northern sky the best-known is probably the binary 61 Cygni, which is only 11 ly distant, with a total apparent brightness of just m_v = 4.8 and the spectral classes K5V and K7V. Therefore, all above listed, bright "highlights" of the K class have been classified much earlier during their former journey on the main sequence. Now they are just temporarily located here on the giant branch, spending a relatively short time here before they go supernova or end up as a white dwarf. Therefore, with exception of the approximate effective temperature, their parameters differ totally from the values for K main sequence stars, presented in Table 10.1.

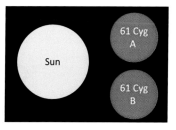

Figure 10.1 Size comparison: binary components of 61 Cygni in contrast to the Sun

Table 10.1 Data for the late to early subclasses, K-class stars

Mass $M/M_\odot$	Lifetime on Main Sequence [Gy]	Effective Temperature [K]	Radius $R/R_\odot$	Luminosity $L/L_\odot$
0.6–0.9	>20	3600–5000	0.65–0.8	0.10–0.42

10.3 Spectral Characteristics of the K Class

The temperature of this class is considerably lower than in the G category, but still approximately as high as within the solar sunspots. Correspondingly, the spectra of the early K class look still very similar as in the whole G category. Figure 10.2 (900 L mm^{-1}) shows in the blue short-wave domain the superimposed profiles of Pollux (blue) and the Sun (green). Fraunhofer himself noticed the striking similarity of the two spectra.

At a glance, the two profiles look indeed very similar, although Pollux (K0 III) is a giant and the Sun (G2V) is still established on the main sequence. Apparently the trend continues that the luminosity-related differences in the profiles become increasingly smaller towards later spectral classes. Figure 10.3 shows a similar comparison between the highly resolved profile cores of the H and K lines of the Sun and Arcturus (SQUES echelle spectrograph). In contrast to the Sun, the profile of the K-type giant (K0 III) typically shows small and slightly asymmetrical double peak emissions in the cores of both lines. The extent of the asymmetry provides information about the differential outflow velocities of the wind within the chromosphere [33]. These features are also used as long term indicators for investigation on periods of variability, mainly connected with magnetic activity. Further, the width of the emission core in the K line allows estimating the absolute magnitude for giant stars of the spectral classes G, K and M by the Wilson–Bappu effect [3].

Within the G and early K subclasses apparently no spectacular changes of spectral characteristics take place. However, considered more in detail, several differences in the line intensity and in the shape of the continuum become recognizable. In the K class the spectral lines are mainly due to neutral atoms or simple diatomic molecules. The H-Balmer series is now very weak and difficult to identify, except for the Hα and the considerably shrunken Hβ line. The significant

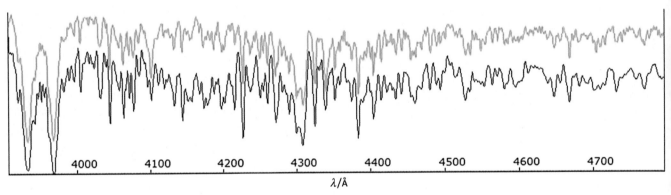

Figure 10.2 Similarities in the spectra of the Sun G2V (green) and Pollux K0 III (blue)

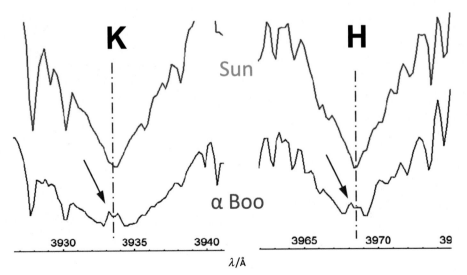

Figure 10.3 Double peak emission in the cores of the H and K lines in the spectrum of Arcturus (blue) due to an outflowing wind from the chromosphere of the giant. At the Sun (green profile) this feature is lacking

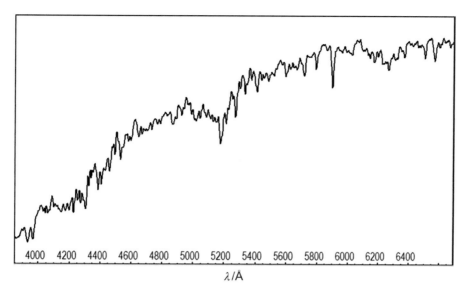

Figure 10.4 Continuum of a synthetic K4 III standard star

intensity loss of the Ca II H and K lines between the classes K0 and K5 is also striking. The magnesium triplet remains still quite intensive.

Within the spectra of the later K subclasses a remarkable break occurs and the similarity to the solar spectrum becomes increasingly lost. Since the class K5 and later, particularly in the long-wave (red) part of the spectrum, titanium monoxide (TiO) bands appear and start to overprint the telluric H_2O and O_2 absorptions. Father Angelo Secchi noticed Aldebaran (K5 III) as a transition star because of the appearance of the impressive TiO band absorptions. The intensity maximum of the real continuum is shifted here into the red region of the visible spectral range. Figure 10.4 shows the theoretical continuum for a synthetic K4 III standard star (Vspec/Tools/Library [7]).

10.4 Comments on Observed Spectra

The line identification is based on SpectroWeb [9] and [2], [10], [29], [34], [35], [36], [37].

Plate 18 Development of Spectral Features within the K Class

Arcturus (α Boo) and Alterf (λ Leo)

Plate 18 demonstrates the development of the K subclasses by a montage of two broadband overview spectra (200 L mm^{-1}) representing an early and a rather late subtype.

Arcturus (37 ly), classified with K0 III [11], is an early K star established on the giant branch of the HRD. However, currently mostly the classification K 1.5 III pe is applied. The

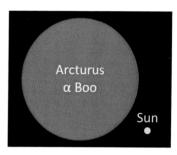

Figure 10.5 Size comparison: Arcturus and the Sun

effective temperature is about 4290 K. It exhibits a relative underabundance of iron. Compared to the Sun the Fe/H ratio of Arcturus is only about 20% [14]. The apparent rotational velocity is 4.2 km s^{-1} [11]. Like Pollux, the spectrum of Arcturus is also very similar to that of the Sun. The main differences also consist of the intensity and shape of individual lines.

Alterf (320 ly) is a K4.5 III giant of the rather late K class. Its effective temperature is 3950 K, some hundred kelvin lower than Arcturus. The apparent rotation speed is <17 km s^{-1} [11]. Particularly the blue, short-wave side of the Alterf profile is still dominated by discrete absorption lines, which are mostly the same, as seen in the spectra of Pollux and Arcturus. However, the shape of the continuum in the yellow-red range is already similar to that of Betelgeuse and Antares (see Plate 22), albeit much less intense. This shows the beginning of the influence of the molecular TiO absorption at this spectral class, already some TiO absorption can be safely identified. The molecular G band remains still recognizable here. The intensity of the neutral metals continues to increase and in the class K5 the neutral calcium Ca I at λ4227 has exceeded the molecular G Band (detail marked in red). The latter now becomes very faint and just after K5 it becomes split into several discrete lines.

Plate 19 Detailed Spectrum of an Early K-Class Star

Pollux (β Gem)

Plate 19 shows Pollux (34 ly) classified as K0 III, an early K star already established on the giant branch of the HRD. Its effective temperature is about 4770 K and the apparent rotation velocity 2.8 km s^{-1} [11]. On this plate Pollux is documented with a broadband overview spectrum (200 L mm^{-1}) and a higher resolved profile in the blue area (900 L mm^{-1}).

Plate 20 Detailed Spectrum of a Late K-Class Star

Aldebaran (α Tau)

Complementary to Plate 19, **Plate 20** shows some higher-resolved spectra of the K star Aldebaran (66 ly) in the green and red spectral domain (900 L mm^{-1}). Aldebaran, with K5 III, represents the later K class, is established on the giant branch of the HRD and classified equally with Alterf. The effective temperature of Aldebaran is about 4010 K, the apparent rotation velocity 4.3 km s^{-1}. In the range of ∼ λ6300 are some

titanium monoxide absorptions, which, however, are barely visible in the vicinity of the still numerous discrete lines (compare with Alterf, Plate 18).

Plate 21 Luminosity Effects on Spectra of the Late K Class

Alsciaukat (α Lyncis) and 61 Cygni B

Plate 21 shows the effect of the luminosity on spectra of the late K class.

Alsciaukat (220 ly) is classified as K7 III [11] and thus a normal giant. Its effective temperature is about 3860 K and the apparent rotational velocity 6.4 km s^{-1} [11]. Its slightly variable brightness suggests that it stands just before a long-period, Mira-type variable stage [14].

61 Cygni B (11 ly) is the weaker of two binary components. With K7 V it is classified as an ordinary dwarf on the main sequence. The effective temperature is some 4120 K, its apparent rotational velocity 3.2 km s^{-1} [11].

The comparison of these spectra shows here a similar profile shape for the different luminosity classes. In each section, however, clear differences and even different lines are visible. The spectrum of Alsciaukat shows at this resolution clearly the splitting of the former Fraunhofer G band into three discrete absorption lines in: Ti I at λλ4301, 4314 and Fe I at λ4308 – according to BSA [10].

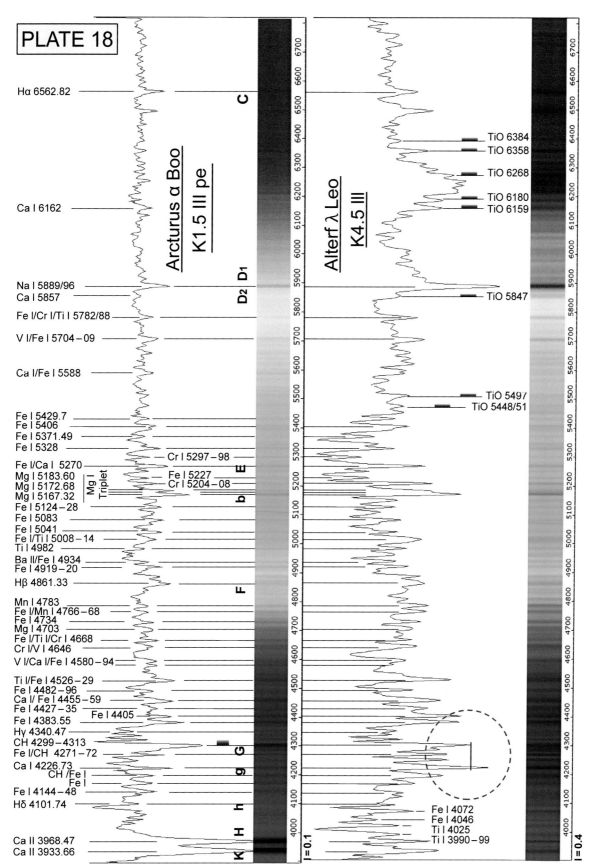

PLATE 18

Hα 6562.82

Ca I 6162

Na I 5889/96
Ca I 5857

Fe I/Cr I/Ti I 5782/88

V I/Fe I 5704—09

Ca I/Fe I 5588

Fe I 5429.7
Fe I 5406
Fe I 5371.49
Fe I 5328

Fe I/Ca I 5270
Mg I 5183.60
Mg I 5172.68
Mg I 5167.32
Fe I 5124—28
Fe I 5083
Fe I 5041
Fe I/Ti I 5008—14
Ti I 4982
Ba II/Fe I 4934
Fe I 4919—20

Hβ 4861.33

Mn I 4783
Fe I/Mn I 4766—68
Fe I 4734
Mg I 4703
Fe I/Ti I/Cr I 4668
Cr I/V I 4646
V I/Ca I/Fe I 4580—94

Ti I/Fe I 4526—29
Fe I 4482—96
Ca I/ Fe I 4455—59
Fe I 4427—35
Fe I 4383.55
Hγ 4340.47
CH 4299—4313
Fe I/CH 4271—72
Ca I 4226.73
CH /Fe I
Fe I
Fe I 4144—48
Hδ 4101.74

Ca II 3968.47
Ca II 3933.66

Arcturus α Boo
K1.5 III pe

Cr I 5297—98

Fe I 5227
Cr I 5204—08

Mg I
Triplet

Fe I 4405

Alterf λ Leo
K4.5 III

TiO 6384
TiO 6358

TiO 6268

TiO 6180
TiO 6159

TiO 5847

TiO 5497
TiO 5448/51

Fe I 4072
Fe I 4046
Ti I 4025
Ti I 3990—99

C

D2 D1

E

b

F

G

g

h

H

K

I = 0.1

I = 0.4

Plate 18 Development of Spectral Features within the K Class

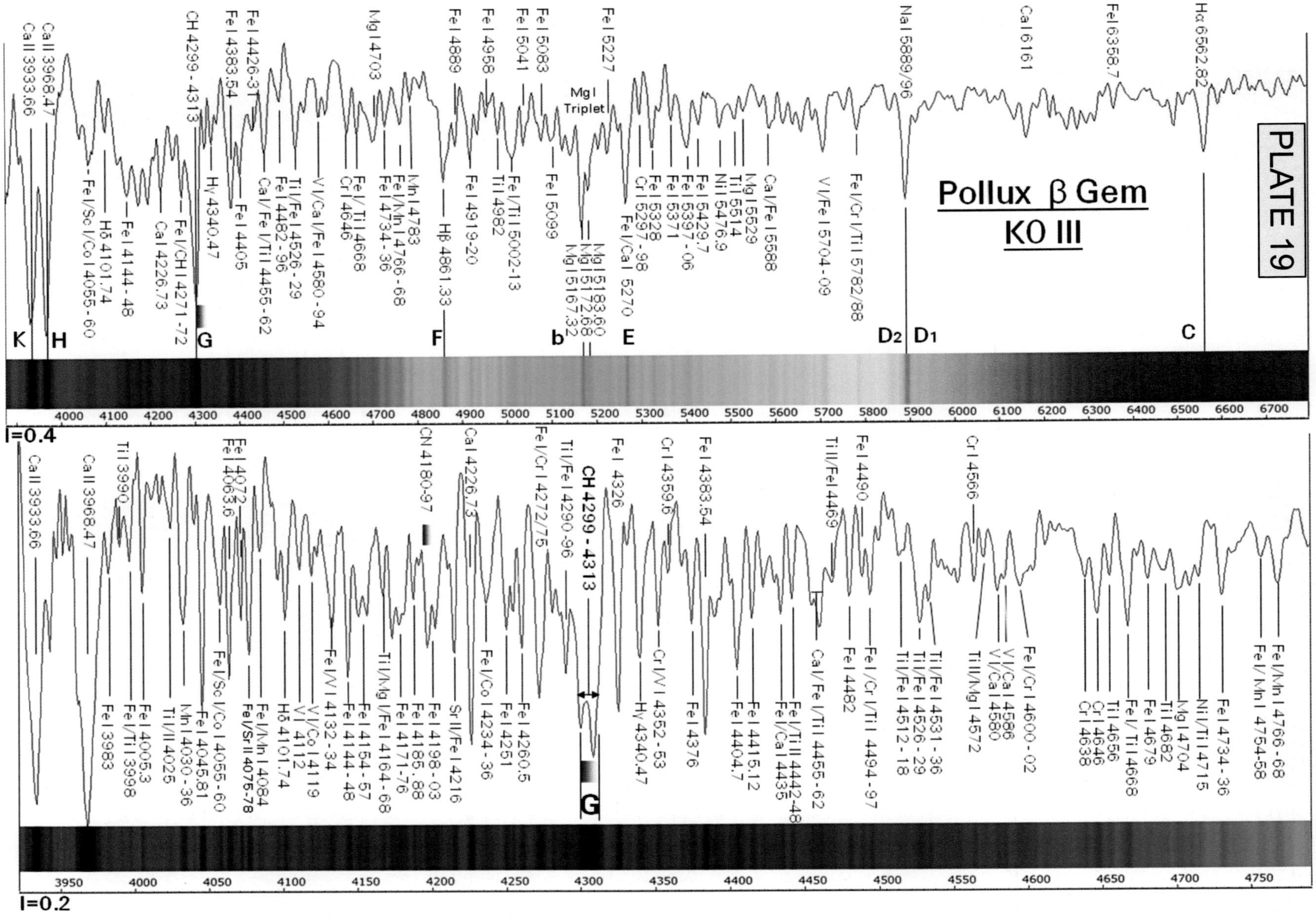
PLATE 19
Pollux β Gem
K0 III
Hα 6562.82
Fe I 6358.7
Ca I 6161
Na I 5889/96
Fe I/Cr I/Ti I 5782/88
V I/Fe I 5704-09
Ca I/Fe I 5588
Mg I 5529
Ti I 5514
Ni I 5476.9
Fe I 5429.7
Fe I 5397-06
Fe I 5371
Fe I 5328
Cr I 5297-98
Fe I/Ca I 5270
Fe I 5227
Mg I Triplet
Mg I 5183.60
Mg I 5172.68
Mg I 5167.32
Fe I 5099
Fe I 5083
Fe I 5041
Fe I/Ti I 5002-13
Ti I 4982
Fe I 4958
Fe I 4919-20
Fe I 4889
Hβ 4861.33
Mn I 4783
Fe I/Mn I 4766-68
Fe I 4734-36
Mg I 4703
Fe I/ Ti I 4668
Cr I 4646
V I/Ca I/Fe I 4580-94
Ti I/Fe I 4526-29
Fe I 4482-96
Ca I/ Fe I/Ti I 4455-62
Fe I 4426-31
Fe I 4405
Fe I 4383.54
Hγ 4340.47
CH 4299-4313
Fe I/CH I 4271-72
Ca I 4226.73
Fe I 4144-48
Hδ 4101.74
Fe I/Sc I/Co I 4055-60
Ca II 3968.47
Ca II 3933.66
C
D1
D2
E
b
F
G
H
K
I=0.4
Fe I/Mn I 4766-68
Fe I/ Mn I 4754-58
Fe I 4734-36
Ni I/Ti I 4715
Mg I 4704
Ti I 4682
Fe I 4679
Fe I/ Ti I 4668
Ti I 4656
Cr I 4646
Cr I 4638
Fe I/Cr I 4600-02
V I/Ca I 4586
V I/Ca I 4580
Ti II/Mg I 4572
Cr I 4566
Ti I/Fe I 4531-36
Ti I/Fe I 4526-29
Ti I/Fe I 4512-18
Fe I 4490
Fe I/Cr I/Ti I 4494-97
Fe I 4482
Ti II/Fe I 4469
Ca I/ Fe I/Ti I 4455-62
Fe I/Ti II 4442-48
Fe I/Ca I 4435
Fe I 4415.12
Fe I 4404.7
Fe I 4383.54
Fe I 4376
Cr I 4359.6
Cr I/V I 4352-53
Hγ 4340.47
Fe I 4326
CH 4299-4313
G
Ti I/Fe I 4290-96
Fe I/Cr I 4272/75
Fe I 4260.5
Fe I 4251
Fe I/Co I 4234-36
Ca I 4226.73
Sr II/Fe I 4216
Fe I 4198-03
CN 4180-97
Fe I 4185.88
Fe I 4171-76
Ti I/Mg I/Fe I 4164-68
Fe I 4154-57
Fe I 4144-48
Fe I/V I 4132-34
V I/Co I 4119
V I 4112
Hδ 4101.74
Fe I/Mn I 4084
Fe I/Sr II 4075-78
Fe I 4072
Fe I 4063.6
Fe I/Sc I/Co I 4055-60
Fe I 4045.81
Mn I 4030-36
Ti I/II 4025
Fe I 4005.3
Fe I/Ti I 3998
Ti I 3990
Fe I 3983
Ca II 3968.47
Ca II 3933.66
I=0.2

Plate 19 Detailed Spectrum of an Early K-Class Star

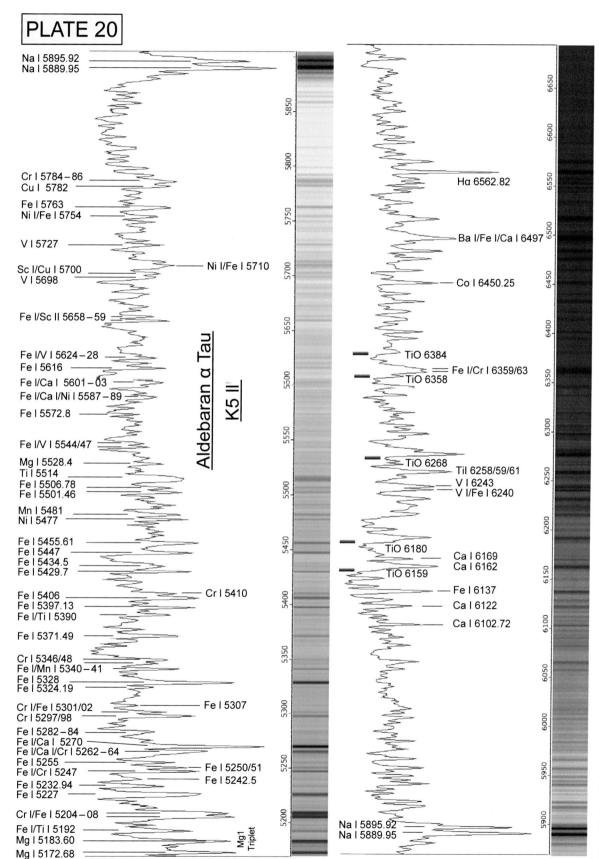

PLATE 20

Na I 5895.92
Na I 5889.95

Cr I 5784−86
Cu I 5782

Fe I 5763
Ni I/Fe I 5754

V I 5727

Sc I/Cu I 5700   Ni I/Fe I 5710
V I 5698

Fe I/Sc II 5658−59

Fe I/V I 5624−28
Fe I 5616
Fe I/Ca I 5601−03
Fe I/Ca I/Ni I 5587−89
Fe I 5572.8

Fe I/V I 5544/47
Mg I 5528.4
Ti I 5514
Fe I 5506.78
Fe I 5501.46
Mn I 5481
Ni I 5477

Fe I 5455.61
Fe I 5447
Fe I 5434.5
Fe I 5429.7

Fe I 5406   Cr I 5410
Fe I 5397.13
Fe I/Ti I 5390

Fe I 5371.49

Cr I 5346/48
Fe I/Mn I 5340−41
Fe I 5328
Fe I 5324.19

Cr I/Fe I 5301/02   Fe I 5307
Cr I 5297/98
Fe I 5282−84
Fe I/Ca I 5270
Fe I/Ca I/Cr I 5262−64
Fe I 5255   Fe I 5250/51
Fe I/Cr I 5247   Fe I 5242.5
Fe I 5232.94
Fe I 5227

Cr I/Fe I 5204−08
Fe I/Ti I 5192
Mg I 5183.60
Mg I 5172.68

Mg1
Triplet

Aldebaran α Tau

K5 II

Hα 6562.82

Ba I/Fe I/Ca I 6497

Co I 6450.25

TiO 6384
Fe I/Cr I 6359/63
TiO 6358

TiO 6268   Ti I 6258/59/61
V I 6243
V I/Fe I 6240

TiO 6180   Ca I 6169
Ca I 6162
TiO 6159
Fe I 6137
Ca I 6122
Ca I 6102.72

Na I 5895.92
Na I 5889.95

Plate 20 Detailed Spectrum of a Late K-Class Star

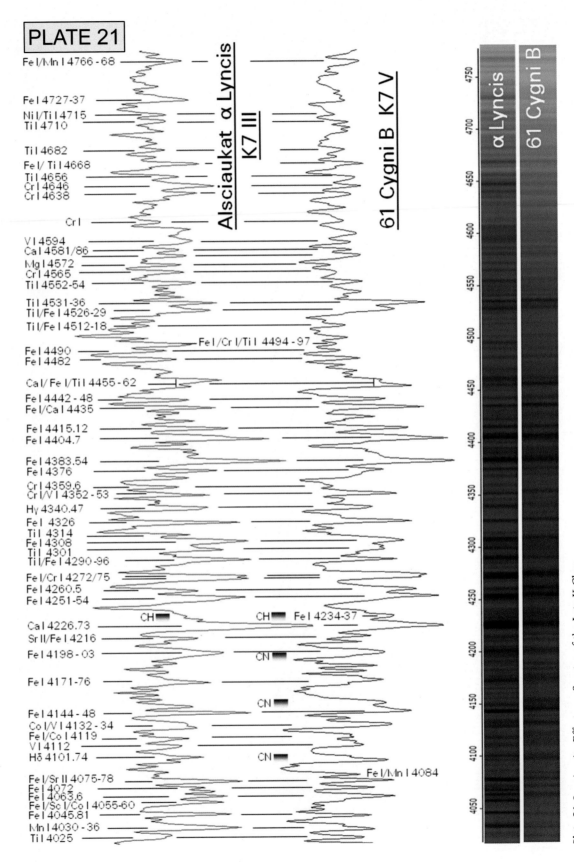

PLATE 21

Fe I/Mn I 4766 - 68

Fe I 4727-37
Ni I/Ti I 4715
Ti I 4710

Ti I 4682
Fe I/ Ti I 4668
Ti I 4656
Cr I 4646
Cr I 4638

Cr I

V I 4594
Ca I 4581/86
Mg I 4572
Cr I 4565
Ti I 4552-54

Ti I 4531-36
Ti I/Fe I 4526-29
Ti I/Fe I 4512-18

Fe I/Cr I/Ti I 4494 - 97
Fe I 4490
Fe I 4482

Ca I/ Fe I/Ti I 4455 - 62

Fe I 4442 - 48
Fe I/Ca I 4435

Fe I 4415.12
Fe I 4404.7

Fe I 4383.54
Fe I 4376
Cr I 4359.6
Cr I/V I 4352 - 53
Hγ 4340.47
Fe I 4326
Ti I 4314
Fe I 4308
Ti I 4301
Ti I/Fe I 4290 -96

Fe I/Cr I 4272/75
Fe I 4260.5
Fe I 4251-54

CH   CH   Fe I 4234-37
Ca I 4226.73
Sr II/Fe I 4216

Fe I 4198 - 03   CN

Fe I 4171-76

CN
Fe I 4144 - 48
Co I/V I 4132 - 34
Fe I/Co I 4119
V I 4112
Hδ 4101.74   CN

Fe I/Sr II 4075-78   Fe I/Mn I 4084
Fe I 4072
Fe I 4063.6
Fe I/Sc I/Co I 4055-60
Fe I 4045.81
Mn I 4030 - 36
Ti I 4025

Alsciaukat α Lyncis
K7 III

61 Cygni B K7 V

4750
4700
4650
4600
4550
4500
4450
4400
4350
4300
4250
4200
4150
4100
4050

α Lyncis

61 Cygni B

Plate 21 Luminosity Effects on Spectra of the Late K Class

CHAPTER

11 Spectral Class M

11.1 Overview

Some of the orange-reddish shining stars of the M class are well-known giants of luminosity classes I–III such as Bctelgeuse, Antares, Mirach (β And), Scheat (β Peg), Ras Algheti (α Her), Menkar (α Cet), Tejat Posterior (μ Gem). One of the probably biggest stars in the Milky Way, VY Canis Majoris, is classified as M3–4 II [38].

11.2 Parameters of the Late to Early M-Class Stars

Table 11.1 shows the data for the late to early subclasses, exclusively for the main sequence stars, compared to the Sun (⊙).

The huge percentage increase of the covered mass range compared to the earlier classes A, F, G and K is very striking here. The "late end" of the M class already touches the range limit of the brown dwarfs, which are separately classified with L, T and Y and are not observable by amateur means. The K, as well as the M dwarfs, stay much longer on the main sequence than the estimated age of the Universe of some 13.7 billion years. This means that after the Big Bang, in the entire Universe, up to the present, not a single main sequence star of the M class has migrated to the giant branch in the HRD! All M-type stars, visible with the naked eye, are former class B to G main sequence stars and spend in this late

spectral class just a relatively short time until the end of their giant stage. Their parameters are therefore totally different to the values in Table 11.1, with exception of the approximate effective temperature.

With the naked eye, no real M-type main sequence stars are visible, although they account for ~76% of the Sun's near neighbors (Table 4.4)! They are also denoted as "red dwarfs." The most famous representative is Proxima Centauri, found in the southern sky, with the spectral class M6 Ve. With a distance of 4.2 ly it is the very nearest neighbor to the Sun, but reaching an apparent magnitude of just $m_v \approx 11$! Thus it was not discovered until 1915. Figure 11.1 shows Proxima Centauri compared in size to the Sun. Its diameter was determined with the HST to about 200,000 km. Its life expectancy on the main sequence is estimated to some 4 trillion years! Whether our Universe will then still exist – in whatever form – must be answered by cosmological models. Besides Proxima Centauri, just a few red dwarfs

Figure 11.1 Size comparison: Proxima Centauri and the Sun

Table 11.1 Data for the late to early subclasses, M-class stars

Mass $M/M_\odot$	Lifetime on Main Sequence [Gy]	Effective Temperature [K]	Radius $R/R_\odot$	Luminosity $L/L_\odot$
0.08–0.6	>100	2600–3600	0.17–0.63	0.001–0.08

Table 11.2 Red dwarfs within reach of amateur spectroscopical equipment

Star	Apparent magnitude [m_v]	Distance from the Sun [ly]	Spectral type	Constellation
Proxima Centauri	11.1	4.2	M6 Ve	Centaurus
Gliese699, Barnard's Star	9.5	5.9	M4.0 V	Ophiuchus
Gliese273, Luyten's Star	9.8	12.4	M3.5 V	Canis Minor
Gliese411, Lalande 21185	7.5	8.2	M2.0 V	Ursa Major
Gliese388, AD Leo	8.1	15.9	M4.5 Ve	Leo
Gliese873, EV Lacertae	8.3	16.6	M4.5 V	Lacerta

are spectroscopically reachable by amateur means, some examples are shown in Table 11.2.

11.3 Spectral Characteristics of the M Class

The Fraunhofer H + K lines of ionized calcium Ca II remain striking in the entire M class. In addition to the Hα line neutral calcium Ca I at λ4227 and in the earlier subclasses the sodium double line at λλ5890/96, are the dominant discrete absorptions, which at this resolution are still visible. The former prominent, molecular G band turns now in to at least three discrete absorption lines [10]. The main features here are undoubtedly the huge absorption bands of titanium monoxide, TiO. Their intensity rises significantly towards the late subclasses, overprinting thousands of neutral, atomic absorption lines which would otherwise be visible here [1]. Although they show up already in the late K classes, they form here the unmistakable "brand" of the M types. In our daily life we find titanium dioxide (TiO_2), for example, as additive in toothpastes and colors. Like all other metals it originates from Sun-type progenitor stars, i.e. red giants which exploded as supernovae or repelled their envelopes as planetary nebulae. Much less frequent and with smaller intensity here we find a few absorption bands of CaH (calcium hydride) and MgH (magnesium hydride) molecules. In the late M subclasses the lower temperature allows the formation of even diatomic molecules like vanadium oxide (VO) and molecular hydrogen (H_2).

Red dwarfs on the main sequence belong to the so called "flare stars" or photometrically to the UV Ceti variables. They may display random outbursts in brightness, often just for a few hours, and be accompanied by intense X-ray radiation, as well as significant variability of spectral features. If appearing as emission lines, this concerns mainly the H-Balmer series. With these small stars this phenomenon is caused by the very deep reaching convection zones, generating a strong magnetic activity. An extreme example, at a distance of 16 ly, is EV Lacertae which became famous by the "super flare" observed

in April, 2008. Remarkable in such spectra is the Balmer decrement, which often appears far too flat or even inverse (Section 28.7 and [3])! Due to the short distance from the Sun the stars listed in Table 11.2 (e.g. Barnard's Star) show a very high apparent proper motion.

The intensity maximum of the real continuum is shifted here into the infrared range of the spectrum. Therefore the telluric H_2O and O_2 absorptions in the red range of the spectrum are almost completely overprinted by the stellar titanium monoxide (TiO) molecular bands. The graph in Figure 11.2 shows the theoretical continuum for a synthetic M5 III star (Vspec/Tools/Library [7]). Due to the specific radiation characteristics, influenced by the massive gaps in the continuum, the M-class stars are considerably different from an ideal blackbody radiator. Further, the course of the continuum is here quite difficult to determine.

11.4 Comments on Observed Spectra

Plate 22 Development of Spectral Features within the M Class

Antares (α Sco) and Ras Algethi (α Her)

Plate 22 demonstrates the development of the M subclasses by two superposed broadband overview spectra (200 L mm^{-1}) representing an early and late subtype.

Antares (450 ly) is a supergiant of the early M class with M0.5 Iab [11] and the dominant component of a binary star system. Its effective temperature is about 3600 K and the apparent rotation velocity <20 km s^{-1} [11].

Ras Algethi (400 ly) is a multiple star system. The spectrum is dominated by the supergiant, classified as M5 Ib-II, representing a later M type. As expected its effective temperature of 3300 K is lower than the earlier classified Antares. Ras Algethi is somewhat smaller than Antares, but also surrounded by a cloud of gas and dust. The line identification is based on SpectroWeb [9] and [2], [10], [29], [34], [35], [37], [39].

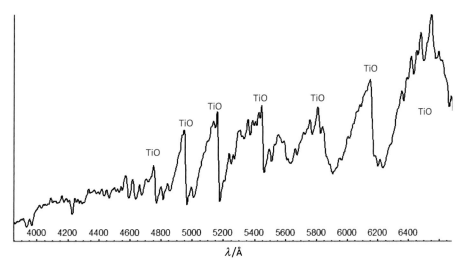

Figure 11.2 Continuum of a synthetic M5 III standard star

Figure 11.3 Size comparison: Antares, Arcturus and the Sun

Plate 23 Red Dwarfs and Flare Stars

Gliese 388 and Lalande 21185

Plate 23 shows a comparison of two flare stars on the main sequence. At the time of recording only Lalande 21185 showed the H-Balmer lines in emission. In contrast to the supergiant Antares on Plate 22, the spectrum of this somewhat later classified main sequence dwarf displays TiO absorption bands far into the short-wave (blue) domain.

Gliese 388, GJ 388, AD Leonis (15.9 ly), located in the constellation Leo, is a red dwarf star, classified on the main sequence as M4.5 Ve [11]. Its effective temperature is about 3390 K. As a flare star it undergoes random outbursts in luminosity and shows the H-Balmer series as emission lines. Its mass is some 40% of the Sun's and the apparent brightness $m_v = 8.1$. On this plate the emission lines of the H-Balmer series are labeled in red.

Gliese 411, Lalande 21185, HD 95735 (8.2 ly) is a red dwarf, classified as M2 V [11] on the main sequence, located in the constellation Ursa Major. Its effective temperature is about 3400 K and its mass some 40% of the Sun's. With $m_v = 7.5$ it is one of the apparently brightest red dwarfs. The recorded spectrum here displays no emission lines.

Plate 24 Effects of the Luminosity on Spectra of the Late M Class

Barnard's Star (Gliese 699) and Ras Algethi (α Her)

Plate 24 shows a montage of two equally normalized broadband overview spectra (200 L mm^{-1}), demonstrating luminosity effects on spectra of the later M class. The slightly later classified supergiant Ras Algethi (α Her) is compared here with Barnard's Star, Gliese 699, a dwarf on the main sequence. Both spectra are mainly dominated by impressive TiO absorption bands. However, in the profile of the red dwarf they appear somewhat "underdeveloped," particularly towards shorter wavelengths. Otherwise, the Na I absorption is here much more intense. To optimize the clarity, the labeling here is limited to some prominent lines.

Barnard's Star, GJ699 (5.9 ly) is a red dwarf on the main sequence, classified as M4.0 V [11], located in the constellation Ophiuchus. Its effective temperature is about 3170 K and the mass, with just some 17% of the Sun's, is very low. With $m_v = 9.5$ the apparent brightness is very faint, considering the very short distance of just 5.9 ly [11]. With the very large ~10.3 arcsec yr^{-1} this star is the "record holder" in respect of proper motion. The recorded spectrum displays no emission lines.

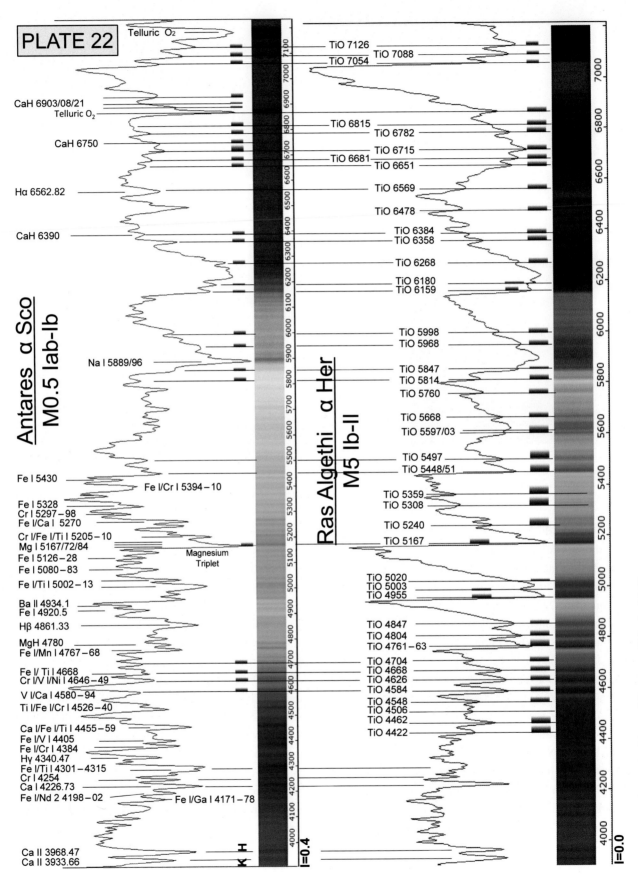

PLATE 22

Telluric O₂

TiO 7126
TiO 7088
TiO 7054

CaH 6903/08/21
Telluric O₂

TiO 6815
TiO 6782

CaH 6750

TiO 6715
TiO 6681
TiO 6651

TiO 6569

Hα 6562.82

TiO 6478

CaH 6390

TiO 6384
TiO 6358

TiO 6268

TiO 6180
TiO 6159

TiO 5998
TiO 5968

Na I 5889/96

TiO 5847
TiO 5814

TiO 5760

TiO 5668
TiO 5597/03

TiO 5497
TiO 5448/51

Fe I 5430
Fe I/Cr I 5394−10

TiO 5359
TiO 5308

Fe I 5328
Cr I 5297−98
Fe I/Ca I 5270

TiO 5240

Cr I/Fe I/Ti I 5205−10
Mg I 5167/72/84

TiO 5167

Fe I 5126−28
Fe I 5080−83

Magnesium
Triplet

Fe I/Ti I 5002−13

TiO 5020
TiO 5003
TiO 4955

Ba II 4934.1
Fe I 4920.5

Hβ 4861.33

TiO 4847
TiO 4804
TiO 4761−63

MgH 4780
Fe I/Mn I 4767−68

TiO 4704
TiO 4668
TiO 4626
TiO 4584

Fe I/ Ti I 4668
Cr I/V I/Ni I 4646−49

V I/Ca I 4580−94
Ti I/Fe I/Cr I 4526−40

TiO 4548
TiO 4506
TiO 4462
TiO 4422

Ca I/Fe I/Ti I 4455−59
Fe I/V I 4405
Fe I/Cr I 4384
Hγ 4340.47
Fe I/Ti I 4301−4315
Cr I 4254
Ca I 4226.73
Fe I/Nd 2 4198−02

Fe I/Ga I 4171−78

Ca II 3968.47
Ca II 3933.66

K H

I=0.4

Antares α Sco
M0.5 Iab-Ib

Ras Algethi α Her
M5 Ib-II

I=0.0

Plate 22 Development of Spectral Features within the M Class

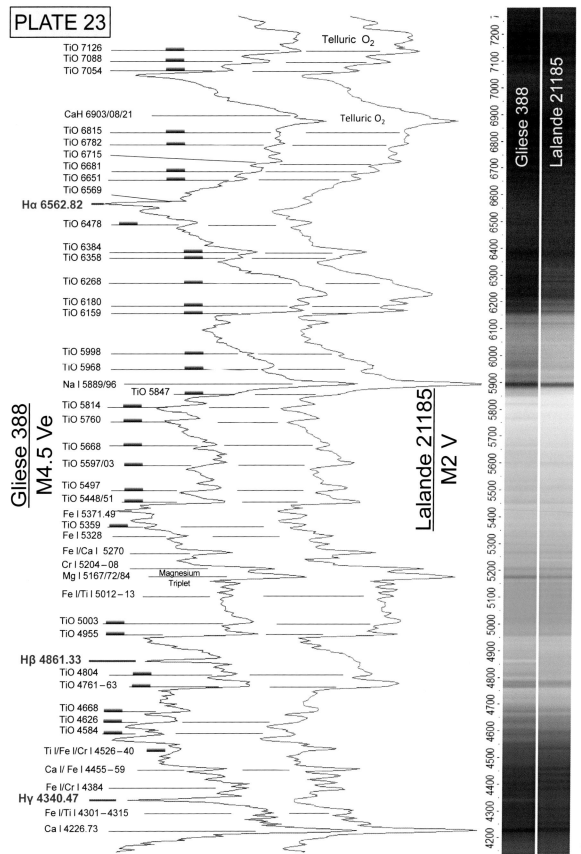

PLATE 23

TiO 7126
TiO 7088
TiO 7054

Telluric O₂

CaH 6903/08/21

Telluric O₂

TiO 6815
TiO 6782
TiO 6715
TiO 6681
TiO 6651
TiO 6569

Hα 6562.82

TiO 6478

TiO 6384
TiO 6358

TiO 6268

TiO 6180
TiO 6159

TiO 5998

TiO 5968

Na I 5889/96
TiO 5847
TiO 5814
TiO 5760

TiO 5668
TiO 5597/03

TiO 5497
TiO 5448/51

Fe I 5371.49
TiO 5359
Fe I 5328

Fe I/Ca I 5270
Cr I 5204−08
Mg I 5167/72/84
Magnesium
Triplet
Fe I/Ti I 5012−13

TiO 5003

TiO 4955

Hβ 4861.33
TiO 4804

TiO 4761−63

TiO 4668
TiO 4626
TiO 4584

Ti I/Fe I/Cr I 4526−40

Ca I/ Fe I 4455−59

Fe I/Cr I 4384

Hγ 4340.47

Fe I/Ti I 4301−4315

Ca I 4226.73

Gliese 388
M4.5 Ve

Lalande 21185
M2 V

Gliese 388

Lalande 21185

7200
7100
7000
6900
6800
6700
6600
6500
6400
6300
6200
6100
6000
5900
5800
5700
5600
5500
5400
5300
5200
5100
5000
4900
4800
4700
4600
4500
4400
4300
4200

Plate 23 Red Dwarfs and Flare Stars

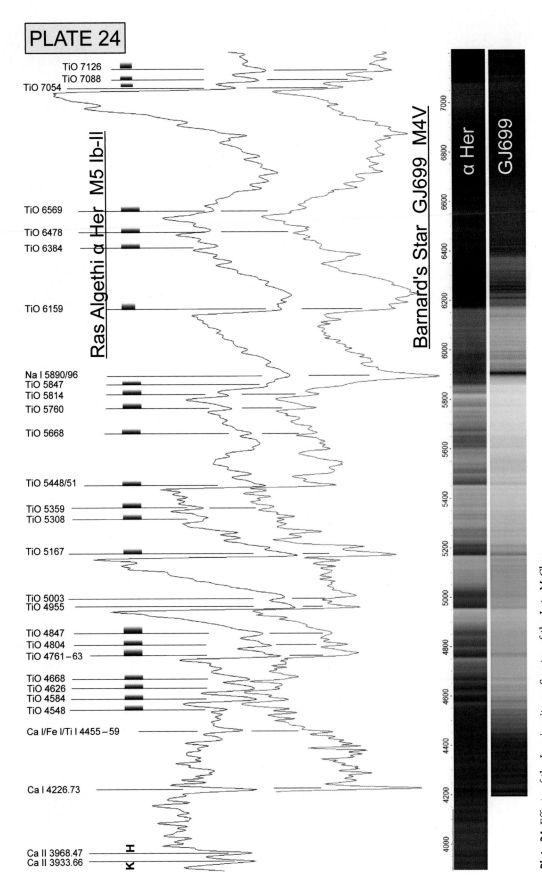

PLATE 24

TiO 7126
TiO 7088
TiO 7054

Ras Algethi α Her   M5 Ib-II

TiO 6569

TiO 6478

TiO 6384

TiO 6159

Barnard's Star GJ699 M4V

α Her

GJ699

Na I 5890/96
TiO 5847
TiO 5814
TiO 5760

TiO 5668

TiO 5448/51

TiO 5359
TiO 5308

TiO 5167

TiO 5003
TiO 4955

TiO 4847
TiO 4804
TiO 4761–63

TiO 4668
TiO 4626
TiO 4584
TiO 4548

Ca I/Fe I/Ti I 4455–59

Ca I 4226.73

Ca II 3968.47   H
Ca II 3933.66   K

7000
6800
6600
6400
6200
6000
5800
5600
5400
5200
5000
4800
4600
4400
4200
4000

Plate 24 Effects of the Luminosity on Spectra of the Late M Class

12 Spectral Sequence on the AGB

12.1 Evolution of Stars in the Post-Main Sequence Stage

The following roughly explained evolutionary pathways are relevant for stars $\lesssim 8\,M_\odot$. Considering the whole course of a star's lifetime on the HRD these just form an important phase at the end, lasting some billion years.

Red Giant Branch

After leaving the main sequence the star moves on to the red giant branch (RGB) towards the upper right corner of the HRD. In this phase, the hydrogen fusion zone grows outwards as a shell around the increasing helium core. Thus the radius and the luminosity of the star dramatically increase, while the density of the atmosphere, the effective temperature and the rotation velocity substantially decrease.

Horizontal Branch

The so-called helium flash stops this ascent abruptly on the RGB. However, it does not destroy the star. This event is triggered by the nuclear ignition of the helium core, which was formed during the RGB phase. From now on, by helium fusion and other highly complex nuclear processes, carbon and oxygen are formed in the core of the star. As a result, the star first moves somewhat to the left and slightly down on the horizontal branch (HB).

Asymptotic Giant Branch

In a later phase, the helium begins to "burn" within a shell around the growing core of carbon and oxygen and in addition triggers hydrogen fusion in a further outer shell of the star. This multiple or alternating shell burning process causes the rise of the expanding and unstable star along the asymptotic giant branch (AGB). In stars of $1\,M_\odot$ the stellar radius reaches about 1 AU and the luminosity is about 10,000 times that of the former main sequence star.

Post-AGB Phase

In the final stage these stars eject their outer shells as planetary nebulae (Section 28.3), which are excited to the emission of light by the remaining, extremely hot core. Later on, they cool down and end as white dwarfs at the stellar "cemetery" in the bottom part of the HRD.

12.2 The Spectral Sequence of the Mira Variables on the AGB

This sequence consists of stars in their unstable AGB phase, limited to the range of about 1–2 $M_\odot$ and the spectral class M as well as to some late K types. For amateur astronomers this category is not only interesting for photometric monitoring, but also in terms of spectral analysis. The astrophysical context is still not fully understood and explanation here is greatly simplified, mainly based on [2].

Once arrived on the AGB the star becomes unstable as a result of the aforementioned complex fusion processes. It now starts pulsating and pushing off a lot of matter. Therefore, most but not all of these stars show the behavior of long-period Mira variables. Current theories postulate that at this stage strong, so-called thermal pulses, trigger deep

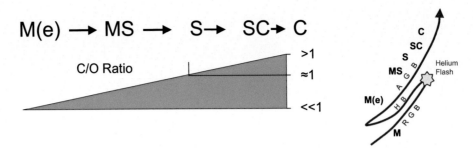

Figure 12.1 During the ascent onto the AGB the abundance of free oxygen in the stellar atmosphere becomes progressively reduced by chemical combination with carbon and other "metals." This process is described by the increasing C/O ratio.

convective "dredge up" processes inside the star. As a result carbon and products of the nuclear s-process (for example, barium, lithium, zirconium, technetium, etc.) are conveyed to the stellar surface.

On this course the oxygen in the stellar atmosphere is not all bound by the formation of metal oxides (e.g. TiO, ZrO, YO). The rest of it combines with the rapidly increasing carbon mainly to CO molecules [2]. Thus, the amount of free oxygen gets continuously reduced. In this context it is expressed as a ratio of carbon to oxygen C/O.

During its ascent on the AGB the star passes the following sequence of spectral classes (Figure 12.1):

Spectral class M(e) In the initial phase, at the lower part of the AGB, the stellar atmosphere is still oxygen-rich with a ratio C/O $\approx$ 0.5. Here, TiO absorptions continue to dominate the profile, usually associated with individual atomic emission lines, mainly of the H-Balmer series. Therefore, in this Atlas the "index" (e) is added to this class, to distinguish it clearly from the relatively stable, exclusively hydrogen fusing giants of spectral types K to M on the RGB.

Spectral class MS Within the transition class MS the stellar atmosphere exhibits an increasing oxygen deficiency. Thus, absorptions of diatomic zirconium monoxide (ZrO) displace more and more the formerly dominating titanium monoxide bands (TiO).

Spectral class S Within the spectral class S, the ratio becomes C/O $\approx$ 1 and therefore the abundance of unbound oxygen in the stellar atmosphere is now significantly reduced.

Spectral classes SC and C Above the spectral class S the ratio becomes C/O $>$ 1. This inevitably creates a carbon excess, which accumulates in a circumstellar cloud and impressively dominates the stellar spectrum. Thus, in the transition class SC and increasingly in the following C class (carbon stars), moderately high resolved spectra just show absorptions of diatomic carbon molecules CH, CN, and C_2. In addition, atomic lines of the s–process products and impressive absorptions of Na I show up here.

After arriving at the top of the AGB the star starts to repel its outer hydrogen shell as a photogenic planetary nebula, triggered by intense thermal pulses and associated shockwaves.

CHAPTER

13 M(e) Stars on the AGB

13.1 Overview

The unstable M(e) stars on the AGB still have an oxygen-rich atmosphere (C/O <1) and are usually of the LPV type (long period variable), also called Mira variables. The period of the brightness variation is in the order of about 100 to 1000 days. The prototype of this class is Mira (o Ceti). The instability leads to shockwaves in the stellar atmosphere, expanding and contracting with the rhythm of the brightness variations (radius up to a factor 2!). Figure 13.1 shows impressively how the star continuously loses matter as a result of this spectacular process. The star on its way through the galaxy leaves a tremendous trace of stellar material, which can be seen in the UV range on a length of about 13 ly.

13.2 Spectral Characteristics of the M(e) Stars on the AGB

Parallel to the "pulsations" of the stellar atmosphere, emission lines of variable intensity appear in the spectrum, predominantly the short-wave II-Balmer lines on the blue side of Hγ. The maximum intensity mostly shows Hδ (Plate 25). Why the observed Balmer decrement (Section 28.7 and [3]) is mostly running inversely in the spectra of AGB stars is subject of several hypotheses. Highly intense TiO absorption bands often seem completely to overprint Hα and Hβ and strongly attenuating Hγ. This is evidence that the emission lines are formed in much deeper layers of the stellar atmosphere as the titanium monoxide absorptions [1]. The spectral type of the

Figure 13.1 Mira (o Ceti) leaving an impressive trace of repelled stellar matter. The picture was recorded in 2006 in the UV range by the GALEX satellite
(Credit:NASA/JPL-Caltech)

star itself is often variable within the range of several decimal subclasses. Apart from the emission lines, in low resolved spectra normally no further differences to a normal M star can be detected.

13.3 Comments on Observed Spectra

Plate 25 Mira Variable M(e), Comparison to a Late-classified M Star

Mira (o Cet) and Ras Algethi (α Her)

Plate 25 shows a montage of two broadband overview spectra (200 L mm^{-1}): a comparison of the M(e) star Mira, o Ceti, M5–9 IIIe and the ordinary M-type star Ras Algethi (α Her, M5 Ib-II), which is still located on the RGB (see Plate 22).

Mira A (300 ly) probably one of the most popular variables, is steadily losing matter to its binary partner, the white dwarf Mira B (VZ Ceti). It shows an impressive brightness variation of $m_v \approx 3$–10, with a period of about 331 days (LPV). In some cases it becomes much brighter than $m_v \approx 2.0$. Its spectacular variability – discovered in 1639 by J. Holwarda – manifests itself by the strongly variable spectral class M5–9 IIIe [11]. Figure 13.2 shows Mira, embedded in an impressive dust and gas cloud.

Apart from some slight differences in the intensity of TiO bands, the emission lines Hδ and Hγ are essentially the only spectral differences that distinguish the AGB star

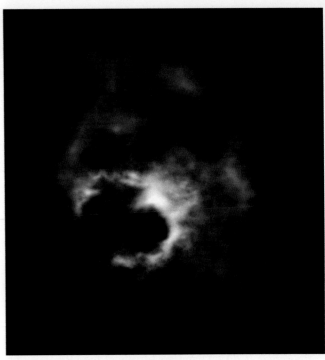

Figure 13.2 Mira, o Ceti, recorded by ALMA Telescope, in Chile (Credit: ESO, S. Ramstedt (Uppsala University, Sweden) and W. Vlemmings (Chalmers University of Technology, Sweden))

Mira from the RGB star Ras Algethi. Here, Hδ appears in fact significantly more intensive than Hγ. The profile was recorded some 60 days after the maximum brightness, $m_v \approx 4.1$ [38].

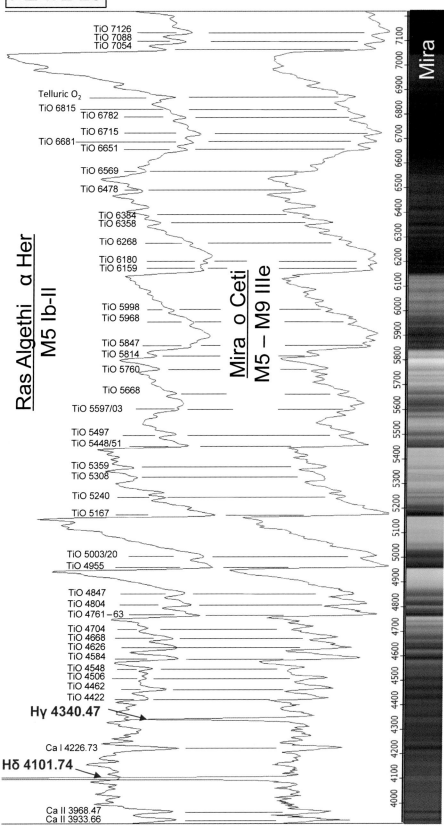

PLATE 25

Ras Algethi α Her
M5 Ib-II

Mira o Ceti
M5 – M9 IIIe

Mira

TiO 7126
TiO 7088
TiO 7054

Telluric O₂
TiO 6815
TiO 6782
TiO 6715
TiO 6681
TiO 6651

TiO 6569

TiO 6478

TiO 6384
TiO 6358

TiO 6268

TiO 6180
TiO 6159

TiO 5998
TiO 5968

TiO 5847
TiO 5814

TiO 5760

TiO 5668

TiO 5597/03

TiO 5497
TiO 5448/51

TiO 5359
TiO 5308

TiO 5240

TiO 5167

TiO 5003/20
TiO 4955

TiO 4847
TiO 4804
TiO 4761 – 63

TiO 4704
TiO 4668
TiO 4626
TiO 4584
TiO 4548
TiO 4506
TiO 4462
TiO 4422

Hγ 4340.47

Ca I 4226.73

Hδ 4101.74

Ca II 3968.47
Ca II 3933.66

Plate 25 Mira Variable M(e), Comparison to a Late-classified M Star

CHAPTER

14 Spectral Class S on the AGB

14.1 Overview and Spectral Characteristics

On its further ascent along the AGB the spectrum of a star becomes mainly formed by the following influences:

- increasing abundance of carbon in the stellar atmosphere
- the consequential reduction of free oxygen
- "dredged up" products of the s-process, for example, zirconium, lithium, technetium.

Thus, at first the spectral type M(e) changes in to the transition class MS. With moderate spectral resolution, if at all, just small differences are detectable in the profile. However, at higher resolution first signs of zirconium monoxide (ZrO) appear within the still dominant titanium monoxide bands. The share of ZrO increases now continuously up to the extreme types of the spectral class S, where ZrO almost entirely replaces the TiO absorptions. This effect is explained with the higher affinity of ZrO to oxygen, and the greater temperature resistance of this molecule [2]. The dissociation energy of ZrO, (as well as lanthanum monoxide, LaO and yttrium monoxide, YO), is >7 eV, in contrast to TiO with <7 eV.

14.2 The Boeshaar–Keenan S-Classification System

The spectral class S and the transition classes MS and SC are specified with the following notation form, corresponding to a temperature sequence, analogously to the ordinary M giants on the RGB. Since all S stars are giants, a specification for the luminosity class is usually omitted.

MXS/n(e), SX/n(e), SCX/n(e)

where X stands for the temperature sequence, analogously to the M class from 1–9, corresponding to a range of ~3600–2500 K, (e) indicates if the spectrum sometimes exhibits emission lines, and n corresponds to a "C/O index" on a scale of 1–10, estimated by the observed relation between the intensities of TiO and ZrO. Table 14.1, after Scalo and Ross [40], provides the C/O index and allows estimation of the C/O ratio.

This classification, mostly indicated with a large range of variability, can be supplemented with characteristic elements or molecules showing up in the spectrum. Additional classification details can be found in [2].

14.3 "Intrinsic" and "Extrinsic" or "Symbiotic" S Stars

This classification system does not explain the entirety of all S class and carbon stars [2]. Many non- or just slightly variable S stars, often staying off the AGB, show no short lived and unstable ^{99}Tc (technetium isotopes in the spectrum) are hence called Tc-poor. This affects at least 40% of all registered MS and S types. In fact, such objects are relatively hot and form components of close binary systems. The analysis of the orbits has shown that the companions must be white dwarfs [42]. These properties, as well as a low variability and the absence of ^{99}Tc in the spectrum, are currently the main distinguishing characteristics for an "extrinsic" S-class star. In such cases the spectral classification is sometimes supplemented by "symbiotic" or "extrinsic." One of the recent, still debated scenarios postulates a mass transfer, bringing the ZrO from a nearby companion star into the stellar atmosphere. Therefore these special types are also

Table 14.1 C/O indices after Scalo and Ross, based on the observed intensity relation of TiO and ZrO [2], [41]. The C/O Index allows estimation of the C/O ratio

Spectral type with C/O index	Criteria for the C/O index	~ C/O ratio	Comment
MXS	TiO still dominates, some signs for ZrO bands are recognizable in highly resolved spectra		Transition class from M(e) to S
SX/1	TiO >> ZrO, YO	< 0.95	Spectral class S
SX/2	TiO > ZrO	0.95	
SX/3	TiO ≈ ZrO, YO intensive	0.96	
SX/4	ZrO > TiO	0.97	
SX/5	ZrO >> TiO	0.97	
SX/6	ZrO very intensive, no TiO recognizable	0.98	
SX/7 = SC X/7	ZrO weaker, intensive Na lines	0.99	
SC X/8	No ZrO recognizable, intensive Na lines	1.00	Transition class from S to the carbon stars C
SC X/9	Very intensive Na lines, C_2 very weak	1.02	
SC X/10 = C–N	Intensive Na lines, C_2 weak	1.1	

called "extrinsic" or "symbiotic." A textbook example is the relatively bright and only slightly variable BD Camelopardalis (Plate 28). For symbiotic (binary) stars refer also to Section 24.4.

In contrast, the S-type stars on the AGB show thermal pulses and are classified as Tc-rich or "intrinsic." Intrinsic means that the elements and molecules responsible for the S-classification, are generated by the star itself. The brightness, some spectral features and even the spectral class of intrinsic S-type stars are usually highly variable.

14.4 Hints for the Observation of S-Class Stars

On the AGB the stellar atmospheres meet the condition $C/O \approx 1$ just for a very short period of time. Therefore S-class stars are very rare. For our Galaxy, just some 1300 are registered. Unfortunately most of the bright S stars exhibit a low C/O index, so, for instance, Chi Cygni is classified with S6-9/1–2e [11]. In low resolved spectra it shows just very weak signs of ZrO. Table 14.2 shows some rather strong S types, with parameters according to [11]. For the planning of the observation, the current apparent brightness for most of the variable S-class stars is available from AAVSO [38].

Table 14.2 Bright intrinsic S-type stars

HD No.	Name	Spectral Class	~m_v max.
4350	U Cas	S4.5-6/3-5e	8.0
53791	R Gem	S3.5-6.5/6e	7.7
51610	R Lyn	S4-6.5/5-6e	7.6
70276	V Cnc	S0-6/6e	7.5
110813	S UMa	S0.5-6/6e	7.1
185456	R Cyg	S4-8/6e	6.1

14.5 Comments on Observed Spectra

Plate 26 Extreme S-Class Star, Comparison to a Mira Variable M(e)

Mira (o Ceti) and R Cygni (HD 185456)

For comparison, **Plate 26** presents a montage of two broad-band overview spectra (200 L mm^{-1}): M(e) star Mira (o Ceti) (Plate 25) and the S-class star R Cygni, S4–8/6e, with a high C/O index. The comparison of the two profiles shows the transition from the spectral type M(e), dominated by TiO bands, to a strong S-type star with a high C/O index, exhibiting mainly ZrO absorptions in the spectrum.

R Cygni is classified with S4–8/6e. The star is strongly variable by $m_v \approx$ 6–14, within a period of ~429 days (LPV). The maximal and minimal brightness may differ considerably, compared between the individual periods. In the oxygen-rich atmosphere of Mira the prominent TiO bands dominate. However, due to the strong lack of oxygen, in the profile of R Cygni just impressive ZrO absorption bands appear, further atomic lines of *s*-process elements, as well as a strong Na I absorption. Due to a number of other physical effects, at the pronounced S class the outer layers of the atmosphere become more "transparent" [2]. Therefore all emission lines of the H-Balmer series show up here, however, Hα is significantly dampened. In the Mira profile only Hα and Hβ are in emission. At this low resolution the identification of the "intrinsic marker" technetium is uncertain, i.e. the ^{99}Tc I lines at λλλ4238, 4262 and 4297. The profile was recorded around the maximum brightness. The detailed line identification for ZrO was chiefly enabled by Wyckoff and Clegg [43].

Observation tip: R Cygni is very easy to find. It is located just nearby the bright Θ Cygni, HD 185395.

Plate 27 Development of Spectral Features within the S Class

On **Plate 27** A montage of three broadband overview spectra (200 L mm^{-1}) demonstrates the spectral development within the *S*-class. To optimize the clarity, the labeling of the lines is limited to some prominent lines.

Omicron^1 Orionis (4 Ori, HD 30959)

This star at a distance of 540 ly and with an apparent brightness $m_v \approx$ 4.7, is just slightly variable. With the classification M3S III [11] it forms the transition between the M(e) and S types. According to the very low C/O index ZrO is barely visible in this low resolved spectrum and TiO bands still dominate the profile. None of the Balmer lines appear in emission here.

Chi Cygni (HD 187796)

Chi Cygni is at 340 ly, with a variable classification of S6–9/1–2e, the low C/O index is here just slightly higher

compared to Omicron^1 Orionis. Thus the ZrO absorptions appear here between the still dominant TiO bands, only within a few small sections of the profile at λλ5298/5305, λλ5375/79, λλ5839/49 and λλ6475/81/94 (blend with TiO λ6478). This star shows with $m_v \approx$ 3.3–14.3, one of the highest brightness variations of all known variables with a period of ~408 days (LPV). The Hγ and Hδ lines appear in emission.

R Cygni (HD 185456)

As already shown in Plate 26, due to the high C/O index, the usual TiO bands have been fully replaced here by the ZrO absorptions.

Plate 28 Comparison of an "Intrinsic" and "Extrinsic" S-Class Star

Plate 28 presents a montage of two broadband overview spectra (200 L mm^{-1}) to compare an "intrinsic" and an "extrinsic" S-class star. Both stars have nearly the same spectral classification. Despite some minor differences the two profiles run almost identically – regardless of the suspected different origin of ZrO in the stellar atmospheres. No emission lines are observable here.

Intrinsic S-class HR Pegasi (HD 216672)

HR Pegasi (970 ly), is just slightly variable $m_v \approx$ 6.1–6.5 within a period of some 50 days (type SR, semiregular). It is classified as S4+/1+ [11] with a very low C/O index.

Extrinsic S-class BD Camelopardalis (HD 22649)

BD Cam (510 ly), is also just slightly variable $m_v \approx$ 5.1. It forms a component of a detached spectroscopic binary system with a period of 596 days, exhibiting a small mass transfer by a weak stellar wind (Section 24.4). It is classified as S3.5/2 "symbiotic" [11]. The recorded spectrum shows no emission lines. The companion might be a white dwarf [44]. This extrinsic S star shows no technetium in the spectrum.

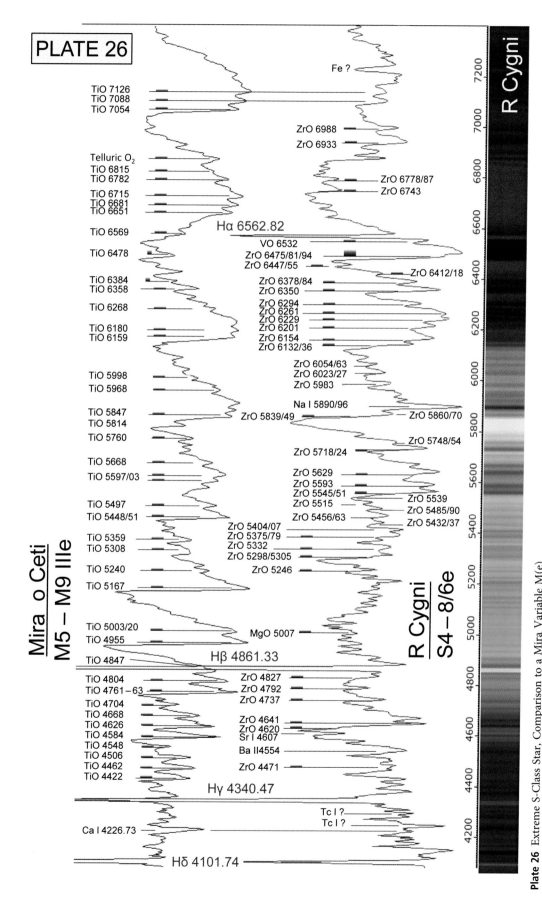

PLATE 26

Mira o Ceti
M5 – M9 IIIe

TiO 7126
TiO 7088
TiO 7054

Telluric O₂
TiO 6815
TiO 6782
TiO 6715
TiO 6681
TiO 6651
TiO 6569
TiO 6478
TiO 6384
TiO 6358
TiO 6268
TiO 6180
TiO 6159
TiO 5998
TiO 5968
TiO 5847
TiO 5814
TiO 5760
TiO 5668
TiO 5597/03
TiO 5497
TiO 5448/51
TiO 5359
TiO 5308
TiO 5240
TiO 5167
TiO 5003/20
TiO 4955
TiO 4847
TiO 4804
TiO 4761 – 63
TiO 4704
TiO 4668
TiO 4626
TiO 4584
TiO 4548
TiO 4506
TiO 4462
TiO 4422

Ca I 4226.73

Fe ?

ZrO 6988
ZrO 6933

ZrO 6778/87
ZrO 6743

Hα 6562.82

VO 6532
ZrO 6475/81/94
ZrO 6447/55
ZrO 6412/18
ZrO 6378/84
ZrO 6350
ZrO 6294
ZrO 6261
ZrO 6229
ZrO 6201
ZrO 6154
ZrO 6132/36
ZrO 6054/63
ZrO 6023/27
ZrO 5983
Na I 5890/96
ZrO 5839/49   ZrO 5860/70
ZrO 5748/54
ZrO 5718/24
ZrO 5629
ZrO 5593
ZrO 5545/51   ZrO 5539
ZrO 5515   ZrO 5485/90
ZrO 5456/63   ZrO 5432/37
ZrO 5404/07
ZrO 5375/79
ZrO 5332
ZrO 5298/5305
ZrO 5246

R Cygni
S4 – 8/6e

MgO 5007

Hβ 4861.33

ZrO 4827
ZrO 4792
ZrO 4737

ZrO 4641
ZrO 4620
Sr I 4607

Ba II4554

ZrO 4471

Hγ 4340.47

Tc I ?
Tc I ?

Hδ 4101.74

R Cygni

Plate 26 Extreme S-Class Star, Comparison to a Mira Variable M(e)

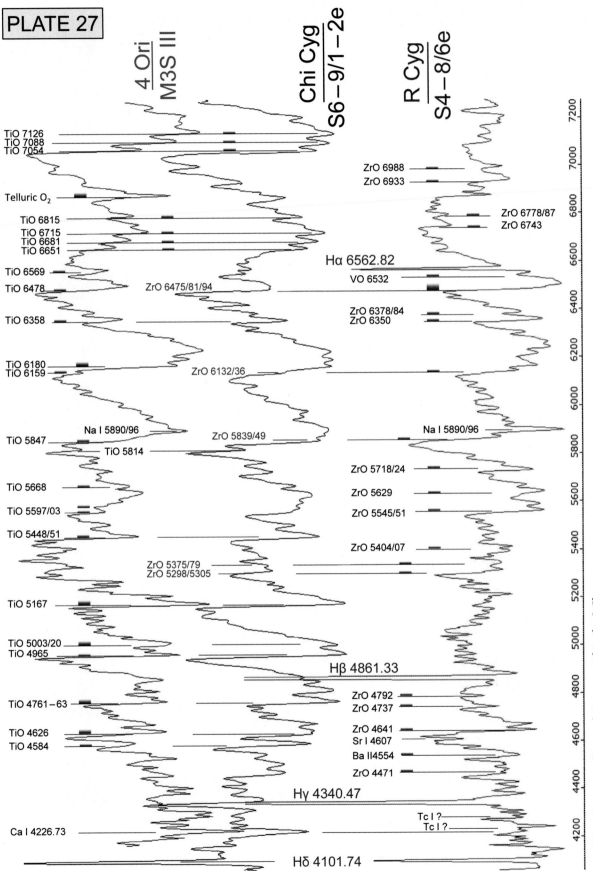

PLATE 27

4 Ori
M3S III

Chi Cyg
S6−9/1−2e

R Cyg
S4−8/6e

TiO 7126
TiO 7088
TiO 7054

ZrO 6988
ZrO 6933

Telluric O₂

TiO 6815
TiO 6715
TiO 6681
TiO 6651

ZrO 6778/87
ZrO 6743

TiO 6569

Hα 6562.82

VO 6532

TiO 6478

ZrO 6475/81/94

TiO 6358

ZrO 6378/84
ZrO 6350

TiO 6180
TiO 6159

ZrO 6132/36

TiO 5847

Na I 5890/96

ZrO 5839/49

Na I 5890/96

TiO 5814

ZrO 5718/24

TiO 5668

ZrO 5629

TiO 5597/03

ZrO 5545/51

TiO 5448/51

ZrO 5404/07

ZrO 5375/79
ZrO 5298/5305

TiO 5167

TiO 5003/20
TiO 4965

Hβ 4861.33

ZrO 4792
ZrO 4737

TiO 4761−63

ZrO 4641
Sr I 4607

TiO 4626
TiO 4584

Ba II4554

ZrO 4471

Hγ 4340.47

Tc I ?
Tc I ?

Ca I 4226.73

Hδ 4101.74

7200
7000
6800
6600
6400
6200
6000
5800
5600
5400
5200
5000
4800
4600
4400
4200

Plate 27 Development of Spectral Features within the S Class

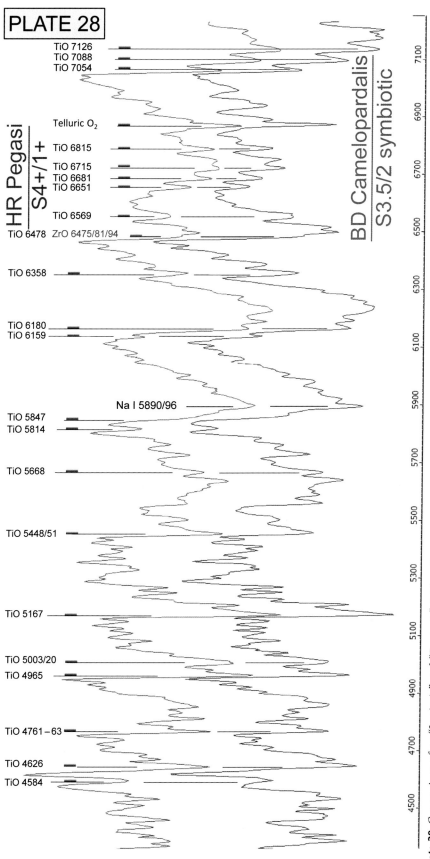

PLATE 28

TiO 7126
TiO 7088
TiO 7054

Telluric O₂

HR Pegasi
S4+/1+

TiO 6815
TiO 6715
TiO 6681
TiO 6651
TiO 6569
TiO 6478   ZrO 6475/81/94

TiO 6358

TiO 6180
TiO 6159

Na I 5890/96

TiO 5847
TiO 5814

TiO 5668

TiO 5448/51

TiO 5167

TiO 5003/20
TiO 4965

TiO 4761–63

TiO 4626

TiO 4584

BD Camelopardalis
S3.5/2 symbiotic

7100
6900
6700
6500
6300
6100
5900
5700
5500
5300
5100
4900
4700
4500

Plate 28 Comparison of an "Intrinsic" and "Extrinsic" S-Class Star

CHAPTER

15 Carbon Stars on the AGB

15.1 Overview and Spectral Characteristics

The final stage of the stellar evolution, at the top of the AGB, is populated by the deeply reddish shining carbon stars, in most of the cases also Mira variables or carbon Miras. Here, in the atmosphere of the star, the C/O ratio becomes C/O > 1. This results in a carbon excess which accumulates in a circumstellar cloud, impressively dominating the star's spectrum. Thus, in the intermediate class SC, and increasingly in the following C class, in moderately high resolved spectra predominantly absorptions of diatomic carbon molecules are recognizable. In addition to CH and CN, due to C_2, the so-called Swan bands appear particularly striking – discovered in 1856 by the Scot William Swan, (see also Plate 102). Further visible are atomic lines of s-process products and impressive absorptions of sodium (Na I). Father Angelo Secchi was the very first to discover that the intensity gradient of the C_2 Swan bands run reversely, compared to other molecular absorptions, such as titanium and zirconium monoxide. For this feature, he created the separate type IV (see Section 4.3). Figure 15.1 shows impressively how the carbon star R Sculptoris ejects thick shells of gas, partly in form of a spiral, probably created by a hidden companion star orbiting the red giant. The picture was recorded by the ALMA Submillimeter Array in Chile.

15.2 Competing Classification Systems

The phenomenon of carbon stars is still far from being fully understood [45]. For the S class, in spite of ongoing disputes, a generally accepted and consistently applied classification system exists. However, for carbon stars the situation is still confusing and unsatisfactory. Obviously the "Revised MK–Carbon Star Classification system 1993," or "Keenan 1993 system," propagated in most of the textbooks, is astonishingly applied very rarely. Its precursor from the 1940s, the "(MK) C-system," however, is very often applied! Surprisingly frequently one can even see classifications according to the much older Harvard Classification System, with the R and N classes, which was further developed by Henry Draper and Edward Pickering, based on Father Secchi's "type IV." Type N corresponds roughly to the "Secchi type IV" and is chiefly limited to the red region of the spectrum, whereas type R, extends into the violet range [2].

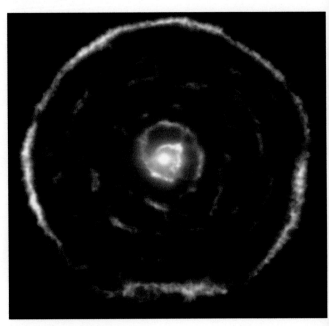

Figure 15.1 R Sculptoris
(Credit: ALMA (ESO/NAOJ/NRAO) M. Maerker et al.)

15.3 The Morgan–Keenan (MK) C-System

Due to difficulties with the N and R classes, this system was replaced in 1941 by Keenan and Morgan by a pure temperature sequence, supplemented just with a single index [46]. The temperatures of the carbon stars have been estimated by measured intensity ratios of certain absorptions like C2 or CN and have afterwards been linked to the classes in the temperature sequence of ordinary, oxygen-rich stars. Anyway, according to [2] and the present state of knowledge the effective temperatures for giants of the spectral classes G4–M4 should be in a range ~3500–5200 K. This simple, old classification system is still very popular, both in many professional publications, as well as in most of the stellar databases. The following format is applied: CX/n. Here X defines, on a scale of 0–7, the position of the star in the temperature sequence, equivalent to the spectral classes from G4 to M4 of ordinary (oxygen-rich) stars. The classification system was extended up to C9 (e.g. WZ Cas). This becomes also evident in the column of the C–N sequence in Table 15.3. Since all C stars are giants, a specification for the luminosity class is usually omitted. And n rates, on a scale of 1–5, the intensity of the C_2 Swan bands. In individual cases, appropriate supplements may be added, for example (e) for emission lines, further also intensive lines of s-process elements.

For example, C 2,5 indicates a stellar temperature, equivalent to the spectral class G9–K0, combined with highly intense C_2 Swan bands.

It is considered as problematic that the entire highly complex class of carbon stars is classified here by a simple, seven stage temperature sequence, supplemented just with one single intensity index!

15.4 The Revised Keenan 1993 System

In 1993, the old "C-system" was changed and expanded to the "Revised Keenan 1993 system" and further adapted to new findings [47]. It comprises five subclasses, subdivided by spectral symptoms – whose astrophysical background, however, remains widely unclear. Perhaps this could be one of the possible reasons why the acceptance of this complex system, even some 20 years after its introduction, seems to be still limited. For the "Keenan 1993 system" the following format is applied:

C Sub X n

Where Sub corresponds to the subclass of the carbon star according to Table 15.2. Where X, according to Table 15.3 is the classification of the star in the temperature sequence of the C class and is equivalent to the spectral classes from G4 to M8.

Table 15.1 Temperature sequence for carbon stars in the (MK) C-system [46]

C0	C1	C2	C3	C4	C5	C6	C7
G4–G6	G7–G8	G9–K0	K1–K2	K3–K4	K5–M0	M1–M2	M3–M4
~4500 K	~4300 K	~4100 K	~3900 K	~3650 K	~3450 K		

Table 15.2 Subclasses of the carbon stars according to the "revised Keenan 1993 system" [2] [53]

Subclass	Supposed status [45]	Criteria, spectral symptoms
C–R	Intrinsic	Intensive Swan bands (C_2) in the blue part of the spectrum. Considerable flux in the blue/violet area – decreasing with lower temperatures. Later types show weaker H-Balmer lines, Flux ratio Hγ/FeI4383 serves as a temperature indicator. s-process elements show average intensity. Intensity of $^{12}C^{13}C$ band head at λ4737 is above average.
C–N	Intrinsic	s-process elements are exceptionally intensive. Early C–N types show a tendency to Merrill–Sanford bands (SiC$_2$ silicon dicarbide). Strong, broad and diffuse absorption, or even a barely detectable flux in the blue range λ < 4400. Swan bands C_2 weaker than with the C–R types.
C–J	Intrinsic	Very intensive Swan bands (C_2) and CN absorptions, further Merrill–Sanford bands (Si C_2) and an infrared excess.
C–H	Extrinsic	Dominant CH absorptions in the blue/violet area. Fraunhofer G band at λ4300 exceptionally and s-process elements above average intensive.
C–Hd	?	Hd indicates hydrogen deficient. Hydrogen lines or CH absorptions weak or even absent. CN and Swan bands (C_2) above average intensive, very often irregular R Crb variables, showing sudden dramatic brightness dips, due to veiling of the photosphere by circumstellar dust clouds.

Table 15.3 Temperature sequence for carbon stars in the "revised Keenan 1993 system" [47]

Temperature sequence of the main sequence	C–R Sequence	C–N Sequence	C–H Sequence
G4–G6	C–R0		C–H0
G7–G8	C–R1	C–N1	C–H1
G9–K0	C–R2	C–N2	C–H2
K1–K2	C–R3	C–N3	C–H3
K3–K4	C–R4	C–N4	C–H4
K5–M0	C–R5	C–N5	C–H5
M1–M2	C–R6	C–N6	C–H6
M3–M4		C–N7	
M5–M6		C–N8	
M7–M8		C–N9	

Table 15.4 Indices with various specifications for the "revised Keenan 1993 system" [2]

Index	Specification
C_2	Intensity of the molecular C_2 Swan bands, scale 1–5.
CH	Intensity of the molecular CH absorption, scale 1–6.
MS	Intensity of the Merrill–Sanford bands (SiC_2), scale 1–5.
J	Intensity ratio of the C_2 molecular absorption with the isotopes $^{12}C^{13}C$ and $^{12}C^{12}C$, scale 1–5.
Elements	In some cases, for strong lithium and sodium lines according index values are specified.

According to [2] the actually accepted, effective temperatures for giants of the spectral classes G4–M8 would be in the range ~2940–5200 K. And n is the index with various specifications according to Table 15.4. Not included in this system are the class of the dC carbon dwarf stars located on the main sequence which are still not understood.

15.5 Connection of the Subclasses to the Evolution of Carbon Stars

Current professional publications provide a rather diffuse picture about the functions of these subclasses within the evolution of carbon stars. Many details, for example, about the "dredge up" processes, are obviously far from being fully understood. Further, the modeling of "carbon atmospheres" $C/O > 1$ seems to be difficult.

Analogously to the S class, carbon stars on the AGB are also referred to as "intrinsic." It is assumed that the subclass C–N, as well as the late representatives of C–R, form the last development stage of the AGB sequence. These subgroups show behavior like M giants with very similar spectra. Furthermore they are also "low mass objects" with $\lesssim 3\ M_\odot$, typically showing a mass loss and a variable brightness.

In contrast, early representatives of C–R type show neither variability nor a significant mass loss. Further, they seem to be not AGB stars but behave rather like K giants. One of the discussed scenarios postulates a star on the horizontal branch (HB) with a helium burning core. However, this would require a strongly instable stage, enabling to dredge up the carbon from the stellar core to the surface. Another hypothesis is that they are former binary systems whose masses have fused into single stars.

The position of the C–J class on the AGB remains unclear. It is argued that these low mass objects could be descendants of the C–R class on the AGB.

The C–H class seems to be "extrinsic." Most representatives seem to be located on the horizontal branch (HB) and components of close binary stars. This suggests a mass transfer scenario similar to the extrinsic S class.

15.6 Merrill–Sanford Bands (MS)

At the beginning of the twentieth century, intensive absorptions in spectra of certain carbon stars in the range of ~$\lambda\lambda 4640$–5200 attracted attention and for a long time could not be interpreted. The most intense band heads are at $\lambda\lambda 4977$, 4909, 4867, 4640 and $\lambda 4581$. These absorption bands are named after the American astronomers Paul W. Merrill (1887–1961) and Roscoe F. Sanford (1883–1958) who first described this absorption in 1926. Not until 1955 would the Swedish physicist Bengt Kleman (1922–2011) prove with laboratory spectra that these bands are caused by silicon dicarbide SiC_2, [49]. For this purpose he heated silicon up to 2400 K in the graphite tube of a King furnace.

P. J. Sarre *et al.* have also shown that the Merrill–Sanford bands are generated in cooler layers of the stellar atmosphere, far beyond the photosphere. Merrill–Sanford bands are the most common in the entire C–J and the early C–N classes [50].

15.7 Comments on Observed Spectra

The majority of carbon stars, documented here with some examples, are part of the subclass C–N. The spectral classes are indicated here in the old (MK) C-system, according to the original paper by Morgan and Keenan from 1941 [46] or, if available, by more recent data from [11] or [56]. If available, the classification corresponding to the Revised Keenan 1993 system is also specified [47].

Plate 29 Comparison of Differently Classified Carbon Stars

Plate 29 presents a montage of three broadband overview spectra (200 L mm^{-1}) to demonstrate the differences between the profiles of differently classified carbon stars.

WZ Cassiopeiae (HD 224855)

WZ Cassiopeiae has a spectral class C9,2 Li [56] or C–N7 III: C2 2, Li 10 [47]. This extremely late classified, cool supergiant is located at a distance of ~2000 ly in the constellation Cassiopeia with a variable brightness $m_v \approx 6.5$–8.5 and a very low effective temperature of just 2500 K. It forms the dominant component of a binary star system, impressively contrasting with its white-bluish shining companion star ($m_v \approx 8$, class A0).

Besides the rather weak Swan bands (here classified with index value 2), this spectrum is dominated by the striking, almost fully saturated Na I line. Further, by the impressive absorption of lithium Li I (λ6708), whose intensity is rated here as 10. Therefore, WZ Cassiopeiae is often called "Lithium star" [51]. This intense Li I absorption line was the first evidence of lithium outside the solar system, found by McKellar in 1941 – a small but interesting detail in the history of science [52]! Further the profile is dominated by CN and C$_2$ absorption bands and the H-Balmer lines are barely recognizable here. The recording information is: DADOS 200 L mm^{-1},

Figure 15.2 Binary WZ Cassiopeiae
(Credit: SIMBAD/Aladin Previewer)

35 μm slit, C8, Atik 314L+, 3×85 s. The line identification is based amongst others on [51], [52], [53], [54], [55].

Z Piscium (HD 7561)

This star, with spectral class C7,3 [46] or C–N5 C2 4 [47], is a supergiant with an effective temperature of some 3000 K and located some 1500 ly distant in the constellation Pisces. With C7,3 it is classified earlier than WZ Cas (C9,2). The variable brightness is reaching a maximum of $m_v \approx 6.8$. The carbon absorptions are here much more intense, anyway the Na I line somewhat weaker, but still very impressive! Instead of the exceptional lithium line (λ6708) the relatively undisturbed CN absorption shows up here at λ6656. Recording information: DADOS 200 L mm^{-1}, 50 μm slit, C8, Atik 314L+, 3×60 s. The line identification is based amongst others on [51], [53], [54], [55].

W Orionis (HD 32736)

W Orionis is of spectral class C5,4 [11]. The luminosity class of this carbon giant (~700 ly distant), is difficult to determine [14]. The variable brightness reaches a maximum of $m_v \approx 5.9$. The Merrill–Sanford bands appear striking here. In the profiles of the other two, much later classified carbon stars, these absorption bands of triatomic SiC$_2$ silicon dicarbide are hardly recognizable. Recording information: DADOS 200 L mm^{-1}, 25 μm slit, C8, Atik 314L+, 5×42 s. The line identification is based amongst others on [47], [50], [51], [53], [54].

Plate 30 Merrill–Sanford Bands: Details at a Higher Resolved Spectrum

W Orionis

Plate 30 shows the higher resolved spectrum for W Orionis (900 L mm^{-1}) in the wavelength domain of the Merrill–Sanford bands. The line identification is additionally based here on [57].

Plate 31 Carbon Star with Hα Emission Lines

R Leporis (Hind's Crimson Star, HD 31996)

Plate 31: R Leporis, with spectral class C7,6e [11] and ~1100 ly, is one of the reddest appearing stars, and probably the most famous representative of the carbon class, discovered in 1845 by John Russell Hind. It is located in the constellation Lepus with a temperature of about 2290 K. It is almost equally classified as Z Piscium, except of the index e, which documents that the Hα line appears in emission here. The variable brightness is in a range of $m_v \approx 5.5$–11 with a period of some 427 days. The spectrum was recorded in a phase of increasing brightness at $m_v \approx 7$.

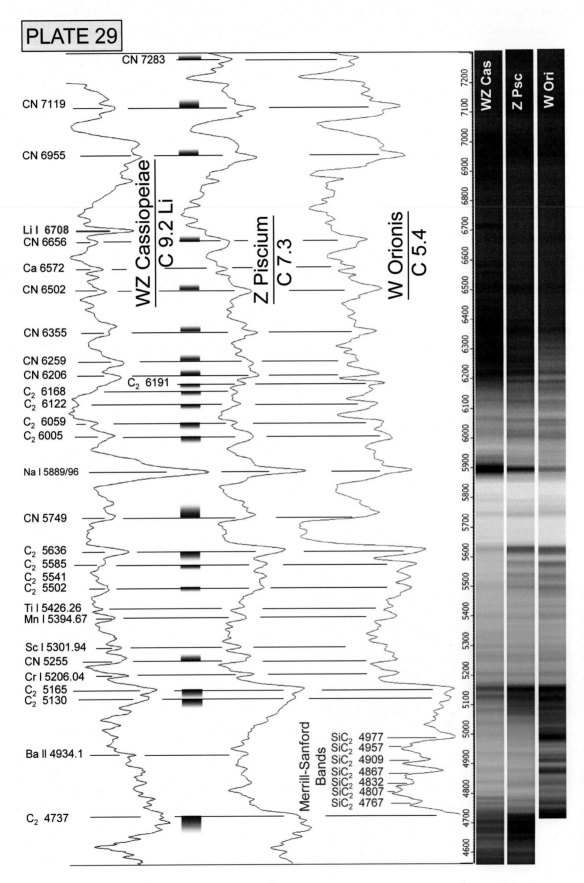

PLATE 29

CN 7283

CN 7119

CN 6955

WZ Cassiopeiae
C 9.2 Li

Z Piscium
C 7.3

W Orionis
C 5.4

Li I 6708
CN 6656

Ca 6572

CN 6502

CN 6355

CN 6259
CN 6206
C₂ 6168   C₂ 6191
C₂ 6122
C₂ 6059
C₂ 6005

Na I 5889/96

CN 5749

C₂ 5636
C₂ 5585
C₂ 5541
C₂ 5502

Ti I 5426.26
Mn I 5394.67

Sc I 5301.94
CN 5255
Cr I 5206.04
C₂ 5165
C₂ 5130

Merrill-Sanford
Bands

SiC₂ 4977
SiC₂ 4957
SiC₂ 4909
SiC₂ 4867
SiC₂ 4832
SiC₂ 4807
SiC₂ 4767

Ba II 4934.1

C₂ 4737

WZ Cas   Z Psc   W Ori

4600 4700 4800 4900 5000 5100 5200 5300 5400 5500 5600 5700 5800 5900 6000 6100 6200 6300 6400 6500 6600 6700 6800 6900 7000 7100 7200

Plate 29 Comparison of Differently Classified Carbon Stars

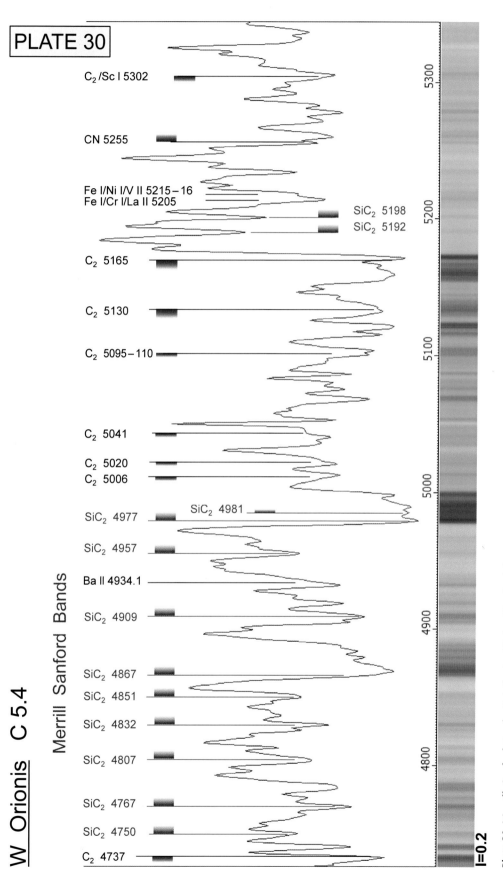

Plate 30 Merrill–Sanford Bands: Details at a Higher Resolved Spectrum

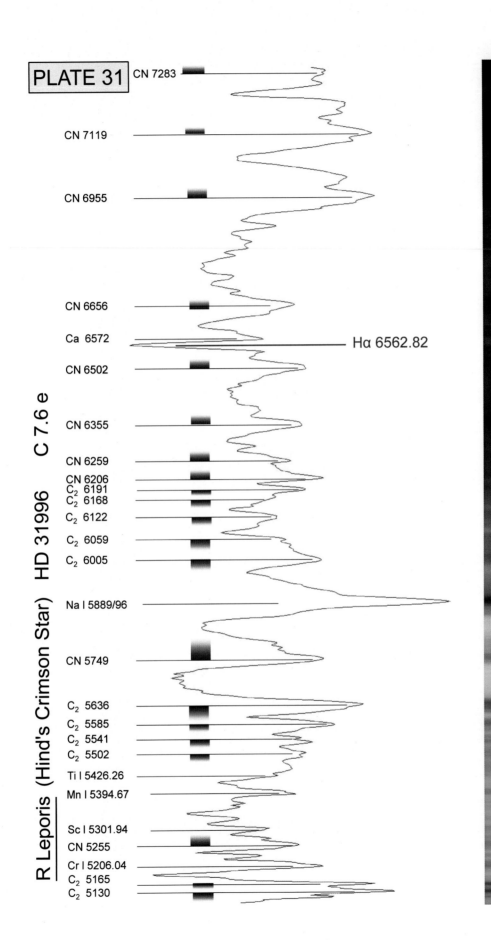

PLATE 31

CN 7283

CN 7119

CN 6955

CN 6656

Ca 6572

Hα 6562.82

CN 6502

CN 6355

CN 6259

CN 6206
C₂ 6191
C₂ 6168

C₂ 6122

C₂ 6059

C₂ 6005

Na I 5889/96

CN 5749

C₂ 5636

C₂ 5585

C₂ 5541

C₂ 5502

Ti I 5426.26

Mn I 5394.67

Sc I 5301.94

CN 5255

Cr I 5206.04
C₂ 5165
C₂ 5130

R Leporis (Hind's Crimson Star)   HD 31996   C 7.6 e

I=0.0

7200
7100
7000
6900
6800
6700
6600
6500
6400
6300
6200
6100
6000
5900
5800
5700
5600
5500
5400
5300
5200

Plate 31 Carbon Star with Hα Emission Line

16 Post-AGB Stars and White Dwarfs

16.1 Position of Post-AGB Stars in Stellar Evolution

This section explains the final stage for stars with $\lesssim 8\,M_\odot$. Their mass is too low to generate a core collapse supernova explosion of type II. Exclusively as a member of a binary system a white dwarf may become a SN type Ia by accretion of matter and finally exceeding its critical Chandrasekhar mass limit (Section 25.4).

16.2 Post-AGB Stars

After the final AGB stage as a carbon star it begins to repel its envelope as a planetary nebula. This stellar stage is called "post-AGB" and includes the central stars of planetary nebulae. Such a very early post-AGB object (HD 44179) is presented in Section 28.13 (Plate 66) as an exciting source of the protoplanetary nebula "Red Rectangle." It is important to note such proto- or pre-planetary nebulae must never be confused with the protoplanetary accretion disks during the birth stage of a star, according to Section 21.1!

16.3 Spectral Features of Post-AGB Stars

As the most important peculiarity during the whole post-AGB phase the spectra of these rather small objects are simulating much higher masses! At the very beginning the expanding shell of gas forms a kind of a pseudo-photosphere, a pretence of a stellar giant, by generating very slim and intense absorptions. Thereby the star becomes increasingly hotter, rising up within a very short time through the whole spectral sequence from K to O [2], mostly exhibiting an extremely low metallicity. Finally, it becomes hot enough to ionize the expanding planetary nebula, with a dramatically increasing temperature in this stage it performs an impressive loop in the upper part of the HRD and passes almost all spectral classes. In extreme cases, its temperature may finally reach far beyond 100,000 K and the dying star may simulate for a very short time a Wolf–Rayet-like spectrum of the type WRPN (see Section 17.5).

16.4 White Dwarfs

By repelling a planetary nebula, the star loses mass, the thermonuclear fusion processes inside the star extinguish and the remnant is finally reduced to an approximately Earth-sized, extremely dense object, with an enormously strong surface gravity [3]. Most of the white dwarfs are composed of a "degenerate" carbon–oxygen core, the products of the former helium fusion, and a thin shell of hydrogen and helium. The spectral class drops within the HRD [3] to the lowest domain of the white dwarfs. The absorption spectrum is produced here just by the residual heat of the very slowly cooling stellar "corpse." Their absolute luminosity is now so low that for amateurs only a few objects are detectable, located in the immediate solar neighborhood, for example within a radius of ~50 ly. A corresponding list can be found in [58]. The nearest and apparently brightest white dwarfs are the companion stars Sirius B and Procyon B. However, due to their close orbits around the much brighter A-components they remain spectroscopically unreachable for amateurs. Easiest to observe, visually and somewhat limited also spectroscopically, is 40 Eridani B ($m_v = 9.5$) as a component of a triple system, not identified as a white dwarf until 1910.

Visually, this object was observed already by William Herschel in 1783. The apparent brightness of the remaining white dwarfs lies already within a range of $m_v \approx 12$–13.

16.5 Spectral Characteristics and Special Features of White Dwarfs

The excessively high surface gravity causes extremely broadened absorption lines – especially due to the Stark effect. This effect forms here the spectral "brand" and affects mainly the hydrogen Balmer series. In addition, also lines of helium and calcium may appear. Due to the low brightness, the display of the finer absorption lines is reserved to high-resolution spectrographs at large professional telescopes.

The surface gravity is so extremely high, that the gravitational redshift, predicted by Einstein's general theory of relativity (GTR), could be relevant even for amateurs [3]. The required work for the light (photons) to escape the gravitational field of a white dwarf, causes at Hα a redshift of slightly < 1 Å, corresponding to a radial velocity of about 20–40 km s^{-1}. This value is star-specifically different and must therefore inevitably be accounted for by measurement of the Doppler shift which, however, is strongly aggravated due to the massively broadened H-lines. For the spectral class DA ~ 30 km s^{-1} can be assumed [59]. However, recent measurements with the HST yielded for Sirius B, class DA 1.9, a significantly higher value of ~ 80 km s^{-1} (solar reference value ~ 600 m s^{-1}).

In astrophysics the surface gravity g is expressed as decimal logarithm, however, not in the otherwise usual SI system [m s^{-2}], but in cgs units [cm s^{-2}]. The cgs system (centimeter–gram–second) is still very common in the field of spectroscopy [3]. The log g values of white dwarfs are in the range of ~ 7–9. For Sirius B it yields $\log g = 8.57$, corresponding to some $371,500,000$ cm s^{-2} or $3,715,000$ m s^{-2} (the terrestrial reference value is $g \approx 9.81$ m s^{-2}).

16.6 Classification System by McCook and Sion

This classification system is among earlier works mainly based on Liebert and Sion [61] and further on McCook and Sion, published in the *McCook–Sion White Dwarf Catalog* [60]. Presented in Table 16.1 is a further updated version by [2]. The first letter of the classification is D, which means "degenerate." The second letter indicates the primary spectroscopic characteristics [2], [61]. Apart from type DA the physical background of most other subclasses is not yet fully understood.

Table 16.1 Updated version of the classification after McCook and Sion [2]

Spectral Class	Characteristics
DA	Only H-Balmer lines; no He I or metals
DB	He I lines; no H or metals present
DC	Continuous spectrum, no lines deeper than 5% in any part of the spectrum
DO	He II strong; He I or H present
DZ	Metal lines only; no H or He lines
DQ	Carbon features, either atomic or molecular in any part of the spectrum
Additional letters and symbols that may specify further effects and features	
P	Magnetic white dwarfs with detectable polarization
H	Magnetic white dwarfs without detectable polarization
X	Peculiar or unclassifiable spectrum
E	Emission lines are present
?	Uncertain assigned classification
V	Optional symbol to denote variability
d	Circumstellar dust
C I, C II, O I, O II	Atomic species in spectra of hot DQ dwarfs

The classification digit, following this letter combination, specifies the effective temperature T_{eff} according to the equation:

$$T_{\text{eff}} = 50,400/\text{classification digit} \qquad \{16.1\}$$

For rarely occurring temperatures above 50,400 K the classification digit becomes <1.0 and is expressed for example as 0.9, 0.8, 0.7 etc.

Finally, sometimes also the log g value of the surface gravity is added. So the classification, for example DA 2.5_7.8 means a spectrum with hydrogen lines, T_{eff} of about 20,000 K and a surface gravity of $\log g = 7.8$.

16.7 Comments on Observed Spectra

Plate 32 White Dwarfs with Different Spectral Classifications

Plate 32 shows a montage of broadband spectral profiles (200 L mm^{-1}) of three differently classified white dwarfs, all

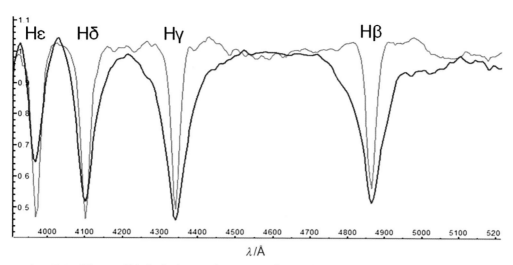

Figure 16.1 Compared to Sirius (blue profile) the hydrogen absorptions of 40 Eridani B (red profile) appear strongly broadened

normalized on the same section of the continuum. Depending on the source, the indicated spectral class may vary considerably [62].

WD 0644 +375 Gliese 246

This white dwarf has spectral class DA 2.3, with m_v = 12.1 and is at a distance of 62 ly [11]. Apart from the highly broadened hydrogen lines no further spectral features are recognizable here. The equivalent width of the Hβ line is EW $\approx$ 65 Å, about 2.4 times as large as the impressive H-absorptions of the main sequence star Sirius, A1Vm (EW $\approx$ 27 Å). The classification digit 2.3 suggests here an effective temperature of about 22,000 K. Recording information: DADOS 50 μm slit, C8, Atik 314L+: 1×1800 s, 2×2 binning.

WD 0413–077 40 Eridani B

This white dwarf has spectral class DA 2.9, with m_v = 9.5 and is at a distance of 16 ly [11]. Here again, the strongly broadened hydrogen lines are the most prominent spectral features. The equivalent width of the Hβ line is here EW $\approx$ 78 Å even about three times as large as the main sequence star Sirius. Compared to WD 0644 +375 this is caused by the lower temperature of 40 Eridani B, favoring this type of absorption and may be estimated by the classification digit, to about 16,000 K. Figure 16.1 shows a montage of the profiles of 40 Eridani B (red) and Sirius (blue). It is striking that the broadening chiefly affects the Hβ and to a lesser extent the Hγ line.

On Plate 32 the Hα line at 40 Eridani B appears strongly deformed. Whether this effect is caused by the strong magnetic fields could not be clarified here. Recording information: DADOS 50 μm slit, C8, 2×1800 s, 2×2 binning. For further info and tips to record this very close binary star component refer to [3].

WD 0046 +051

Known as Van Maanen's Star, this has spectral class DZ 8, is at a distance of 14 ly, with m_v = 12.4 [11]. Adriaan Van Maanen discovered this object in 1917 as the first standalone white dwarf. This object has already cooled down to ~6000 K, which is a similar temperature range to the solar photosphere. Its surface is possibly strongly contaminated with interstellar and planetary particles (metals) [63]. Therefore, at this resolution just the two intense and strongly broadened Fraunhofer H and K lines of ionized calcium Ca II can be seen. Recording: DADOS 50 μm slit, C8, 1 × 1800 s, 2 × 2 binning.

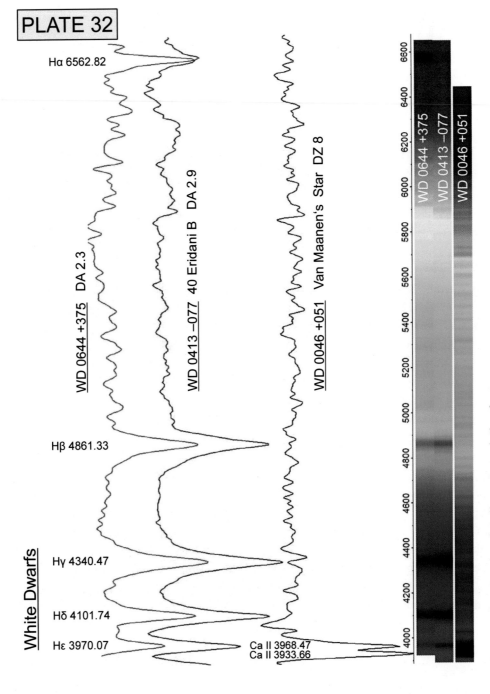

PLATE 32

White Dwarfs

Hα 6562.82

Hβ 4861.33

Hγ 4340.47

Hδ 4101.74

Hε 3970.07

Ca II 3968.47
Ca II 3933.66

WD 0644 +375   DA 2.3

WD 0413 –077   40 Eridani B   DA 2.9

WD 0046 +051   Van Maanen's Star   DZ 8

WD 0644 +375
WD 0413 –077
WD 0046 +051

4000 4200 4400 4600 4800 5000 5200 5400 5600 5800 6000 6200 6400 6600

Plate 32 White Dwarfs with Different Spectral Classifications

CHAPTER

17 Wolf–Rayet Stars

17.1 Overview

The French astronomers Charles Wolf and Georges Rayet discovered in 1867 very rare stars, whose spectra stand out by massively broadened, intense helium emission lines, combined with an almost complete absence of hydrogen. These objects are grouped in an extra class, marking the final stage of massive O type stars with $\gtrsim 25\ M_\odot$. Below this mass limit of ~25 $M_\odot$, it is assumed that the SN explosion takes place directly at the end of the red giant stage, without passing through a WR phase.

In the northern hemisphere mainly in the constellation Cygnus, a concentration of such extreme stars can be found which are members of the Cygnus OB associations. Here, some 23 WR stars are located, ~14 are classified as WN-type, ~8 as WC, and just one as WO type (WR142). The two brightest of them reach an apparent magnitude in the range of $m_v \approx 6$–7 and are therefore well accessible with a slit spectrograph – even for moderately sized amateur telescopes. By chance they just represent the two main types, WN and WC.

At the beginning of the WR phase, the star blasts away its entire outer hydrogen shell resulting in a huge stellar wind with velocities of up to 2000 km s^{-1}. Further on the path to the SN of the category Ib or Ic (Section 25.4), the layers of the stellar core are repelled from the top down, similar to the peeling of an onion. This causes an annual mass loss rate of about 10^{-5} to $10^{-4}\ M_\odot$ [70]. On the surface of the star the extremely hot, former nuclear H- and He fusion zones become exposed where, in addition to helium, the metals C, N, and O have been formed. Depending on the source, the entire WR stage is estimated to last about 200,000–500,000 years.

17.2 Spectral Characteristics and Classification

The spectrum differs completely now from the previous O class. The broad lines of ionized helium He II appear together with the emissions of highly ionized C, N, or O – depending on the "currently" exposed surface layer of the former H- and He fusion layers [70]. These metals determine the three subclasses of the WR stars WN, WC and WO and their high stage of ionization is an indicator of the extremely high temperatures and the corresponding excitation energies (Appendix F). Similar to the other spectral types the subclasses of the WR stars are further subdivided into decimal sub-subclasses (see Table 17.1).

Limited to the very early stages of the WR evolution, the H-Balmer series of the not yet completely repelled hydrogen shell can still be detected exclusively within blends at the very late WN (WNL) types (see Table 17.2), [2], [70].

Table 17.1 Subclasses of WR stars

WR Subclass	Spectral Features in the Optical Range
WN	WR stars with nitrogen emissions:
	Late subtypes WNL: (WN11, WN10) exhibit H lines (Table 17.2)
	Early subtypes WNE: without any H lines [70]
WC	WR stars with carbon emissions (by partial helium fusion)
WO	WR stars with oxygen emissions (by complete helium fusion, very rare)

Table 17.2 Smith classification system for WR stars [65]

WN	Nitrogen Emission Lines	Other Emission Criteria	T_{eff} [K]
WN11	N II ≈ He II, N III weak or absent	Balmer lines, He I P-Cyg	WN11: ~30,000
WN10	N III ≈ N II	Balmer lines, He I P-Cyg	WN8: ~40,000
WN9	N III > N II, N IV absent	He I P-Cyg	WN2: ~100,000
WN8	N III ≫ N IV, N III ≈ He II 4686	He I strong P-Cyg	[66]
WN7	N III > N IV, N III < He II 4686	He I weak P-Cyg	
WN6	N III ≈ N IV, N V present but weak		
WN5	N III ≈ N IV ≈ N V		
WN4.5	N IV > N V, N III weak or absent		
WN4	N IV ≈ N V, N III weak or absent		
WN3	N IV ≪ N V, N III weak or absent		
WN2.5	N V present, N IV absent		
WN2	N V weak or absent	He II strong	

WC	Carbon Emission Lines	Other Emission Criteria	T_{eff} [K]
WC9	C III > C IV	C II present, O V weak or absent	WC9: ~50,000
WC8	C III > C IV	C II absent, O V weak or absent	WC8: ~70,000
WC7	C III < C IV	C III ≫ O V	WC4: ≫100,000
WC6	C III ≪ C IV	C III > O V	[66]
WC5	C III ≪ C IV	C III < O V	
WC4	C IV strong, C II weak or absent	O V moderate intensity	

WO	Oxygen Emission Lines	Other Emission Criteria	T_{eff} [K]
WO4		C IV ≫ O VI, C III absent	
WO3	O VII weak or absent	C IV ≈ O VI, C III absent	
WO2	O VII < O V	C IV < O VI, C III absent	
WO1	O VII ≥ O V, O VIII present	C III absent	~200,000

In the later stages of WR development, the "early" WC subclasses show, though still weakly, the highly ionized oxygen O VI doublet at λλ3811 and 3834.

In the final WO phase, this O VI doublet appears strongly developed, together with other, highly ionized oxygen emissions (Plate 34). Pure WO stars are extremely rare and the corresponding final phase is probably very short – in addition associated with a significant X-ray radiation.

17.3 Classification System for WR Stars in the Optical Spectral Range

Table 17.2 shows the Smith classification system (1968, according to [65]) for the decimal subclasses based on the

ionization stage and the intensity of dedicated emission lines. The higher the subclass, the higher the ionization stage. This system was amended by many authors, including K. van der Hucht *et al.* [65]. The subtypes are here sorted so that the age and temperature of the individual WR stages increase from top to bottom. Within each WR subclass the development takes place against the "semantic logic," i.e. from "late" to "early."

Classification lines in the optical spectral range:

WN Stars: He I λ3888, He I λ4027, He I λ4471, He I λ4921, He I λ5875, He II λ4200, He II λ4340, He II λ4541, He II λ4686, He II λ4861, He II λ5411, He II λ6560, N II λ3995, N III λλ4634–4641, N III λ5314, N IV λ4058, N V λ4603, N V λ4619 and N V λλ4933–4944.

WC Stars: C II λ4267, C III λ5696, C III/C IV λ4650, C IV λλ5801–12 and O V λλ5572–98.

WO Stars: C IV λλ5801–12, O V λλ5572–98, O VI λλ3811–34, O VII λ5670 and O VIII λ6068.

17.4 The WR Phase in Stellar Evolution

Crowther *et al.* have proposed three different evolutionary scenarios for WR stars, as a function of the initial stellar mass M_i, and these are shown in Table 17.3 [66].

According to [66] the evolutionary role of the LBV phase (Chapter 18) is not yet clear and may possibly be skipped in some cases. It also unclear in which cases a WO stage is passed before the onset of the final SN explosion.

Figure 17.1 shows the star WR124 (2200 ly) located in the constellation Sagitta. With the very late spectral class WN8 it is still at the very beginning of the WR stage. Just about

Table 17.3 Evolutionary scenarios for WR stars after Crowther *et al.* [66].

$M_i > 75\,M$	O → WNL → LBV → WNE → WC → SN Type Ic
$40\,M < M_i < 75\,M$	O → LBV → WNE → WC →WO →SN Type Ic
$25\,M < M_i < 40\,M$	O → LBV/RSG → WNE→ WC → SN Type Ib

Figure 17.1 Young WR124 (WN8), repelling its hydrogen envelope (Credit: ESA/Hubble and NASA; J. Schmidt)

10,000 years ago, it started to repel its hydrogen envelope with a stellar wind of about 2000 km s^{-1}.

In the southern sky, some 500 ly distant, the famous γ Velorum, alias WR11, classified as WC8, is much further developed than WR124. A very long time ago it repelled its hydrogen envelope, and is therefore no longer visible as a WR nebula. With $m_v \approx 1.8$ it is by far the apparently brightest representative of all WR stars.

Such "stellar monsters" with originally about 25–80 $M_\odot$ and effective temperatures of 30,000–100,000 K, will end up in a final cataclysmic SN explosion and the small remnant will most likely implode and end as a black hole. This event will probably be accompanied by a high-energy gamma-ray burst, ejected in both directions of the stellar rotation axis (Section 25.6).

17.5 Analogies and Differences to the Central Stars of Planetary Nebulae

The far smaller central stars of planetary nebulae (PN) also repel their shells (Section 28.3). During their final stage, they simulate similar spectra, and are also classified as WR types, denoted as WRPN. However, their absolute magnitude and mass is significantly lower and they finally end up, much less spectacular, as white dwarfs (Section 16.4).

17.6 Comments on Observed Spectra of the WR Classes WN, WC and WO

Plate 33 Wolf–Rayet Stars: Types WN and WC

Plate 33 shows a montage of two broadband overview spectra (200 L mm^{-1}) of the WR subtypes WN and WC.

WR133 (HD 190918)

Some 6500 ly distant, with a spectral class WN5 and $m_v \approx 6.8$, the very well observed WR133 in the constellation Cygnus hides itself inconspicuously within the central members of the open cluster NGC 6871. Despite the use of "GoTo" telescopes there is a certain risk of recording the wrong star – however, that will be instantly recognizable in the spectrum. Particularly the bright He II "emission knots" can hardly be overlooked (Figure 17.2).

As a spectroscopic binary (SB2) WR133 orbits with an O9 I supergiant within a period of about 112 days. Therefore, we see a composite spectrum, which is clearly dominated by the emission lines of the WR star (labeled red). The numerous lines of differently high ionized nitrogen, responsible for the classification WN are striking here. However, one single faint carbon emission shows up, C IV at λ5801–12 [67]. This spectral feature is significantly weaker, compared to the WC type WR140.

Figure 17.2 Unprocessed raw picture of WR133, DADOS 200 L mm^{-1}

Figure 17.3 Unprocessed raw picture of WR140, DADOS 200 L mm^{-1}

The very close O9 companion star contributes some absorptions of helium- and H-Balmer lines (labeled black in Plate 33). Due to the extremely high temperature T_{eff} the sodium double line here is inevitably of interstellar origin [82]. In Section 28.13 the very similarly classified WR7 is presented as ionizing source of NGC 2359, showing a "pure" WN4 spectrum (Plate 75). Compared to WR140 (presented later), the continuum is here relatively intense. The strong He II emission at λ6560 poses a considerable risk to be misinterpreted as Hα line at λ6562. The most intense line here is clearly He II at λ4686. The terminal velocity of the stellar wind can be estimated from the FWHM values of the two intense He II emissions. Corrected by the instrumental broadening and inserted into the basic Doppler equation:

$$v_\infty = \frac{\text{FWHM}}{\lambda_0} c \qquad \{17.1\}$$

where λ_0 is the rest wavelength of the analyzed line, c is the speed of light in a vacuum, 300,000 km s^{-1}, the terminal stellar wind velocities are found to be $v_\infty \approx 1800$ km s^{-1} at λ4686 and $v_\infty \approx 1500$ km s^{-1} at λ6560.

In the order of magnitude, these figures agree quite well with values of $v_r < 2000$ km s^{-1} found in the literature. The spectrum was recorded with the 36 inch CEDES Cassegrain Telescope in Falera – exposure: 4×30 s. The line identification is based amongst others on [65], [67], [68], [69], [70].

WR140 (HD 193793)

Also located in the constellation Cygnus and some 8900 ly distant, WR140 is a member of a spectroscopic binary system (SB2) with an O5 V main sequence star in a highly eccentric orbit with a period of some 2900 days. It has spectral class WC7 and $m_v \approx 6.9$. During the periastron passages this binary system usually attracts the worldwide attention of professional and amateur astronomers, mainly observing effects caused by the colliding stellar winds (colliding wind binary) [71], [72].

This composite spectrum is clearly dominated by the WR star. One of the reasons is presumably the time of the last periastron passage of January 2009. At the time of recording it dated back more than 1.5 years. In contrast to the very close binary system WR133, absorption lines are barely visible here, besides the well known telluric lines and the interstellar Na I absorption. The numerous different lines of highly ionized carbon show clearly that WR140, with the classification WC7, is a representative of the carbon type among the WR stars. Nitrogen is not detectable. The intense C III/C IV emission at λ4650 is blended with the He II line at λ4686. This feature is the most striking of the spectrum, followed by the C IV emission line at λ5801–12 and the C III "hump" at λ5696. The He II emission at λ6560 is relatively weak. The spectrum was recorded with Celestron C8/Meade DSI III, 10× 90 s, 1×1 binning. The line identification is based amongst others on [65], [67], [68], [69], [70].

Plate 34 Wolf–Rayet Stars: Types WN and WO

Plate 34: The montage of the two broadband overview spectra (200 L mm^{-1}) allows for the WR sequence the direct comparison of the early stage WN with the final stage WO, probably shortly before the onset of the final SN explosion.

WR136 (HD 192163)

At a distance of about 4700 light years, WR136 is located in the constellation Cygnus. It has spectral class WN6 and $m_v = 7.5$ and belongs to the Cygnus OB1 association. With $T_{eff} \approx 55,000$ K [73], it forms the origin and ionizing source of the elliptical shaped Crescent emission nebula NGC 6888 (Plate 74). Within about 30,000 years this repelled hydrogen shell has expanded to ~16 × 25 ly [74] and is still visible. Therefore WR136 is still at the beginning of the WR sequence. It is clearly further developed than WR124, classified with WN8. The velocity of the stellar wind is

Table 17.4 Pickering and Balmer series

Balmer H I [λ]	Pickering He II [λ]
6563	6560
–	5412
4861	4859
–	4542
4340	4339
–	4200
4102	4100

$v_r \approx 1700$ km s^{-1} [65]. It generates the visible shockwave within NGC 6888 by collision with the interstellar matter, which still propagates with about 75 km s^{-1} [75]. WR136 has no confirmed companion star.

The spectrum is dominated by numerous striking He II emissions. Most, but not all of the more intense of them belong to the Pickering series, which was discovered in 1896 by E. Pickering (Table 17.4 and Plate 34). Towards shorter wavelengths it shows a similar decrement, i.e. intensity loss, like the H-Balmer series (Section 28.7 and [1]). Further some emissions of the He II Pickering series, generated by electron transitions with base level $n = 4$, are also very close to the H-Balmer lines because the much heavier nucleus of the He II ion is orbited just by one electron on the innermost shell. If hydrogen is present in the spectrum, it becomes evident in the blends with He II lines, whose peaks clearly exceed the decrement line of the Pickering series [2]. At WR136, hydrogen is theoretically detectable at $\lambda \leq H\beta$ [73]. The spectrum was recorded with Celestron C8/Atik 314L+, 6 × 120 s. The line identification is based here on [65], [68], [69], [70].

WR142

Also known as Sand 5 [76] or ST3 [77], WR142 has a spectral class WO2 and $m_v \approx 13.8$ [65] or 13.4 [78].

The 4000 ly distant WR142 in the constellation Cygnus, is inconspicuously embedded in the open star cluster Berkeley 87. It is currently one of four oxygen types of the WO stage, detected in the Milky Way. In addition to WR142 and a recently discovered specimen in the Scutum arm of the Milky Way [79], there is also WR102 ($m_v = 15.8$). All are very probably close to a final explosion as a SN. In the Magellanic Clouds three other WO stars have been detected. Like most of the WO class stars also WR142 is an active X-ray source [81]. Five of these WO stars are numbered with Sand 1–5, named after Nicholas Sanduleak, who, in the 1970s was searching with objective prism for these "exotic stars." These extremely rare and highly interesting objects are well researched and documented.

Compared to WR136, WR142 shows mainly extremely broad and highly ionized emissions of oxygen. These obviously excessive amounts of energy, generated in the final stage of a WO star, can very roughly be estimated with help of the generated ionization stages. The required energy to ionize oxygen to the stage O VI is 113.9 eV (Appendix F). This is as much as 4.6 times the energy necessary to generate He II and 8.3 times the amount required to ionize hydrogen, H II. In addition to the extremely high temperatures, this may also be caused due to the photoionization by X-ray sources [80].

The Doppler analysis of the line widths yield here stellar wind velocities of 3600 km s^{-1} up to >5000 km s^{-1}, which is some 30–50% of the radial velocity of a typical SN explosion (Chapter 25) or ~1.6% of the speed of light! The unblended shape of the O VI line at $\lambda 5290$ is suitable for amateurs to measure the FWHM, enabling the estimation of the stellar wind velocity by Doppler analysis.

With $m_v = 13.8$, WR142 is by far the most apparent brightest WO representative and for sure the only one, which can be recorded with a C8/DADOS setup. The spectrum was obtained with Celestron C8/Atik 314L+, 4 × 1300 s, 2 × 2 binned. The line identification is based here on [2], [79], [81].

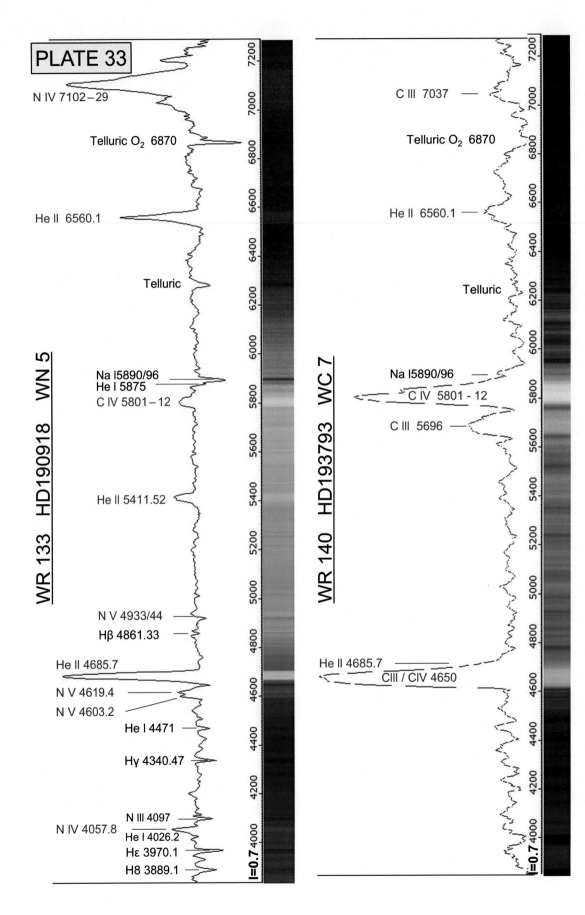

PLATE 33

N IV 7102–29

Telluric O₂ 6870

He II 6560.1

Telluric

Na I5890/96
He I 5875
C IV 5801–12

He II 5411.52

N V 4933/44

Hβ 4861.33

He II 4685.7

N V 4619.4

N V 4603.2

He I 4471

Hγ 4340.47

N III 4097

N IV 4057.8

He I 4026.2

Hε 3970.1

H8 3889.1

WR 133   HD190918   WN 5

I=0.7

C III 7037

Telluric O₂ 6870

He II 6560.1

Telluric

Na I5890/96

C IV 5801 - 12

C III 5696

He II 4685.7

CIII / CIV 4650

WR 140   HD193793   WC 7

I=0.7

Plate 33 Wolf–Rayet Stars: Types WN and WC

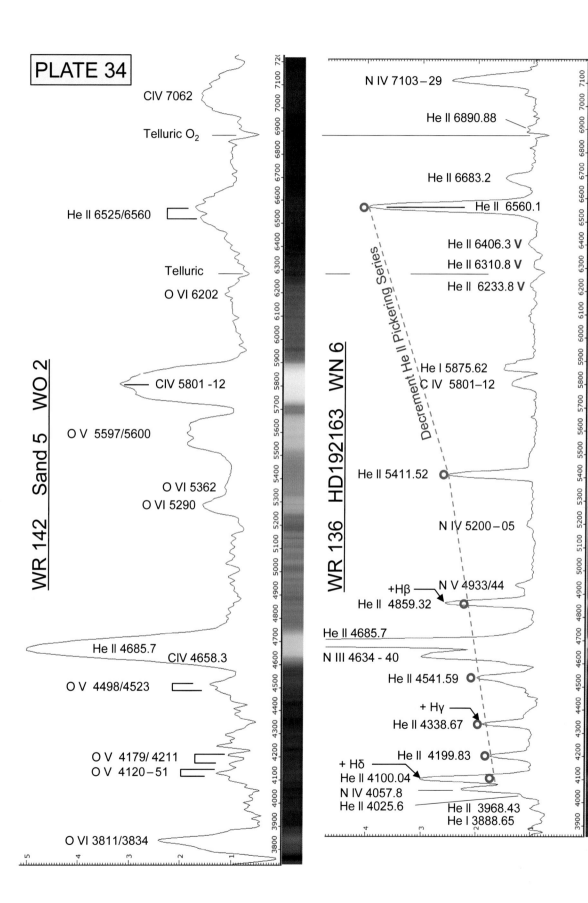

PLATE 34

WR 142   Sand 5   WO 2

CIV 7062

Telluric O₂ —

He II 6525/6560

Telluric —

O VI 6202

CIV 5801 -12

O V 5597/5600

O VI 5362

O VI 5290

He II 4685.7

CIV 4658.3

O V 4498/4523

O V 4179/ 4211

O V 4120 – 51

O VI 3811/3834

WR 136   HD192163   WN 6

N IV 7103 – 29

He II 6890.88

He II 6683.2

He II 6560.1

He II 6406.3 **V**

He II 6310.8 **V**

He II 6233.8 **V**

He I 5875.62

C IV 5801–12

He II 5411.52

Decrement He II Pickering Series

N IV 5200 – 05

+Hβ —
He II 4859.32   N V 4933/44

He II 4685.7

N III 4634 - 40

He II 4541.59

+ Hγ
He II 4338.67

He II 4199.83

+ Hδ
He II 4100.04

N IV 4057.8

He II 4025.6   He II 3968.43
                     He I 3888.65

Plate 34 Wolf–Rayet Stars: Types WN and WO

CHAPTER

18 LBV Stars

18.1 Overview

P Cygni is one of the "luminous blue variables" (LBV), probably progenitors of WR stars. It is an unstable and variable supergiant of spectral class B1–2 Ia–0 pe [11], with an effective temperature of about 19,000 K. The distance to P Cygni is estimated to be some 6000 ly [11]. What we analyze here with our spectrographs, apparently in "realtime," therefore took place at the end of the Neolithic period! At the beginning of the seventeenth century, it showed a tremendous outburst, which is known as Nova Cygni 1600. After about six years as a star of the third magnitude, the brightness decreased to weaker than $m_v \approx 5$. Apart from a few minor episodic outbursts, the luminosity increased again in the eighteenth century, until it reached the current, slightly variable value of $m_v \approx 4.8$. Further examples of LBV stars are S Doradus, Eta Carinae and AG Carinae (Figure 18.1). Depending on the source, the duration of the entire LBV phase is estimated to be approximately 20,000–40,000 years.

18.2 Spectral Characteristics of LBV Stars

P Cygni is the eponymous star to generate so called P Cygni profiles, already presented in Plate 3 (Alnitak) and Plate 6 (Alnilam). Such spectral features can be found in nearly all spectral classes and are a reliable sign of a huge amount of matter radially ejected by a stellar wind. Not surprisingly, this phenomenon sometimes appears in the spectra of novae and supernovae. In general, spectra of LBV stars are highly variable, particularly the intensity of the emission lines. In addition to the H-Balmer series mainly He, N and Fe appear here in various stages of ionization. Figure 18.1 shows the expanding shell of LBV star AG Carinae (~20,000 ly), recorded with the Hubble Space Telescope (HST).

For the case of expansion Figure 18.2 shows schematically the formation of the P Cygni profiles. A small section of the expanding shell is moving exactly in our line of sight, becoming quickly cooler. Thus, considered against the light of the radiation source, an absorption trough is generated, appearing strongly blueshifted by the Doppler effect.

The hot, sideward expanding sections of the shell generate a fairly symmetric emission line which may appear to be slightly redshifted. The addition of both spectral features results in the "P Cygni profile." In the case of expansion the blueshifted absorption part continuously transits into the emission peak. However, in the case of contraction the absorption part appears in the "inverse P Cygni profiles" to be redshifted [3].

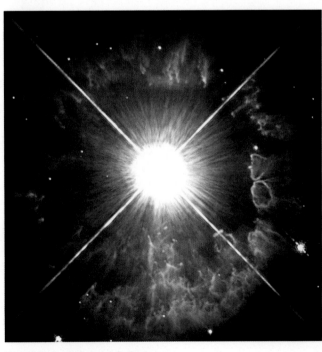

Figure 18.1 LBV star AG Carinae
(Credit: ESA/Hubble and NASA)

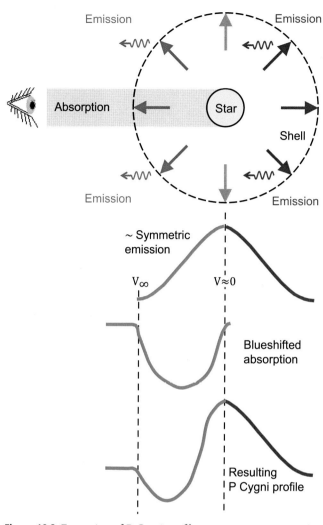

Figure 18.2 Formation of P Cygni profiles

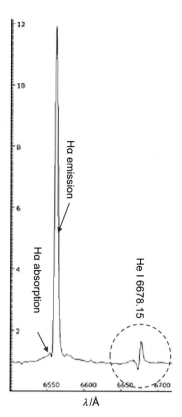

Figure 18.3 P Cygni Hα and He I emission

The specific shapes of the individual P Cygni profiles may differ considerably and depend on a number of factors, for example the density and transparency of the expanding shell as well as the intensities of the absorption and emission parts.

Figure 18.3 shows the highly impressive Hα emission line of P Cygni (900 L mm⁻¹). Therefore, the other spectral features appear to be strongly compressed, so they are barely recognizable. The blue absorption part of the Hα line is here impressively stunted. However, the helium emission at λ6678 shows here, and in the lower profile of Plate 35, a textbook-like P Cygni profile. The same applies to Hδ in the upper profile of Plate 36.

In this low resolved profile the difference of $\Delta\lambda \approx 4.4$ Å between the absorption and emission peak of the line, the basic spectroscopic Doppler equation enables the simplistic estimation of the terminal velocity v_∞ of the expanding shell to some 200 km s⁻¹.

$$v_\infty = \frac{\Delta\lambda}{\lambda_0} c \qquad \{18.1\}$$

where $\lambda_0 = 6563$ Å is the rest wavelength and $c = 300{,}000$ km s⁻¹ is the speed of light. In highly resolved profiles and on a professional level this calculation is mostly based on the width of the absorption trough [3].

18.3 Comments on Observed Spectra

Plate 35 LBV Star, Early B Class, P Cygni Profiles

Plate 35 shows for P Cygni the broadband overview spectrum (200 L mm⁻¹) and a higher resolved profile around the Hα line (900 L mm⁻¹). The line identification is based on [82]. To show the other lines, here overprinting the telluric absorptions, the profiles are strongly zoomed in the vertical intensity axis. Thus, the upper part of the Hα emission appears to be cropped. Within an area of some 100 Å around Hα, the continuum level is significantly elevated.

Plate 36 Detailed Spectrum of LBV Star, Early B Class, P Cygni Profiles

Plate 36 shows a higher resolved profile sections (900 L mm⁻¹) in the green and blue area of the spectrum. The line identification in the range >λ4100 is based on [82], in the range of <λ4100 according to [83], [84]. In these profiles most of the absorptions turn out as P Cygni profiles, if recorded at a higher resolution!

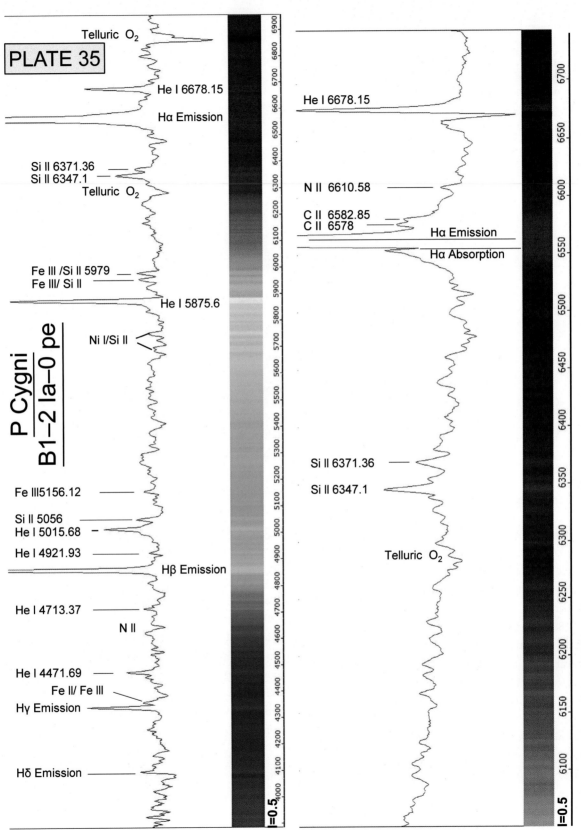

PLATE 35

Telluric O₂

He I 6678.15

Hα Emission

Si II 6371.36
Si II 6347.1

Telluric O₂

Fe III /Si II 5979
Fe III/ Si II

He I 5875.6

Ni I/Si II

P Cygni
B1–2 Ia–0 pe

Fe III5156.12

Si II 5056
He I 5015.68

He I 4921.93

Hβ Emission

He I 4713.37

N II

He I 4471.69

Fe II/ Fe III

Hγ Emission

Hδ Emission

He I 6678.15

N II 6610.58

C II 6582.85
C II 6578

Hα Emission

Hα Absorption

Si II 6371.36

Si II 6347.1

Telluric O₂

I=0.5

I=0.5

Plate 35 LBV Star, Early B Class, P Cygni Profiles

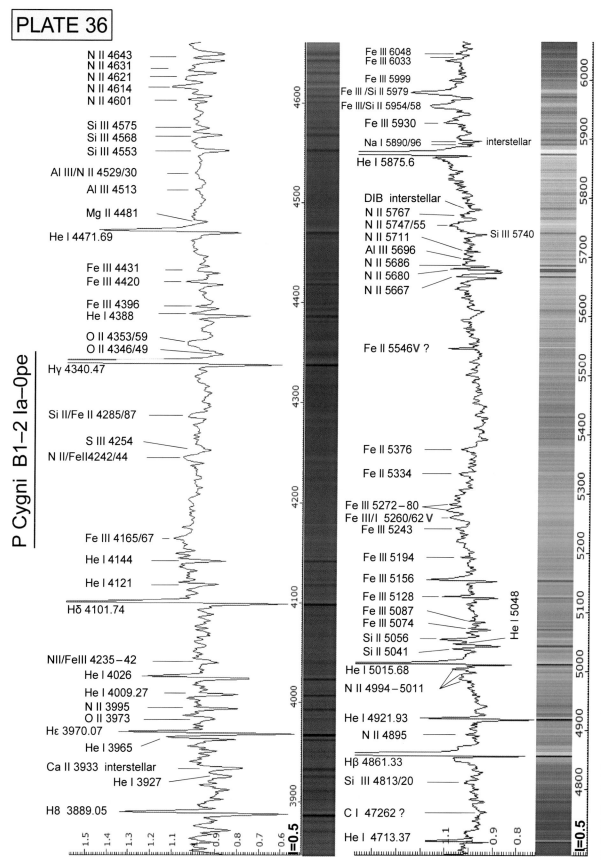

PLATE 36

P Cygni B1–2 Ia–0pe

N II 4643
N II 4631
N II 4621
N II 4614
N II 4601

Si III 4575
Si III 4568
Si III 4553

Al III/N II 4529/30

Al III 4513

Mg II 4481

He I 4471.69

Fe III 4431
Fe III 4420

Fe III 4396
He I 4388

O II 4353/59
O II 4346/49

Hγ 4340.47

Si II/Fe II 4285/87

S III 4254
N II/FeII4242/44

Fe III 4165/67

He I 4144

He I 4121

Hδ 4101.74

NII/FeIII 4235–42
He I 4026

He I 4009.27
N II 3995
O II 3973
Hε 3970.07
He I 3965

Ca II 3933 interstellar
He I 3927

H8 3889.05

4600
4500
4400
4300
4200
4100
4000
3900

1.5
1.4
1.3
1.2
1.1
1.0
0.9
0.8
0.7
0.6
I=0.5

Fe III 6048
Fe III 6033

Fe III 5999
Fe III /Si II 5979
Fe III/Si II 5954/58

Fe III 5930

Na I 5890/96   interstellar

He I 5875.6

DIB interstellar
N II 5767
N II 5747/55
N II 5711   Si III 5740
Al III 5696
N II 5686
N II 5680
N II 5667

Fe II 5546V ?

Fe II 5376

Fe II 5334

Fe III 5272–80
Fe III/I 5260/62 V
Fe III 5243

Fe III 5194

Fe III 5156

Fe III 5128
Fe III 5087
Fe III 5074   He I 5048
Si II 5056
Si II 5041

He I 5015.68

N II 4994–5011

He I 4921.93
N II 4895

Hβ 4861.33

Si III 4813/20

C I 47262 ?

He I 4713.37

6000
5900
5800
5700
5600
5500
5400
5300
5200
5100
5000
4900
4800

1.1
1.0
0.9
0.8
I=0.5

P Cygni Profiles

Plate 36 Detailed Spectrum of LBV Star, Early B Class, P Cygni Profiles

CHAPTER

19 Be Stars

19.1 Overview

The Be stars form a large subgroup of spectral class B. They all have in common that they can form, at least temporarily, an equatorial decretion disk, fed by the mass loss of the star (Figure 19.1). The first Be star discovered was γ Cassiopeiae in 1868, by Father Angelo Secchi, who wondered about the "bright lines" in its spectrum. Here follow some parameters about such objects, based on lectures by A. Miroshnichenko [85], [86], and further publications:

- Of the 240 brightest Be stars 25% have been identified as components of binary systems.
- Most Be stars are still on the main sequence of the HRD, with spectral classes O1–A1 [86], or according to other sources in the range O7–F5 (F5 for shell stars). Only a few Be stars have reached the giant stage with an upper limit of luminosity class III.
- Be stars consistently show high rotational velocities, up to $v \sin i > 400$ km s^{-1}, in some cases even close to the "break up limit." The different observed inclination angles of the stellar rotation axes are assumed as the main cause for the spread of these values.
- The cause for the formation of the circumstellar disk and the associated mass loss is still not fully understood. In addition to the strikingly high rotational velocity, it is also attributed to non-radial pulsations of the star or the passage of a close binary star component in the periastron of the orbit.
- Such disks can arise within a short time but also quickly disappear this way. This phenomenon may pass through three stages: Common B star, Be star and Be shell star. A classical example for that behavior is Pleione (28 Tau)

in the Pleiades cluster M45, which has passed all three phases within a few decades (Section 27.6).
- Due to its viscosity, during the rotation the disk material moves outwards [86].
- With increasing distance from the star, the thickness of the disk is growing and the density is fading.
- If the mass loss of the star exceeds that of the disk, the material is collected close to the star. In the opposite case it may form a ring.
- The X-ray and infrared radiation are strongly increased.

19.2 Spectral Characteristics of Be Stars

In contrast to the stars showing P Cygni profiles due to expanding matter (Chapter 18), Be stars develop, often just episodically, a rotating circumstellar disk of gas in the equatorial plane. The mechanism of formation is not yet fully understood. This phenomenon is accompanied by H I and He I emissions in the spectrum, as well as strong infrared and X-ray radiation. Outside such episodes, the star can seemingly

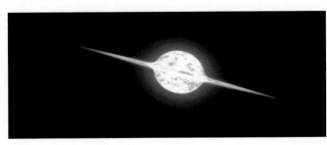

Figure 19.1 Artist's impression: "edge-on" view of a Be star with ionized gas disk
(Credit: NASA/ESA/G. Bacon (STScI))

spend a normal "life" in the B class. Probably due to the sometimes just temporary Be stage, the suffix "e" for emission lines is missing in some stellar databases, which should indicate that emission lines may occur.

If a B-class star mutates within a relatively short time to a Be star (e.g. δ Scorpii), the Hα absorption from the stellar photosphere changes to an emission line, generated in the circumstellar disk. At the same time it becomes the most intense spectral feature representing the kinematic state of the ionized gas disk.

19.3 A Textbook Example: δ Scorpii

The approximately 730 ly distant δ Scorpii (Dschubba), dominant member in a binary system, has mutated since about 2000 from spectral type B0.2 IV to a Be star and developed a typical circumstellar disk of gas. The apparent rotation velocity of the central star is approximately 165 km s^{-1} [11].

The spectra on Plate 37 show the state of the gas disk at different dates. The Hα emission line was taken on August 18, 2009. Its intensity is significantly lower than P Cygni's, but still so high that the interesting much weaker double peak profile of the helium line at λ6678 may be overlooked. To make such details visible, the profiles in Plate 38 are strongly zoomed in the intensity axis. In [3] it is demonstrated how this double peak arises due to a combination of Doppler and perspective effects. The intensities of the V and R peaks as well as the distance between the two peaks, inform us about the apparent state of the disk.

The main interest of the research seems to be focused on these spectral features but even more remarkable effects can be seen. Also, Hβ shows an emission line developing in the middle of a broad photospheric absorption sink, generated by the central star [87]. Such spectral features are therefore called emission or shell cores [3]. However, at this resolution Hγ, Hδ and Hε appear as "normal" absorption lines, a suggestion for amateurs to observe these lines with higher resolutions. Further, at about λ3700 the profile shows a kind of double peak emission.

19.4 Classification System for Be Stars

Table 19.1 presents the seven stage classification system according to Janet Rountree-Lesh (1968) and [2]. For the range of λ3800–5000 the following criteria are applied – mainly based on the number and intensity of the concerned H-Balmer, as well as Fe II lines, appearing in emission.

19.5 Comments on Observed Spectra

Plates 37 and 38 Be Star, Early B Class

Dschubba δ Sco

Plate 37 shows a broadband profile (200 L mm^{-1}) of the Be star Dschubba δ Sco (730 ly), classified as B0.2 IV e [11]. Further, it displays higher resolved profiles (900 L mm^{-1}) around the Hα and Hβ lines and a zoomed detail of the helium double peak at λ6678. The analysis of such spectral features is explained in [3]. The Hβ line exhibits here a textbook-like "shell core" emission (estimated classification e_{1+}).

Tsih, γ Cassiopeiae

Plate 38 displays a broadband overview spectrum (200 L mm^{-1}) of the Be star γ Cassiopeiae (300 ly), spectral class B0.5 IV pe. The apparent rotation speed of the central star is typically

Table 19.1 Classification system for Be stars, according to Janet Rountree-Lesh and [2]

Classification	Standard Star	Classification Criteria
e_1	66 Oph	No obvious H emission, weak "filling-in" in the core of Hβ
e_{1+}	48 Per	Hβ shows a narrow emission core, but remains predominantly an absorption line (example see δ Sco, Plate 37)
e_2	Ψ Per, 120 Tau,	Hβ in emission, Hγ shows weak "filling-in," the higher Balmer lines are not affected
e_{2+}	HD 45995	Hγ shows a weak emission core, the higher Balmer lines show a weak filling-in
e_3	11 Cam	Complete H-emission spectrum, Fe II emissions are present
e_{3+}	HD 41335	Fe II lines are prominent, intensity of the H emissions is lower compared to e_4 (for example see γ Cas, Plate 38)
e_4	X Oph	Extreme Be star, high intensity of all H emissions, highly intensive Fe II lines

very high, ~295 km s^{-1}. The spectrum was recorded on April 9, 2011. In contrast to δ Sco many more lines are in emission here. Some of them are caused by Fe I/Fe II (estimated classification e$_{3+}$). The He I line at λ6678 is hidden within a broad absorption and barely visible in this low resolved profile. The red profile represents the actual intensity ratios with the dominant Hα and Hβ emission. The blue one is strongly zoomed to make the fine lines more visible.

Pleiades

For comments on the spectra of the Be stars in the Pleiades cluster see Section 27.6, Plates 61 and 62.

PLATE 37

Dschubba δ Sco B0.2 IV e

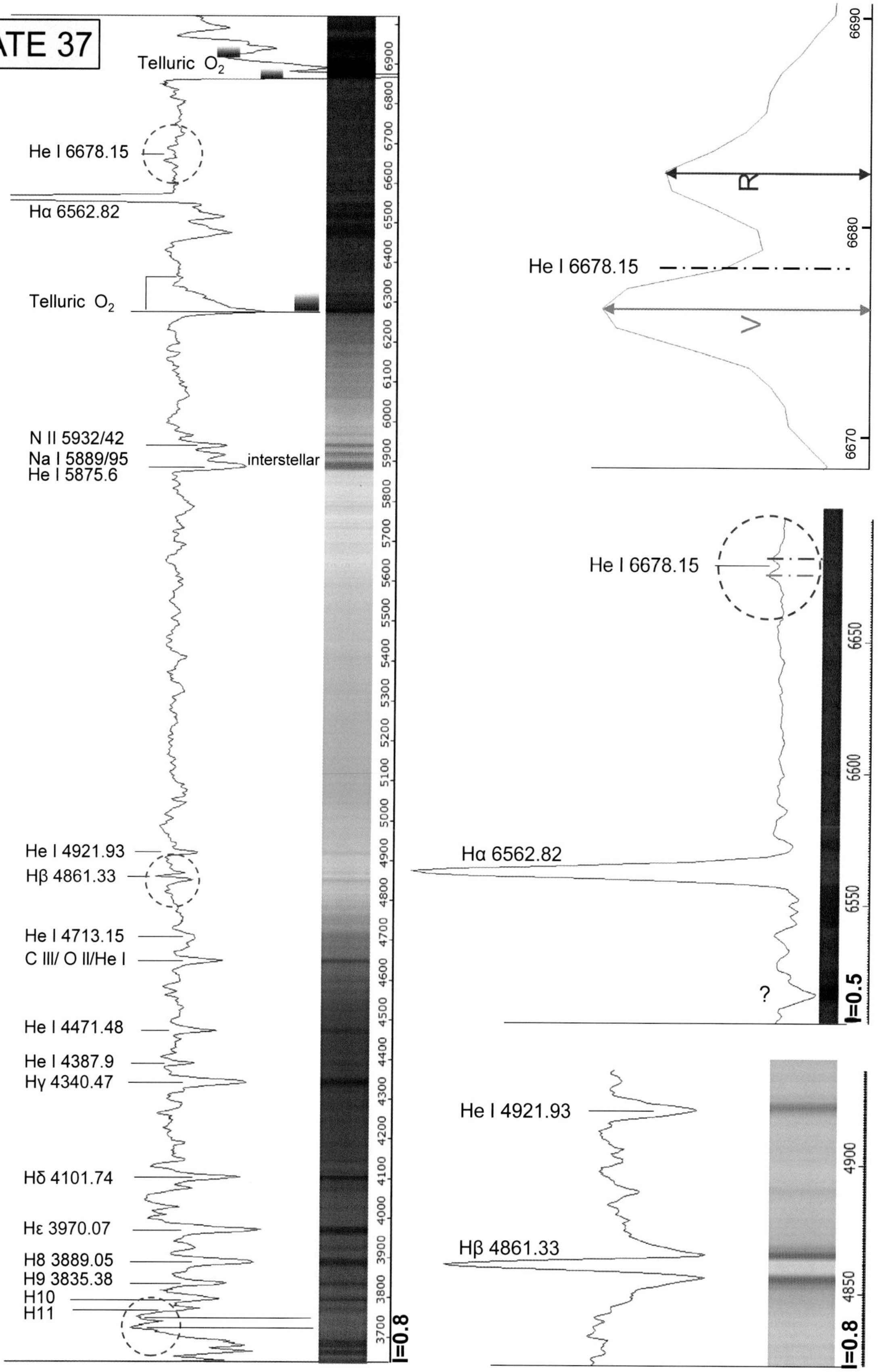

Telluric O₂

He I 6678.15

Hα 6562.82

Telluric O₂

N II 5932/42
Na I 5889/95
He I 5875.6
interstellar

He I 4921.93
Hβ 4861.33

He I 4713.15
C III/ O II/He I

He I 4471.48

He I 4387.9
Hγ 4340.47

Hδ 4101.74

Hε 3970.07

H8 3889.05
H9 3835.38
H10
H11

I=0.8

He I 6678.15

R

V

He I 6678.15

Hα 6562.82

?

I=0.5

He I 4921.93

Hβ 4861.33

I=0.8

Plates 37 and 38 Be Star, Early B Class

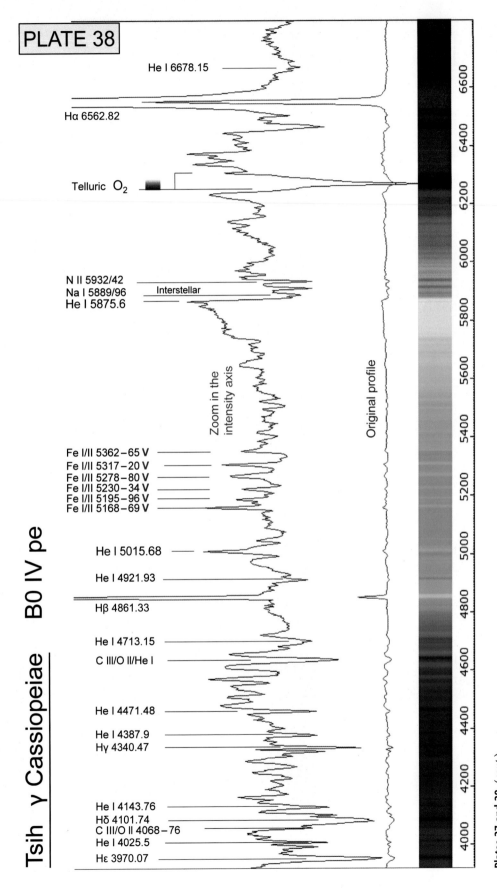

PLATE 38

Tsih γ Cassiopeiae   B0 IV pe

He I 6678.15

Hα 6562.82

Telluric O₂

N II 5932/42
Na I 5889/96   Interstellar
He I 5875.6

Zoom in the intensity axis

Original profile

Fe I/II 5362–65 V
Fe I/II 5317–20 V
Fe I/II 5278–80 V
Fe I/II 5230–34 V
Fe I/II 5195–96 V
Fe I/II 5168–69 V

He I 5015.68

He I 4921.93

Hβ 4861.33

He I 4713.15

C III/O II/He I

He I 4471.48

He I 4387.9
Hγ 4340.47

He I 4143.76
Hδ 4101.74
C III/O II 4068–76
He I 4025.5
Hε 3970.07

6600
6400
6200
6000
5800
5600
5400
5200
5000
4800
4600
4400
4200
4000

CHAPTER

20 Be Shell Stars

20.1 Overview

During the Be phase a star may in some cases pass through a so called shell stage. The slim absorption lines in the recorded spectra suggest that the star forms a kind of a large scale, low density shell or "pseudo-photosphere" [2], which has similar characteristics to the thin atmospheres of supergiants with a luminosity class of approximately I–II. However, the details of this process are widely unknown and subject of numerous hypotheses.

20.2 Spectral Characteristics of Be-Shell Stars

Only few emission lines are observed and many more very narrow but strikingly intense absorption lines but with relatively low FWHM values. The prototype for this star category is ζ Tauri. Also, 28 Tauri (Pleione) undergoes this extreme shell stage from time to time (Section 27.6, Plate 61). Sometimes the suffix "pe" is added to the spectral classification of such stars.

20.3 Comments on Observed Spectra

Plate 39 Be-Shell Star, Comparison with an "Ordinary" Be Star

Plate 39 shows a montage of broadband overview spectra (200 L mm^{-1}), presenting a comparison of the Be-shell star ζ Tauri and Be star δ Scorpii.

Zeta Tauri (~400 ly) B1 IV pe [11], shows an effective temperature, T_{eff}, of some 22,000 K and an apparent rotational velocity $v \sin i$ ~125 km s^{-1}. It is the main star of a binary system with a much smaller B component of spectral class G8III. The orbital period is 0.36 years and the distance between the components ~1 AU. The behavior of the spectral lines is sometimes very volatile. According to [13] the width of the H lines may change within just some 10 minutes, this is not unusual for Be shell stars! In contrast to δ Scorpii, the profile of ζ Tauri displays just Hα as an emission line. Figure 20.1 shows this feature as an asymmetrical double peak emission line – a highly rewarding monitoring project for amateur astronomers!

Unlike δ Scorpii, the spectrum of ζ Tauri shows the He I line at λ6678 and the Hβ line at λ4861 not as emission but as relatively intensive absorption lines (marked in Plate 39 with red ellipses). Furthermore, all absorption lines appear much more intensive compared to δ Scorpii, corresponding approximately to those of early B giants. Striking, however, is the strong absorption around He I at λ5016. Due to the strong zoom in the vertical intensity axis, the peaks of the Hα emissions here appear to be cropped. Their real intensity is similar in both spectral profiles. At these early spectral types the Na I lines are always of interstellar origin.

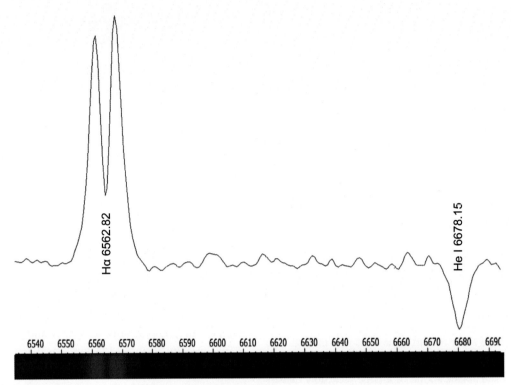

Figure 20.1 ζ Tauri: Hα emission line and He I absorption (900 L mm^{-1})

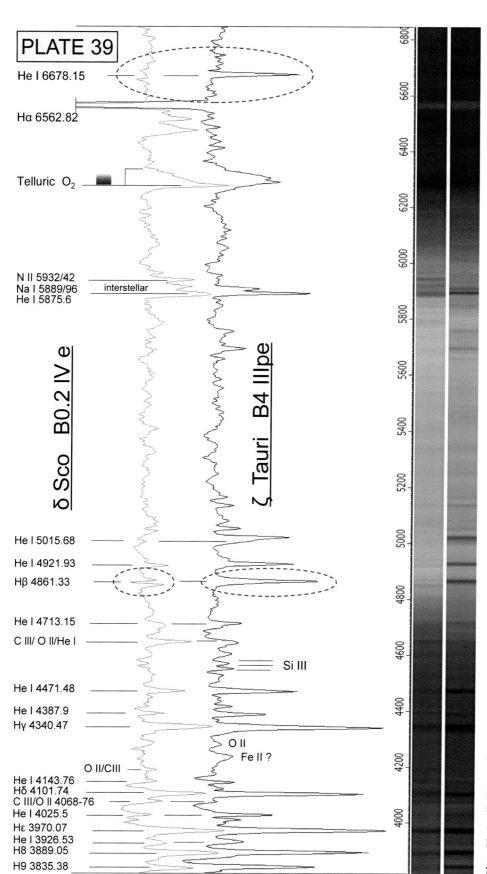

PLATE 39

He I 6678.15

Hα 6562.82

Telluric O₂

N II 5932/42
Na I 5889/96   interstellar
He I 5875.6

δ Sco B0.2 IV e

ζ Tauri B4 IIIpe

He I 5015.68

He I 4921.93

Hβ 4861.33

He I 4713.15

C III/ O II/He I

Si III

He I 4471.48

He I 4387.9

Hγ 4340.47

O II
Fe II ?

O II/CIII

He I 4143.76
Hδ 4101.74
C III/O II 4068-76
He I 4025.5
Hε 3970.07
He I 3926.53
H8 3889.05
H9 3835.38

6800

6600

6400

6200

6000

5800

5600

5400

5200

5000

4800

4600

4400

4200

4000

Plate 39 Be-Shell Star, Comparison with an "Ordinary" Be Star

CHAPTER

21

Pre-Main Sequence Protostars

21.1 Overview

These objects represent the stellar birth stage. These young stellar objects (YSO) are formed from contracting gas and dust clouds. Due to the gravitational process the temperature in the center rises until the onset of hydrogen fusion. Finally, the star begins to shine. From now on, within the rough time frame of several hundred thousand to millions of years, the protostar approaches the main sequence of the HRD from the top downward, until it is ultimately established. Therefore, these objects are also called PMS stars (pre-main sequence) [2]. The highly irregular optical brightness variations proof, that the PMS stars are still very unstable. In a later phase planets are formed from the residual, protoplanetary disk material. Figure 21.1 shows a highly resolved, interferometrically obtained image of a T Tauri object, recorded with the ALMA Submillimeter Array in Chile. It displays HL Tauri (450 ly) with its protoplanetary disk, spanning about 1500 light-minutes across. It consists of concentric, reddish rings, separated by small gaps, which may indicate possible planet formation. The age of the star is estimated to be ~1 My (NASA). On this infrared image the polar jet is not visible.

In this phase, the material forms a rotating accretion disk, at least temporarily veiling the central star. Figure 21.2 schematically shows how a part of the so-called accretion flow does not reach the emerging protostar but is rather deflected, possibly bundled by strong magnetic fields (still a matter of debate), and ejected on both sides of the disk, forming a cone-shaped, bipolar nebula [88]. At some distance from the protostar, this jet may collide with interstellar matter, forming rather short-lived, nebulous structures, so-called Herbig–Haro objects. These are named after George Herbig and

Guillermo Haro. A detailed and illustrated presentation of these effects can be found in [88]. It is important to note that these protoplanetary accretion disks must never be confused with protoplanetary (or pre-planetary) nebulae, ejected at the end of the stellar evolution (Section 28.4).

21.2 Herbig Ae/Be and T Tauri Stars

The Herbig Ae/Be objects represent the birth stage of the stellar spectral classes A–O. For the later F–M types these

Figure 21.1 HL Tauri with protoplanetary accretion disk, ALMA Array
(Credit: ALMA (ESO/NAOJ/NRAO))

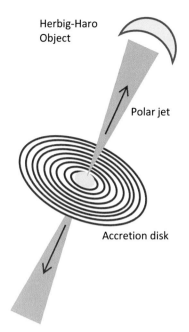

Herbig-Haro
Object

Polar jet

Accretion disk

Figure 21.2 A PMS star with accretion disk and bipolar jet

are the T Tauri stars, which are still subdivided into classical T Tauri stars (CTTS) with intense emission lines and weak T Tauri stars (WTTS), predominantly exhibiting absorption lines. The limit is determined here by the equivalent width of the Hα emission line. According to [2] the object is a CTTS-type if $EW > |10\,\text{Å}|$ and according to other sources if $EW > |5\,\text{Å}|$.

21.3 Spectral Characteristics of PMS Stars

The striking instability also becomes evident in the spectrum. Typically, it shows emission lines of the H-Balmer series, as well as of the H and K (Ca II) Fraunhofer lines. Depending on the state of the accretion disk, and our perspective of the object, in addition numerous Fe I and Fe II lines as well as Ti II may show up [89]. Mainly late (WTTS) T Tauri stars sometimes generate more or less pure absorption spectra, which enable a fairly accurate classification of the star. Gray and Corbally [2] present their own classification system for Herbig Ae/Be stars with the following main criteria:

- The presence and intensity of emission lines of the H-Balmer series.
- The presence and a possible λ shift of shell cores [3], mainly observed at the Balmer series. These are emission lines, raising up from rotationally broadened, photospheric absorption sinks.
- The presence of emission lines of ionized metals, particularly the Fe II (42) multiplet.

- The appearance of the Fe II (42) multiplet as emission lines, absorption lines, or P Cygni profiles.
- The inclination of the Balmer decrement (Section 28.7 and [3]).

21.4 The FU Orionis Phenomenon

FU Orionis is the eponymous star of this phenomenon, which occupies a special position in the birth phase of stellar evolution. In astronomical jargon such objects are also called "FUors." During 1936–1937 FU Ori increased its apparent brightness within 120 days by about six magnitudes from $m_v \approx 16$ up to $m_v \sim 9$ [90]. At first an ordinary nova eruption was suspected. However, the subsequently expected decrease in brightness did not take place, it has remained, with slight fluctuations, more or less on this level until today. In the 1970s, V1057 Cygni showed a similar behavior. Today, it is believed that the long-lasting increase in brightness of T Tauri stars is caused in their late phase by a dramatic increase of matter descending from the accretion disk down to the protostar [90]–[94]. Thereby, the spectrum is changing dramatically and it resembles that of a supergiant of the spectral classes F–K (for details see the comment on Plate 42). This effect could possibly be caused by a temporarily formed "pseudo-photosphere," similar to the Be shell stars (Chapter 20). Further, it is believed that this stage may be passed through several times and is further limited to stellar masses $\lesssim 1\,M_\odot$, i.e. restricted to the T Tauri class [90]. While the T Tauri stage contraction movements are dominating, recognizable by the inverse P Cygni profiles, this feature appears at the Hα lines of "FUors" typically in "expansion mode," i.e. with blueshifted absorption part.

21.5 Comments on Observed Spectra

Plate 40 Herbig Ae/Be Protostar

NGC 2261 (Hubble's Variable Nebula) with R Monocerotis

Plate 40: R Monocerotis (~2500 ly) is a Herbig Ae/Be star. From our perspective, only one of the two conical halves of its bipolar nebula is visible as the nebula NGC 2261 (Figure 21.3). The star itself and the second symmetrical half of the nebula remain hidden behind the accretion disk. The brightness of R Monocerotis varies in the range $m_v \approx 10$–12. Quite recently it became clear that this well-hidden object is a Herbig Ae/Be star of spectral type ~B0. Like most other representatives of early spectral classes, R Monocerotis is also orbited by a smaller companion. In the 1970s it was still considered by many authors as a T Tauri object of middle spectral class.

The spectrum in Plate 40 was recorded by the 50 μm slit of DADOS in the brightest tip of the nebula, near R Mon which is not directly visible. Recording information: C8/Atik 314L+, 4 × 780 s in 2 × 2 binning. The most striking features are the strong H-Balmer lines, which are seen here in emission, appearing superposed on a significant continuum. The Balmer decrement in this profile yields $I_{H\alpha}/I_{H\beta} \approx 5$. For such objects, this selective reddening of the profile is significantly influenced by the density of the circumstellar dust cloud and even poses an important classification criterion (see above).

The high degree of dust occultation is probably one reason why almost exclusively emission lines can be seen here. C. A. Grady *et al.* [95] noted the correlation between the intensity of the Fe I and Fe II emissions, and the degree of dust occultation, combined with the associated infrared excess. In this profile the Fe II (42) multiplet at $\lambda\lambda\lambda4923$, 5018 and 5169 appears striking [2]. This feature can also be found in the spectra of novae and supernovae as well as in active galactic nuclei (AGN). Herbig already noted in the 1960s that the emission of the Fraunhofer K line mostly appears more intense than those of the H line.

Spectra taken in a wider area of the NGC 2261 nebula look similar to those recorded very close to R Monocerotis, proving in this way that NGC 2261 is obviously a reflection nebula. The line identification is among others based on [2], [89].

This highly interesting object is also very popular amongst astrophotographers recording the short-term brightness variations in spectacular series of images. On January 26, 1949 NGC 2261 became famous, because Edwin Hubble himself photographed this nebula as the official first light object of the new 5 m Hale Telescope at Mount Palomar Observatory!

Plate 41 Prototype of the T Tauri Stars

T Tauri (HD 284419)

Plate 41: T Tauri (~460 ly), is a classical T Tauri star (CTTS) with highly intense emission lines appearing superposed on a significant continuum. This star is classified as F8Ve–K1IV-Ve (T) [96]. The CTTS status is clearly demonstrated by the value of $EW_{H\alpha} \approx -87$ Å. At first glance, the two profiles of Plates 40 and 41 look very similar. The Fe II multiplet, however, is here much weaker, but the Ca II emission significantly stronger. T Tauri is roughly classified in a very broad range, much later than spectral class A, the lower limit to the earlier Herbig Ae/Be stars. However, in this profile this can only be recognized at the CH absorption band at ~$\lambda4300$. Highly interesting details here are the forbidden [O I] and the strikingly intense sulfur lines [S II], as well as the Hγ emission, which at the time of recording appeared as inverse P Cygni profiles (see Chapter 18)! The

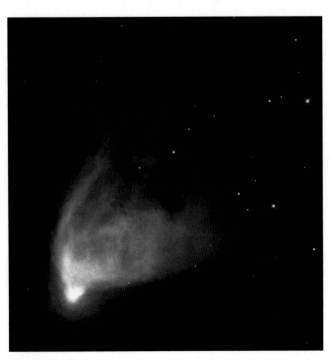

Figure 21.3 NGC 2261 and R Monocerotis with cone-shaped jet (Credit: NASA/ESA/HST Heritage Team (Aura/STScI))

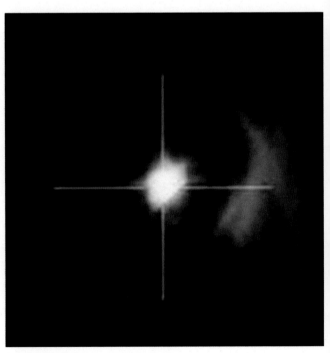

Figure 21.4 T Tauri with NGC 1555 (Credit: Simbad Aladin Previewer, DSS)

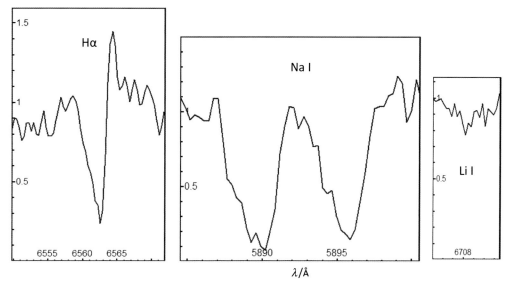

Figure 21.5 FU Orionis, Hα appearing as P Cygni profile and massively broadened, fully saturated Na I lines – clear evidences for a strong outflowing wind. Li I absorption is evidence for a very young object, SQUES echelle spectrograph. Hα and Na I lines, SQUES, slit width 70 μm, 2 × 3600 s, 2 × 2 binning. Li I line, SQUES, slit width 85 μm, 2 × 3600 s, 3 × 3 binning

redshifted absorption part here indicates large-scale contraction movements within the accretion disk, headed towards the star – this in contrast to the normal P Cygni profiles, which are always a reliable sign for an expansion. The Doppler analysis according to Equation {18.1} yields contraction velocities of some 600 km s^{-1}.

According to AAVSO the apparent brightness of the object, at the time of recording, was $m_v \approx 10$. At T Tauri, as a byproduct of this star birth, a closely neighboring Herbig–Haro object is generated. Here it is the variable nebula NGC 1555, named after John Russell Hind (1823–1895), which is visible in large telescopes next to T Tauri (Figure 21.4). Recording info: C8/DADOS 200 L mm^{-1}, Atik 314L+, 3 × 720 s, 2 × 2 binning.

Plate 42 FU Orionis Protostar: Comparison with K0 Giant

FU Orionis and Algieba γ Leonis
Plate 42 shows a spectral comparison of the protostar FU Orionis (~1300 ly and $m_{v\ var} \approx 9.6$) with a composite spectrum of the late classified binary giants of Algieba γ Leonis (~126 ly and K1 III+G7 III [11]). Recording data with C8/DADOS/Atik 314L+, 25 μm slit, 8 × 600 s, 2 × 2 binning.

In the short-wave range the "FUor" spectra look generally like those of F–G supergiants, but in the near-infrared rather like those of K giants. In comparison to γ Leonis, this is easily recognizable here. The two profiles show the following striking differences:

- At FU Orionis the Hα line forms a distorted P Cygni profile, indicating at this stage an intensive outflowing wind of more than 100 km s^{-1}. Typically, the redshifted emission part appears here strongly stunted (Figure 21.5), [91], [92].

- Further evidence for this wind is the impressively widened, fully saturated sodium D1 and D2 absorption (Figure 21.5).

- The here somewhat noisy absorption of lithium Li I at λ6708 shows that FU Orionis must necessarily be a very young object. In lower mass stars ($\lesssim 1M_\odot$) with deep reaching convection zones, lithium gets mixed into the stellar interior where the temperatures are high enough to reduce this element to helium with time [1].

- With the exception of the Hα line the H-Balmer series of FU Orionis appears here very slim but intensive. The profile shows striking similarities with those of γ Leonis.

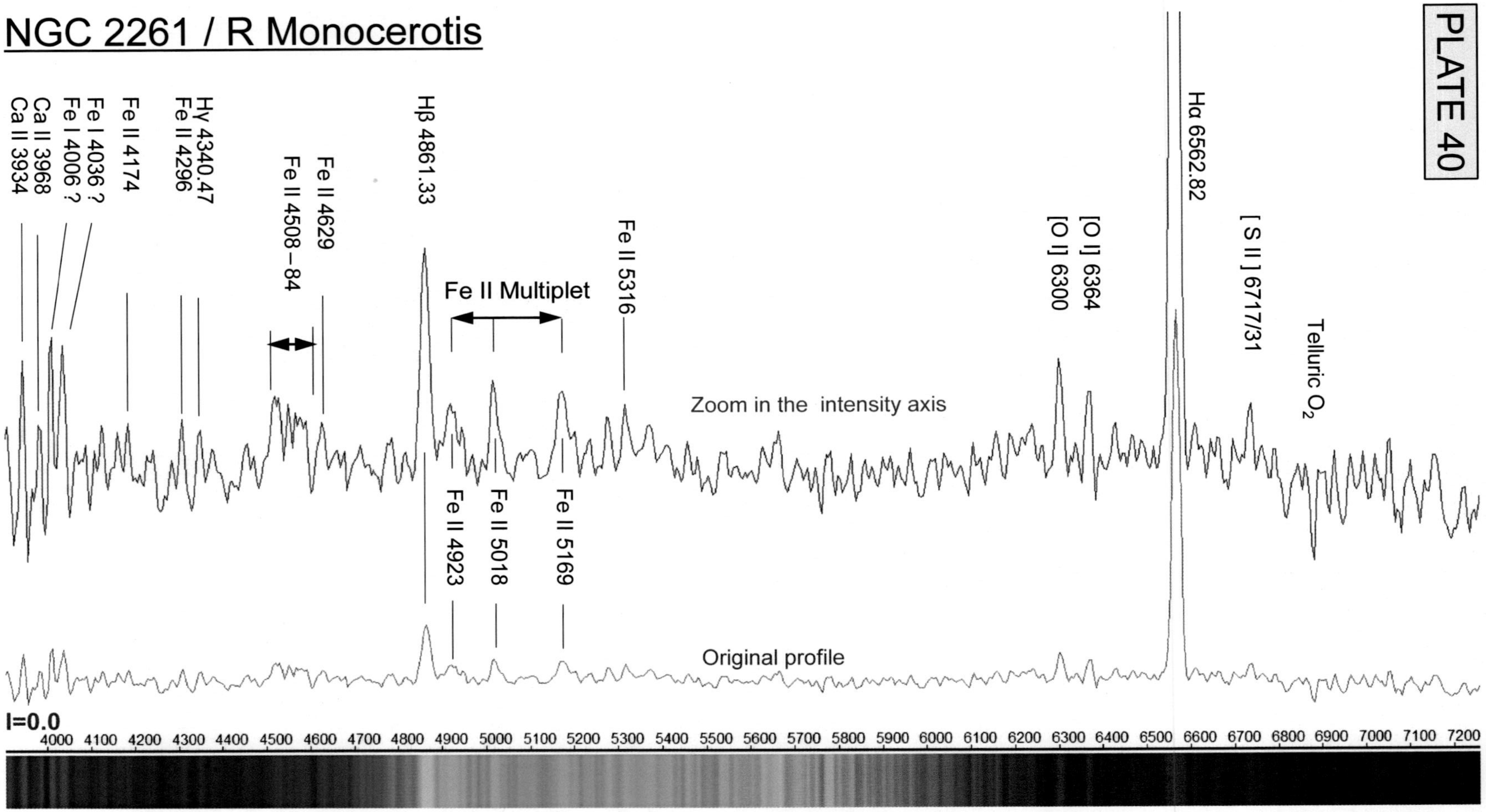

PLATE 40

Telluric O$_2$

[S II] 6717/31

Hα 6562.82

[O I] 6364
[O I] 6300

Zoom in the intensity axis

Original profile

Fe II 5316

Fe II 5169

Fe II Multiplet

Fe II 5018

Fe II 4923

Hβ 4861.33

Fe II 4629

Fe II 4508 – 84

Hγ 4340.47
Fe II 4296

Fe II 4174

Fe I 4036 ?
Fe I 4006 ?

Ca II 3968
Ca II 3934

NGC 2261 / R Monocerotis

I=0.0

4000 4100 4200 4300 4400 4500 4600 4700 4800 4900 5000 5100 5200 5300 5400 5500 5600 5700 5800 5900 6000 6100 6200 6300 6400 6500 6600 6700 6800 6900 7000 7100 7200

Plate 40 Herbig Ae/Be Protostar

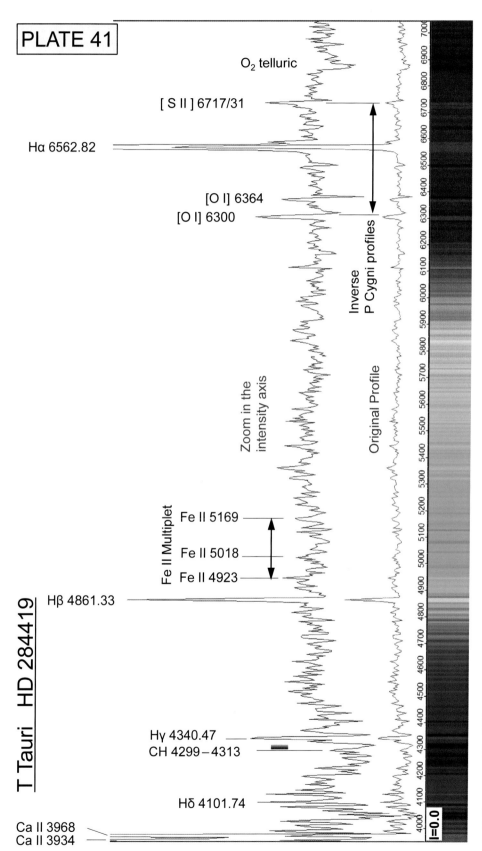

PLATE 41

T Tauri HD 284419

Ca II 3968
Ca II 3934

Hδ 4101.74

Hγ 4340.47
CH 4299 – 4313

Hβ 4861.33

Fe II Multiplet

Fe II 5169

Fe II 5018

Fe II 4923

Zoom in the intensity axis

Original Profile

Inverse P Cygni profiles

[O I] 6364
[O I] 6300

Hα 6562.82

[S II] 6717/31

O₂ telluric

I=0.0

Plate 41 Prototype of the T Tauri Stars

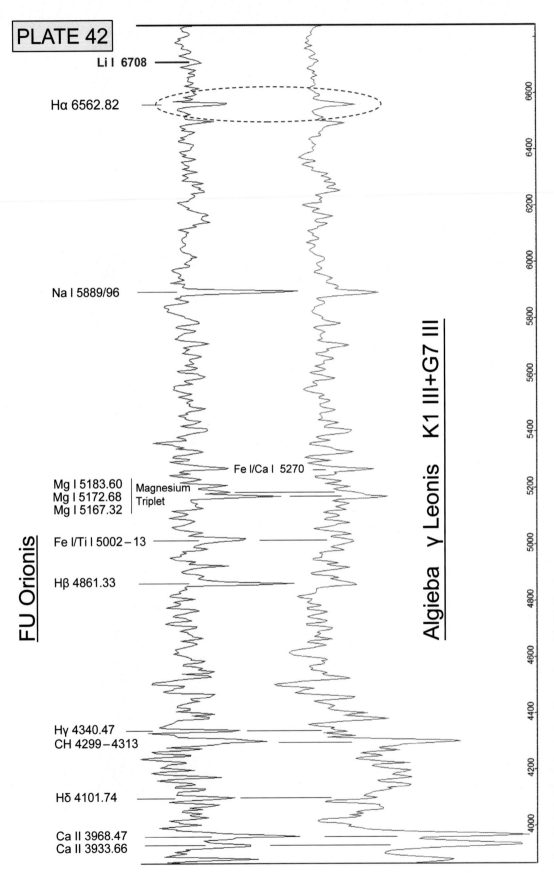

PLATE 42

Li I 6708

Hα 6562.82

Na I 5889/96

Fe I/Ca I 5270

Mg I 5183.60
Mg I 5172.68   Magnesium
Mg I 5167.32   Triplet

Fe I/Ti I 5002–13

Hβ 4861.33

Hγ 4340.47
CH 4299–4313

Hδ 4101.74

Ca II 3968.47
Ca II 3933.66

FU Orionis

Algieba γ Leonis   K1 III+G7 III

Plate 42 FU Orionis Protostar: Comparison with K0 Giant

CHAPTER

22 Chemically Peculiar (CP) Stars

22.1 Overview

Peculiar or chemically peculiar (CP) stars differ from the other representatives of the early B to early F classes by several of the following characteristics:

- Abnormal abundance of certain chemical elements on the stellar surface, indicated by the intensity of the corresponding spectral lines.
- Differently strong magnetic fields with variable intensity, except for the Am–Fm class.
- Photometrically and spectroscopically detectable variations or "spectrum variables."
- Slow stellar rotation, with correspondingly sharp spectral lines, with exception of the λ Bootis and the He-rich class.
- Some representatives show rapid brightness oscillations with low amplitude and short period lengths of just several minutes to hours.

The spectral classification of peculiar stars is usually denoted by the suffix p for "peculiar," possibly followed by the abnormally abundant chemical elements. Thus, for example,

the brighter binary component of α^2 CVn (Cor Caroli) is classified as A0 II-IIIp SiEuCr [2], [11].

For amateurs it primarily remains to reproduce the above listed, distinctive features, for which mostly high-resolution spectrographs ($R \gtrsim 10{,}000$) are required. Another option may be the monitoring of spectrum variables.

22.2 Classification of the CP Stars

Table 22.1 shows according to Smith the allocation of CP stars to six subclasses, according to their spectral features [97]. It includes also the four basic CP classes after Preston [98].

22.3 λ Bootis Class

Stars of this special class are classified according to their H-Balmer lines approximately on the main sequence within the spectral types F0–A0. Striking here is a generally weak metal-line spectrum with the main feature of the Mg II absorption at λ4481. The intensity ratio Mg II λ4481/Fe I λ4383 is here

Table 22.1 Allocation of CP stars to six subclasses according to Smith [97]

Classical Designation	Preston Classification	Spectral Features	Spectral Types	Temperature Range [K]
λ Bootis	–	Mg II weak, metals generally weak	F0–A0	7500–9000
Am–Fm	CP1	Ca II K line and Sc II weak, other metals enhanced	F4–A0	7000–10,000
Ap–Bp	CP2	Sr, Cr, Eu, Si enhanced, partly strong magnetic fields	F4–B6	7000–16,000
Hg–Mn	CP3	Hg II, Mg II enhanced	A0–B6	10,500–16,000
He weak	CP4	He I weak	B8–B2	14,000–20,000
He rich	–	He I enhanced, partly strong magnetic fields	B2	20,000–25,000

significantly smaller than in spectra of normal A-class stars. The distribution of rotational velocities approximately corresponds to that of the ordinary A class. Interestingly, in open star clusters λ Bootis stars are almost completely missing. Confirmed so far are only about 50 stars, of which, in addition to λ Bootis, 29 Cygni is a relatively bright representative of this class. The astrophysical reasons for the spectral peculiarity of the λ Bootis class seem to be still unclear.

22.4 Am–Fm Class

These "metal-line stars" were discovered by Antonia Maury and Annie J. Cannon from E. Pickering's staff. In 1940 this star category was integrated into the MK classification system with the specific suffix "m." The main feature is the Ca II K line, which appears significantly too weak, regarding the spectral type. It corresponds to a star that is classified at least five subclasses earlier [2]. The abundance of calcium (Ca) and scandium (Sc) is here too low, by a factor of ~5–10 [97]. However, the other metal lines are much stronger than usual for this spectral class. The abundance of iron is usually too high, by a factor of ~10.

Shorlin *et al.* [99] were not able to detect any magnetic fields in a sample of 25 Am stars. Compared to equally classified stars on the main sequence, the luminosity is higher by approximately 0.8–1.3 mag (Allen [100]). This effect could still not satisfactorily be explained. The most prominent representative of the Am star is Sirius A [101], [102]. Further examples are 63 Tau, δ Nor and τ UMa.

For the classification of the Am–Fm stars a special notation system was developed. The following lowercase letters correspond to a separate spectral classification:

Fraunhofer K line = k
hydrogen lines   = h
other metal lines = m

According to [11], for example τ UMa is classified as: kA5 hF0 mF5 II, where

kA5:   The Ca II Fraunhofer K line corresponds to a significantly earlier type A5.
hF0:   According to the H-Balmer series the classification is F0.
mF5:   The other metal lines correspond to a significantly later classified type F5.

The luminosity classification II meets here the criteria of the F class. In many of the well-known databases this notation system is not consistently applied – sometimes just a part of the classification is used, frequently the k or the m rating.

All currently known stars of this class are to be found in the range of spectral types F4–A0 and are, like Sirius, members of

relatively close binary systems. The binary system generates strong tidal forces, significantly slowing down the stellar rotation. Thus, the power relations inside the star are changed, favoring the rise of specific elements up to the surface, contributing there to the observed "chemical separation" [2].

22.5 Ap–Bp Class

As the first star of this class α² CVn, Cor Caroli, was discovered by Antonia Maury in 1897. While in the Am–Fm stars the abundance of all metals, except Ca II, appears to be enhanced, this applies here selectively to just a few elements [2]. Accordingly, this class was divided by several authors into subclasses, for example Si, Sr, Si–Cr–Eu or Sr–Cr–Eu.

Analogously to the Am–Fm stars the suspected reason is here a "chemical separation," which is additionally influenced here by very strong magnetic fields, whose causes are the subject of several hypotheses. Their flux density ranges from several hundred gauss up to several kilogauss (kG). Thus, the increased abundance of elements is not able to spread evenly over the entire photosphere, but accumulates in rather spot-like structures. Absorption lines can therefore vary with the rotation period of the star, so it is designated photometrically as "rotational variable" and spectroscopically as a "spectrum variable." Particularly interesting is the rare subclass "roAP" (rapid oscillating Ap stars) exhibiting extremely rapid brightness oscillations in the frequency range of a few minutes but with a very low amplitude of just a few thousandths of a magnitude (e.g. 33 Lib (HD 137949)). Ap–Bp stars are to be found in the spectral classes B6–F4.

For classification purpose the following lines in the short-wave optical range are applied [2]:

Si II:	λλ4128, 4131 (close, strong doublet), λ4200 (in hotter Si II stars), further λλ3856, 3862, 4002, 4028, 4076
Cr II:	λ4172 (strongest absorption), λ4111 (blend with Hδ), further λλ3866, λ4077
Sr II:	λ4077 (strongest Sr II absorption, possible blend with Si II / Cr II), further λ4216
Eu II:	λλ4205, 4130 (possible blend with Si II)

22.6 Mercury–Manganese Class

In 1906 Lockyer and Baxendall noticed stars of this class for the very first time, because the spectra contained absorptions at λλ3944, 3984, 4136, 4206, 4252, which could not be identified at that time. Until 1914 these turned out as ionized manganese (Mn II), with the exception of a strong line at λ3984, which was identified in 1961 by Bidelman as ionized mercury Hg II. These absorptions are defined today as the peculiar Hg–Mn class. These stars are all classified as late B types, which rotate slowly, but in contrast to the Ap–Bp class, exhibit weaker magnetic fields with a flux density of at most several tens gauss. Typical representative of this class is Alpheratz, α Andromedae, with the classification B8 IV–V Hg–Mn [11]. A further example is Gienah Corvi (γ Crv), B8IIIp Hg–Mn [13] (see Section 6.4). This class usually exhibits neither a spectral nor photometric variability.

22.7 Helium-weak Stars

In the 1960s, B-class spectra were detected, which for their class exhibited significantly weaker helium lines, sharp absorptions of P II, Ga II and of individual noble gases. In addition, such stars appear to be photometrically too blue [2]. In this class, weak magnetic fields can also be detected.

Here, subclasses have been formed, according to the occurrence of the metal lines and in some cases even based on isotopes. Typical representatives are 3 Centauri A, B5IIIp [13] and α Sculptoris, B7 IIIp [13]. Today it is assumed that this class probably forms the "hot end" of the magnetic Ap–Bp stars [2].

22.8 Helium-rich Stars

A small number of stars of the early B type exhibit for their class significantly too strong helium lines, combined with a C II absorption at λ4267. The prototype is the rotational variable σ Orionis E, B2 Vp [11], which displays in addition a quite strong mean magnetic field modulus of maximum ~3 kG (see Equation {22.1}). This becomes, however, clearly surpassed by the 44 kG of HD 37776. Magnetic fields are very common in this class but never a general characteristic feature. Some representatives show a variable He line, which is also here a possible indication for an uneven distribution of this element on the stellar surface.

22.9 Subdwarf Luminosity Class VI

This very rare and low luminosity category of "cool subdwarfs" does not belong to the systematics of peculiar CP stars. However, since the strong metal-underabundance at these old population II stars is the cause for their relatively low luminosity, this class is briefly presented in this section. These stars of later spectral types G, K, M are located in the HRD approximately 1.5–2 magnitudes below the main sequence, appearing on the branch of luminosity class VI. They are labeled with the prefix "sd" (subdwarf). The underabundance of metals enhances the transparency in the interior of the star, which reduces the outwardly directed radiation pressure. This results finally in a smaller stellar radius. For amateurs,

these objects are of little interest because the few accessible, mostly faint specimens are distinguished by small spectral nuances from their cousins on the main sequence, so, for example Gliese191, Kapteyn's Star, is classified as sD M1 [11].

This category must not be confused with the just partly understood "hot subdwarfs" of spectral types O and B where this effect is presumably caused by a significant mass loss after the helium flash.

22.10 Comments on Observed Spectra

Plate 43 Comparison of Metallicity

Spectral Differences between Vega (α Lyr) and Am-Star Sirius A (α CMa)

Characteristically for Am stars, **Plate 43** displays the deficit of calcium (Ca) as well as the excess of iron (Fe) in the spectrum of Sirius. The comparison of the "metallicity" [3] between Vega and Sirius has been the subject of numerous professional studies in the past, such as [101], [102]. The metal abundance in the photosphere of Sirius is about three times higher compared to the Sun. However, the ordinary main-sequence star Vega (A0 V) exhibits a "metallicity" of ~30% compared to the solar value.

Besides the different metal abundance both stars are almost equally classified and also display a similarly low apparent rotational velocity $v \sin i$. Differences in the spectrum must therefore primarily be caused by the different metal abundance, which predestines these bright stars for such investigations.

Even at this modest resolution (DADOS: 900 L mm^{-1}) it can immediately be noticed that the Ca II line λ3933 in the Vega spectrum is much stronger, even though the star is classified slightly earlier than Sirius. In the profile of Sirius most other metal absorptions appear significantly stronger. Within a certain gas mixture the equivalent width of an unsaturated spectral line behaves nearly proportional to the number of involved atoms of an element. Further details and analysis options are outlined in [3].

Plate 44 Comparison of Spectral Classes Ap–Bp with Hg–Mn and the Reference Profile of Vega

Alpheratz (α And), Vega (α Lyr) and Cor Caroli (α² CVn)

Plate 44: Alpheratz α And (97 ly), the prototype of the Hg–Mn class, is compared here with Cor Caroli α² CVn (110 ly), a typical representative of the Ap–Bp stars. In addition as a reference, the profile of Vega A0 V is displayed here. To show these chemical peculiarities a high resolution spectrograph is required (SQUES echelle).

Alpheratz (B8 IV–V Hg–Mn) forms here with a significantly weaker A3 V component a spectroscopic binary system with an orbital period of ~96 days. With a slightly variable brightness of $m_v \approx 2.1$ it is apparently the brightest representative of a mercury–manganese star and shows in this wavelength range the characteristic, strong lines of singly ionized manganese, Mn II. The apparent rotational velocity of $v \sin i \approx 50$ km s^{-1} is not unusual for this spectral class. The absorptions appear remarkably flat. As a late B type, Alpheratz is classified slightly earlier than Vega.

The brighter component of the binary star Cor Caroli (A0 II–IIIp SiEurCr) with a combined brightness of $m_v \approx 2.8$ shows the specified elements of its classification and additionally Sr II at $\lambda4216$ in above average intensity. Especially impressive here are the strong Si II absorptions at $\lambda\lambda4128/4131$, which in 1897 led to the discovery of this class by Antonia Maury. The apparent rotational velocity of $v \sin i \approx 20$ km s^{-1} is typically rather low, compared with ordinary, early classified A stars. Compared to Vega, with the same main classification A0, all the metal lines in the profile of Cor Caroli are much stronger. The flux density of the mean magnetic field modulus fluctuates periodically within ~ 5.5 days between -4 and $+5$ kG [103]. To prove this period, spectrographs with a very high resolution power of R> 50,000 would be required.

Plate 45 Estimation Flux Density of Mean Magnetic Field Modulus

Babcock's Star (HD 215441)

Plate 45 shows several cuttings with a montage of highly resolved profiles from HD 215441 (~2500 ly, $m_v \approx 8.8$, spectral class B9p Si) and, except for the Ap–Bp peculiarity, the very similarly classified Vega. In addition, it is attempted to estimate here the flux density B of the mean magnetic field modulus.

This star belongs to the very small subset within the Ap–Bp class, exhibiting extraordinary strong magnetic fields of several kG (kilogauss). With B ~34 kG it generates one of the strongest mean magnetic field moduli of all non-degenerate stars [104], [105], [106]. Apart from that, the object shows the characteristic symptoms as a relatively low apparent rotational speed of $v \sin i \approx 5$ km s^{-1} and very sharp and – compared to Vega – numerous metal lines. Further, the determination of the spectral class is here not easy. In earlier

times HD 215441 was classified as ~B4–B5. Today the spectral class has been established as B9p Si or A0p [11].

The extraordinarily strong magnetic field generates here a split up of some specific absorption lines. The corresponding theory is outlined in [3]. In the recorded profile of $\lambda\lambda4500$–6500, approximately 90 lines can be counted, clearly exhibiting differently strong Zeeman splits. Six of them, with a measurable split up by $\Delta\lambda$, and known Landé factors g_{eff}, have been selected to estimate the mean magnetic field modulus B. In Equation {22.1} the dimensionless factor g_{eff} compensates for the different sensitivities of the individual spectral lines to magnetic fields. In this spectral class the usable values are within a limited range of only 1.2–1.9. The calculation of the flux density B is simple and based on the measured Zeeman split $\Delta\lambda$:

$$B = \frac{\Delta\lambda}{4.67 \times 10^{-13}\, g_{eff}\, \lambda^2} \qquad \{22.1\}$$

where B is the mean magnetic field modulus, λ is the wavelength of the analyzed line in [Å], $\Delta\lambda$ is the measured Zeeman split of the line in [Å] and g_{eff} is the Landé factor.

If appropriate and possible, a Gaussian fit is applied to the individual splits of the absorptions. In other cases, the cursor position of the maximum intensities was evaluated. So the measured mean value of ~33 kG appears to be consistent with the literature value of ~34 kG. For comparison, the extremely dense, degenerate neutron stars exhibit magnetic fields of B ~10^{12} G, magnetar (neutron star with a rotation period of <10 ms): ~10^{15} G. Recording information: SQUES echelle,

Table 22.2 Ap–Bp stars with mean magnetic field modulus $B \gtrsim 10$ kG

Star	m_v	B [kG]
HD 215441, Babcock's Star	8.8	34
HD 75049	9.1	30
HD 137509	6.9	29
HD 119419	6.4	26
HD 145708	8.1	25
HD 175362	5.4	21
HD 47103	9.2	17
HD 318107	9.3	14
HD 65339, 53 Cam	6.0	13
HD 126515, Preston's Star	7.1	13
HD 70331	8.9	12

slit width 70 μm, 7 inch Aries Fluorite Refractor/Atik314L+: 2 × 1200 s, 2 × 2 binning.

The current "magnetic" record holder with 44 kG would be the helium-rich HD 37776. However, its impressive rotational broadening of $v \sin i \approx 145$ km s^{-1} prevents a determination of the Zeeman split by amateur means. Very few of these stellar magnetic fields are so strong ($B \gtrsim 10$ kG) that they can be analyzed even by amateurs with high-resolution spectrographs. Table 22.2 shows a compilation from various sources.

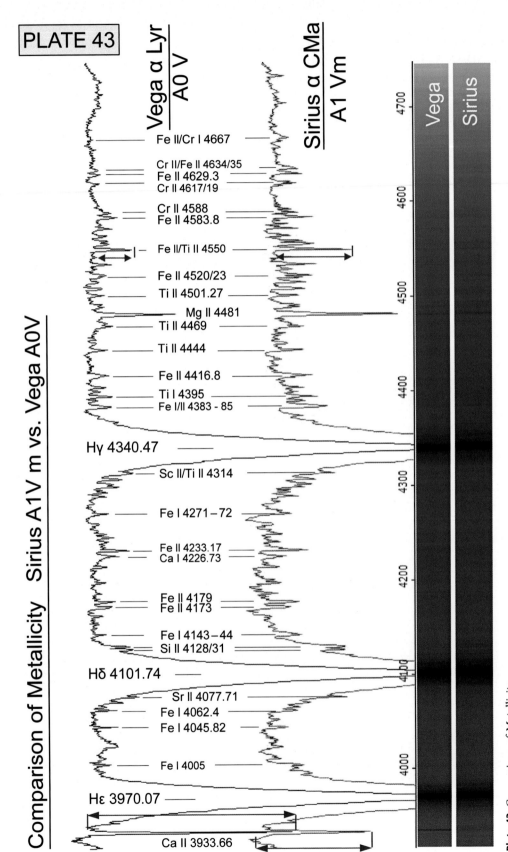

PLATE 43

Comparison of Metallicity   Sirius A1V m vs. Vega A0V

Vega α Lyr
A0 V

Sirius α CMa
A1 Vm

Vega

Sirius

Fe II/Cr I 4667

Cr II/Fe II 4634/35
Fe II 4629.3
Cr II 4617/19

Cr II 4588
Fe II 4583.8

Fe II/Ti II 4550

Fe II 4520/23

Ti II 4501.27

Mg II 4481

Ti II 4469

Ti II 4444

Fe II 4416.8

Ti I 4395
Fe I/II 4383 - 85

Hγ 4340.47

Sc II/Ti II 4314

Fe I 4271−72

Fe II 4233.17
Ca I 4226.73

Fe II 4179
Fe II 4173

Fe I 4143−44
Si II 4128/31

Hδ 4101.74

Sr II 4077.71
Fe I 4062.4
Fe I 4045.82

Fe I 4005

Hε 3970.07

Ca II 3933.66

4700

4600

4500

4400

4300

4200

4100

4000

Plate 43 Comparison of Metallicity

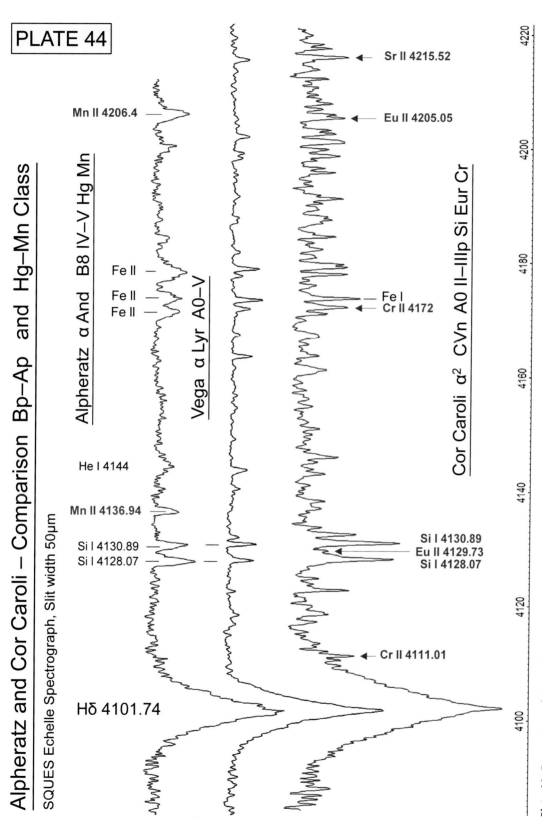

Alpheratz and Cor Caroli – Comparison Bp–Ap and Hg–Mn Class

SQUES Echelle Spectrograph, Slit width 50μm

PLATE 44

Alpheratz α And   B8 IV–V Hg Mn

Vega α Lyr A0–V

Cor Caroli α² CVn A0 II–IIIp Si Eur Cr

Mn II 4206.4

Fe II —
Fe II —
Fe II —

He I 4144

Mn II 4136.94

Si I 4130.89
Si I 4128.07

Hδ 4101.74

Sr II 4215.52

Eu II 4205.05

Fe I
Cr II 4172

Si I 4130.89
Eu II 4129.73
Si I 4128.07

Cr II 4111.01

4220   4200   4180   4160   4140   4120   4100

Plate 44 Comparison of Spectral Classes Ap–Bp with Hg–Mn and the Reference Profile of Vega

PLATE 45

Babcock's Star HD 215441　Mean Magnetic Field Modulus

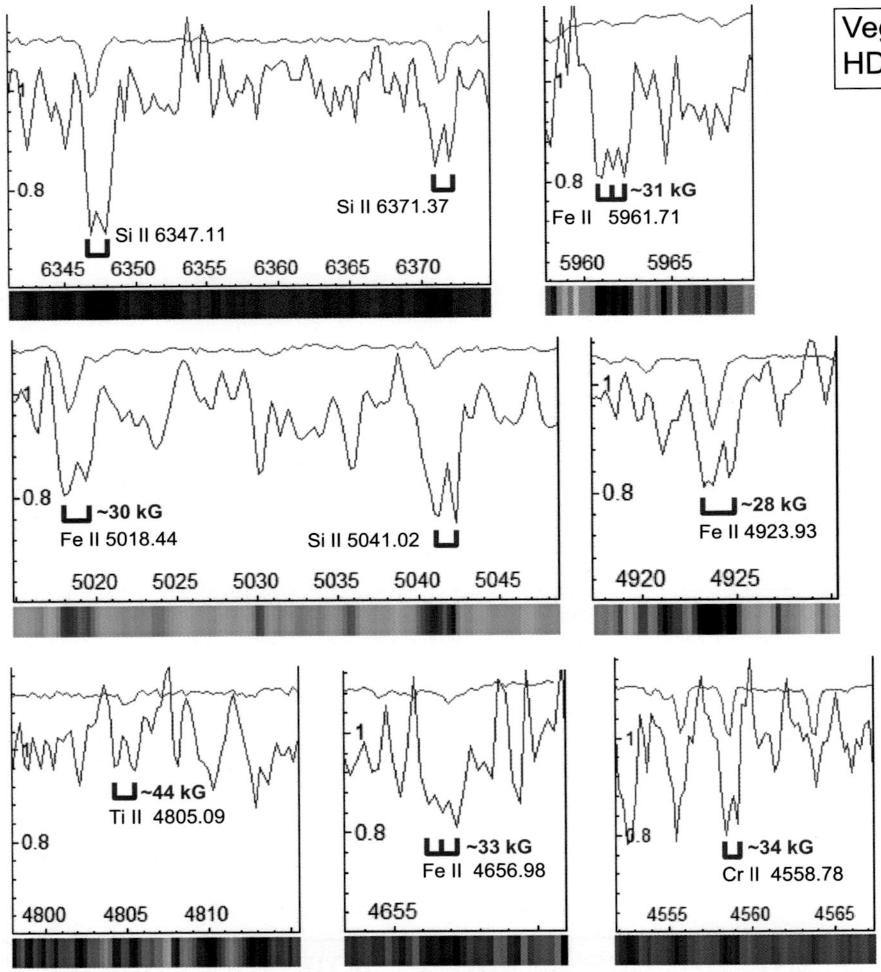

Vega　　　A0V
HD215441　B9p Si

SQUES Echelle Spectrograph
slit width 70µm,　JD2456984.4

Estimation of the flux density B of the
mean magnetic field modulus from
the Zeeman splittings of HD 215441:

Ion	Landé Factor	B
Fe II λ5962	1.2	31 kG
Fe II λ5018	1.9	30 kG
Fe II λ4924	1.7	28 kG
Ti II λ4805	1.2	44 kG
Fe II λ4657	1.7	33 kG
Cr II λ4559	1.2	34 kG

Measured mean value:	~33 kG
H. Babcock, 1960:	~34 kG

Plate 45　Estimation Flux Density of Mean Magnetic Field Modulus

CHAPTER

23 Spectroscopic Binaries

23.1 Short Introduction and Overview

Many more than 50% of the stars in our Milky Way are gravitationally bound components of double or multiple systems. They are primarily concentrated within the spectral classes A, F and G. For astrophysics these objects are of special interest because some of them allow the determination of stellar masses independently of the spectral class. Soon after the invention of the telescope, visual double stars were observed by amateur astronomers. Today, spectroscopy has opened up for us the interesting field of spectroscopic binaries. These objects are orbiting each other very closely around a common gravity center, so they cannot be optically resolved even with the largest telescopes in the world. In contrast to the visual double stars they reveal their binary nature just by the periodic change in spectral characteristics caused by the Doppler effect. For such close orbits the three Kepler laws enforce very short orbital periods combined with high-track velocities, both significantly facilitating the spectroscopic observation of these objects. In this Chapter just the spectroscopically visible effects are discussed. The analysis of such spectra and the rough estimation of the orbital elements and stellar masses with amateur means are presented in [3].

In the special case of eclipsing binaries (e.g. Algol and β Lyrae) the orbital plane is aligned very closely with our line of sight. This additionally allows the photometric observation of the periodic variations in the light curve due to mutual eclipses of the two components.

23.2 Impact on the Spectral Features

Spectroscopic binaries can be subdivided into five types: detached binaries, semi-detached binaries, contact binaries, semi-detached binaries with white dwarfs and semi-detached binaries with neutron stars or black holes (see Figure 23.1).

Detached Binaries

In this case both stars remain within their teardrop-shaped "Roche lobes," defining the gravitational influence zones. The lobes meet at the inner Lagrangian point L1, where the gravitational forces of the two stars cancel each other. Depending on the luminosity difference we record either a

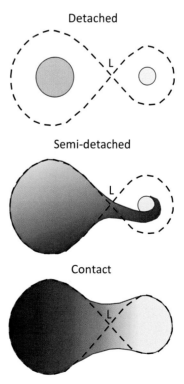

Figure 23.1 Detached, semi-detached and contact systems

composite spectrum or the much brighter component more or less dominates the spectral features in the profile. The periodically changing Doppler shift of the spectral lines due to high orbital velocities is measurable here, for example Spica (α Vir), Mizar A (ζ UMa A) and β Aurigae. In very close but still detached systems a weak mass transfer by a stellar wind may occur, generating some emission lines.

Semi-Detached Binaries

If one of the stars leaves the main sequence and expands past its Roche lobe, then matter can escape from the gravitational influence and transfer to the companion star, forming there a rotating accretion disk (these are so called mass transfer binaries). A part of the accreting matter may be ejected as jets along the rotational axis of the mass gaining star. In such cases the spectra may exhibit highly complex and intense emissions lines, for example Sheliak (β Lyr). In binary orbits with a high eccentricity this stage may be just passed near the periastron (e.g. VV Cephei).

Contact Binaries

Here both stars expand past their Roche lobes, orbiting each other within a common envelope, sometimes within just a few hours.

Semi-Detached Binaries with White Dwarfs

In such close binary systems matters become even more complex if one of the components is a white dwarf, with a

typically excessive high surface gravity. In this case we may observe a so called cataclysmic variable, generating different types of nova eruptions (Chapter 24), or in very rare cases even a final supernova explosion of type Ia (Chapter 25).

Semi-Detached Binaries with Neutron Stars or Black Holes

The very rare combination of mostly massive stars (type O or B), orbiting within a few days extremely dense neutron stars or black holes, generate a huge transfer of matter resulting in so called X-ray binaries or even "microquasars" blasting out quasar-like jets with relativistic velocity. Famous examples are Sco X-1 ($m_v \approx 11.1$), SS 433 ($m_v \approx 13.0$) and Cyg X-1 ($m_v \approx 8.9$) [11]. Such objects exhibit rapidly changing spectral features, particularly in the X-ray domain.

The high mass X-ray binary Cygnus X-1 (~8000 ly), with an apparent brightness of $m_v \approx 8.9$ is spectroscopically accessible also by amateurs, unfortunately limited to the less interesting, optical part of the spectrum. The supergiant HDE 226868 is classified as O9.7 Ia bpe var. With a period of just 5.6 days, it is in orbit with a black hole with estimated 7 $M_\odot$. Figure 23.2 shows a montage, comparing profiles from Cygnus X-1 (HDE 226868) with the similarly classified reference star Alnitak (ζ Ori). Therefore both profiles look indeed quite similar. However, due to the huge mass transfer, combined with an accretion disk, some absorptions of HDE 226868 appear to be deformed by small emission peaks.

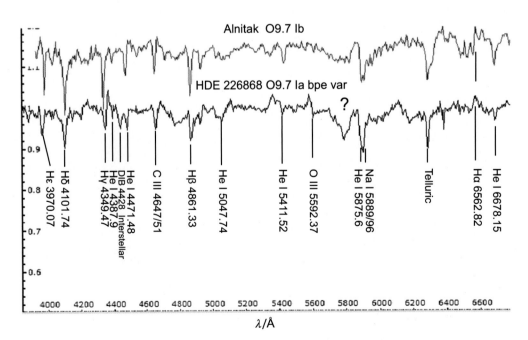

Figure 23.2 Montage, comparing profiles from HDE 226868 with the similarly classified Alnitak

Table 23.1 Simplified overview of the different possible phenomena occurring in interacting binary systems

Separation		Very close but still detached binary system	Semi-detached binary system	Extremely close semi-detached binary system
Type of mass transfer		Weak mass transfer by stellar wind	Significant mass transfer by Roche lobe overflow, causing accretion disks and ejected jets	
Type of mass gaining object spectral features	Main-sequence star	Symbiotic star, some emission lines may show up (Section 24.4)	Highly complex emission spectra e.g. β Lyrae	
	White dwarf		Classical novae, highly complex emission spectra	Dwarf novae, e.g. SS Cygni, AM CVn, short eruption periods
	Neutron star or black hole		X-ray binaries or microquasars ejecting jets with relativistic speed, highly variable X-ray spectrum e.g. Cygnus X-1	

Particularly the barely recognizable Hα as well as the Hγ line, even show P Cygni-like profiles. Recording information: C8, DADOS 50 µm slit, Atik 314L+: 5 × 420 s, 2 × 2 binning.

Table 23.1 displays a rough and simplified overview of the different possible cases and occurring phenomena in interacting binary systems, mainly depending on the type of the mass gaining object and the separation between the components.

23.3 SB1 and SB2 Systems

To understand the influence on the spectra of binaries we have to distinguish between SB1 and SB2 Systems (Figure 23.3). Depending on the apparent brightness difference Δm_v and the resolution power of the spectrograph, we can observe the Doppler shift either just of the brighter component (single lined SB1 system) or in a composite spectrum of both components (double lined SB2 system). For low resolution spectroscopy (R <1000) the upper limit to record an SB2 system has turned out to be approximately $\Delta m_v \approx 1^m$. In higher resolved profiles this allowed luminosity difference may reach up to several magnitudes. Figure 23.3 shows the generation of the Doppler shift for the cases SB1 and SB2 due to different radial velocities, observed in our line of sight. In the SB2 case the luminosity difference is only small and we see a split up of the line.

In SB1 systems just the absorption line of the brighter component is visible, shifting around a neutral position λ_{r0}. Extreme cases are orbits around entirely invisible black holes or extrasolar planets, shifting the spectrum of the orbited star by just a few tens of m s^{-1}!

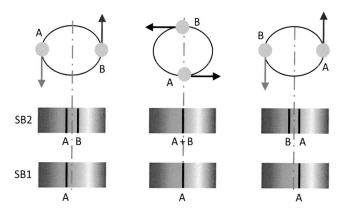

Figure 23.3 Doppler-shift-induced effects in spectra of SB1 and SB2 systems

23.4 Comments on Observed Spectra

Detached SB1 System: β Scorpii A

The following example, recorded with DADOS 900 L mm^{-1}, shows this effect at the spectroscopic A components within the quintuple system β Scorpii (400 ly). Impressively visible in Figure 23.4 is the Hα shift of the brighter component, recorded within three days. The line shifts are plotted here relative to the neutral reference point λ_{r0}. This is the rest wavelength of the spectral line, heliocentrically adjusted from the radial system velocity v_{rSyst} of the binary star's barycenter [3]. So here v_{rSyst} of β Scorpii A yields just +1 km s^{-1} [107]. The x-axis is scaled in Doppler velocities and allows a rough and immediate reading of the measured radial velocity. The $\Delta\lambda$ values have been determined with heliocentrically corrected radial velocities, measured in the profile by Gaussian fits.

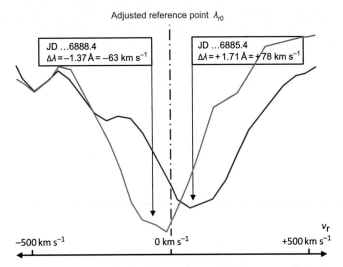

Figure 23.4 Doppler shift of the Hα line in the SB1 system β Sco A

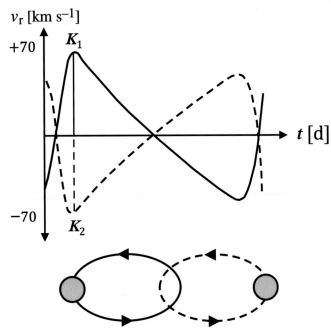

Figure 23.5 Components of Mizar A, variation of the radial velocities

Here follow some orbital parameters of β Scorpii A according to an earlier study by Peterson *et al.* [107]. These values are based on the measured maximum radial velocities K_1 and K_2, obtained from the composite spectrum of both components (see Figure 23.5 for Mizar A). With the large telescopes involved in this study, the spectrum of the weaker component could still be recognized, but analyzed with substantial difficulties [107]. Even in the profiles of Figure 23.4 the weaker B component shows up in the Hα line but just by a very small kink, appearing in the wing which is located opposite to the deflection.

The terms $\sin i$ in the data list and $\sin^3 i$ for the masses, demonstrate that the inclination i was unknown here and therefore the values remained uncertain by this factor [3]. The epoch time is conventionally indicated as a Julian date (JD) showing the number of days after January 1, 4713 BCE. It allows an easy calculation of even larger time differences without bothering with calendar days.

Plate 46 Spectroscopic Binaries: Detached SB2 Systems

Spica (α Vir) and Mizar A (ζ UMa A)

Plate 46 shows a montage of highly resolved spectra, recorded with the SQUES echelle spectrograph, presenting the Doppler shifted Hα and He line (λ6678) of the spectroscopic binaries Spica, α Vir and Mizar A, ζ Uma A. The telluric H_2O lines around Hα have been removed manually (Vspec/Interpolate zone), except for absorptions at both edges of the graph, in order to serve as unshifted optical references. The profiles are labeled with the corresponding orbital phase φ, which is expressed as a fraction of one full orbit (see sketch on the plate). To determine the phase φ the difference in days (d) must be calculated first between

	$β^1$ **Scorpii Aa**	$β^1$ **Scorpii Ab**
Spectral class of the components:	B1V	B2V
Stellar masses:	$M_1 \sin^3 i = 12.5\, M_\odot$	$M_2 \sin^3 i = 7.7\, M_\odot$
Max. recorded radial velocities:	$K_1 = 121$ km s^{-1}	$K_2 = 198$ km s^{-1}
Semi-major axes of orbital ellipse:	$a_{M1} \sin i = 1.09 \times 10^7$ km	$a_{M2} \sin i = 1.78 \times 10^7$ km
$β^1$ **Scorpii A system**		
Epoch periastron time T_0:	JD2418501.531	
Orbital period T (d):	6.8282	
Mass ratio:	$M_1/M_2 = 1.63$	

the observation time T_{Obs} and the epoch time T_0. This difference must then be divided by the orbital period T. The total number of completed orbits since T_0 is mostly not of interest here but the figures after the decimal point indicate the phase φ as a fraction of 1. The indicated epoch time T_0 (orbital phase $\varphi = 0$) corresponds here to the periastron passage.

Spica, α Vir (260 ly) is a spectroscopic binary system with an orbital period of 4.01 days. Here, two giants of the early B class orbit each other at an average distance of only 0.12 AU

unshifted. Even with the high resolution of the SQUES echelle spectrograph just the helium line at $\lambda6678$ shows a real SB2 spectrum with two clearly separated and analyzable absorptions. The Hα line exhibits the weaker B component just by a small kink, appearing in the wing which is located opposite to the deflection.

For comments on the low resolution spectrum of Spica refer to Section 6.4. Presented next are some of the orbital parameters according to [108]. In contrast to β Scorpii the inclination i is approximately known here.

	Spica A	Spica B
Spectral class of the components:	B1.5 IV-V	B3V
Stellar masses:	$M_1 = 10.9\,M_\odot$	$M_2 = 6.8\,M_\odot$
Inclination:	$i = 66°$	
Eccentricity:	$e = 0.146$	
Radial velocities:	$v_1 \sin i = 161$ km s^{-1}	$v_2 \sin i = 70$ km s^{-1}
Spica system		
Semi-major axes of orbital ellipse:	$a_{M1} \sin i = 1.61 \times 10^7$ km	$a_{M2} \sin i = 1.66 \times 10^7$ km
Epoch periastron time	T_0: JD2440678.09	
Orbital period T [d]:	4.014597	

	Mizar Aa	Mizar Ab
Spectral class of the components:	A2V	A2V
Stellar masses:	$M_1 \sin^3 i = 1.68\,M_\odot$	$M_2 \sin^3 i = 1.63\,M_\odot$
Eccentricity:	$e = 0.529$	
Max. recorded radial velocities:	$K_1 = 67.258$ km s^{-1}	$K_2 = 69.179$ km s^{-1}
Mizar A system		
Semi-major axes of orbital ellipse:	$a_{M1} \sin i = 1.61 \times 10^7$ km	$a_{M2} \sin i = 1.66 \times 10^7$ km
Epoch Periastron time T_0:	JD2438085.7	
Orbital Period T (d):	20.5385	

($\sim18 \times 10^6$ km)! The spectral class of the A component is already moving towards the giant branch. The somewhat later classified B component is still established on the main sequence and thus its absolute luminosity is approximately eight times weaker. The systemic radial velocity v_{rSyst} of the entire binary system is just 1 km s^{-1} [11]. Plate 46 shows the Doppler shift of the Hα and He line during a full orbit. As expected the telluric H$_2$O absorptions around Hα remain

Mizar A, ζ Uma A (78 ly) was the first spectroscopic binary ever discovered. In 1890, Antonia Maury, a niece of Henry Draper and a member of Pickering's legendary female staff, analyzed spectral photographic plates of Mizar A. She recognized soon, that the K line on some of the pictures was clearly seen to be double. Not until many years later did it become clear that even the B component is a spectroscopic binary. Mizar A is a nearly symmetrical binary system with two main sequence stars of the early A class. The systemic radial velocity v_{rSyst} of the entire system is –5.6 km s^{-1} [11]. Plate 46 shows the Doppler shift of the Hα line near the maximal and minimal splitting. Presented here are some of the orbital parameters according to [109].

Figure 23.5 roughly shows the variation of the radial velocities for Mizar A, plotted over a full orbit of about 20.5 days.

Plate 47 Semi-Detached System and Eclipsing Binary

Sheliak, β Lyr A

Plate 47 shows a broadband overview spectrum (200 L mm^{-1}) recorded on September 5, 2014. The line identification is based mainly based on [110], [111], [112], [113]. The indicated epoch time T_0 (phase $\varphi = 0$) corresponds here to the primary minimum in the light curve of the eclipsing binary (AAVSO).

Sheliak, β Lyrae A (960 ly), is a semi-detached and also an eclipsing SB2 system. Photometrically it belongs to the Beta Lyrae variables and spectroscopically to the group of W Serpentis stars [38]. In addition, two more distant components belong to the system. The orbit with a period of some 12.9 days is nearly circular and the separation of the components just 38–42 × 10^6 km. According to present knowledge it is a giant of the estimated B6–B8 II class [112] which has left the main sequence and expanded past its Roche lobe. This has caused an overflow of matter to the much earlier classified, hotter companion star, estimated to be B0.5 V. As a dwarf the mass-gaining star still remains on the main sequence and is fully hidden within an opaque accretion disk. This enigmatic system is far from being fully understood and belongs to the best studied objects in astronomic science history [111]. The American Association of Variable Star Observers (AAVSO) expresses this as follows: "In the more than two centuries since its discovery, β Lyrae has played a game of cat and mouse with astronomers attempting to unlock its secrets" [38]. Today, many amateurs are involved in projects contributing to a better understanding of this puzzling object.

The spectra are highly complex and volatile, chiefly reflecting the appropriate radial velocities but in a rather "scrambled" way [110], [113]. Here, just a rough overview can be provided. The mass transfer is estimated to be impressive 1.58 × 10^{-5} $M_\odot$ yr^{-1}, which increases the orbital period of 12.9 days by a rate of about 19 s yr^{-1}. A part of this mass gets lost by a bipolar jet of gas, ejected on both sides and perpendicular to the disk [111]. The strong H and He emission lines are generated mainly by the large scale mass transfer. The origin of the absorption lines is consistently attributed to the B7 giant, because the other component remains fully hidden. There are intense but slim lines which are typical for the thin atmospheres of giants. Depending on the actual recorded phase, even the absorptions sometimes display a double peak, comparable to the H and He emissions. The strikingly slim Na I lines sometimes also appear split, becoming absorbed in the very distant outer gas structures of the system and further by interstellar matter [113]. The photosphere of the B7 giant enables the generation of the quite strong Ca II line at λ3934. However, it would be far too hot to generate any Na I absorptions.

Plate 48 Semi-Detached System and Eclipsing Binary, Spectral Details

Plate 48 shows higher resolved details, recorded with SQUES echelle spectrograph, slit width 50 μm. Here, the complexity of the spectrum becomes obvious as well as the impressive dependency on the recorded phase of the binary system. These effects are difficult to interpret. The shapes and amplitudes of the Hα, Hβ and the numerous He lines do not behave consistently.

Some of the orbital parameters according to [38], [112], [109]:

	Sheliak, β Lyrae Aa	Sheliak, β Lyrae Ab
Spectral class of the components:	B6–B8 II	B0.5V
Stellar masses:	$M_1 \sin^3 i = 2.99\,M_\odot$	$M_2 \sin^3 i = 13.1\,M_\odot$
Eccentricity:	$e = 0.04$	
Max. recorded radial velocities:	$K_1 = 188$ km s^{-1}	$K_2 = 42$ km s^{-1}
Semi-major axes of orbital ellipse:	$a_{M1} \sin i = 3.34 \times 10^7$ km	$a_{M2} \sin i = 7.48 \times 10^6$ km
Sheliak, β Lyrae A system		
Epoch primary minimum T_0:	JD2456897.16	
Orbital period T (d):	12.94062	

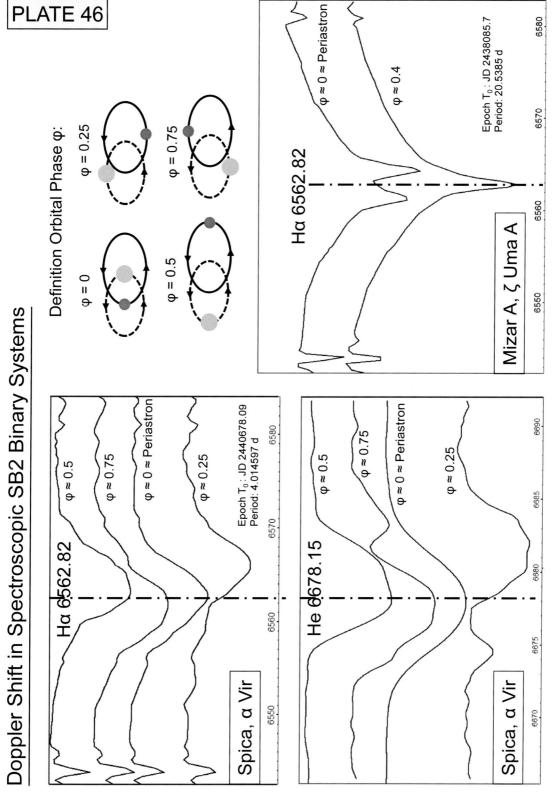

Plate 46 Spectroscopic Binaries: Detached SB2 Systems

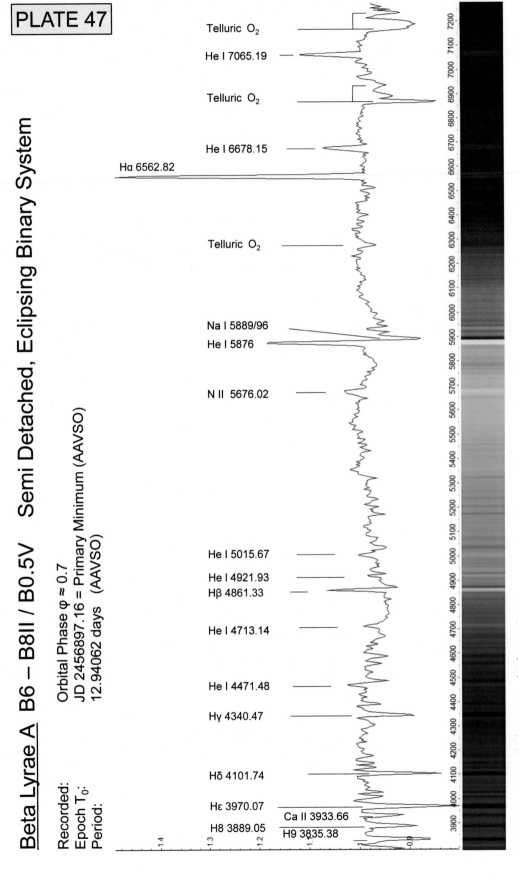

PLATE 47

Beta Lyrae A B6 – B8II / B0.5V Semi Detached, Eclipsing Binary System

Recorded:
Epoch T_0:
Period:

Orbital Phase $\varphi \approx 0.7$
JD 2456897.16 = Primary Minimum (AAVSO)
12.94062 days (AAVSO)

Telluric O$_2$

He I 7065.19

Telluric O$_2$

He I 6678.15

Hα 6562.82

Telluric O$_2$

Na I 5889/96
He I 5876

N II 5676.02

He I 5015.67

He I 4921.93
Hβ 4861.33

He I 4713.14

He I 4471.48

Hγ 4340.47

Hδ 4101.74

Hε 3970.07

Ca II 3933.66

H8 3889.05

H9 3835.38

Plate 47 Semi-Detached System and Eclipsing Binary

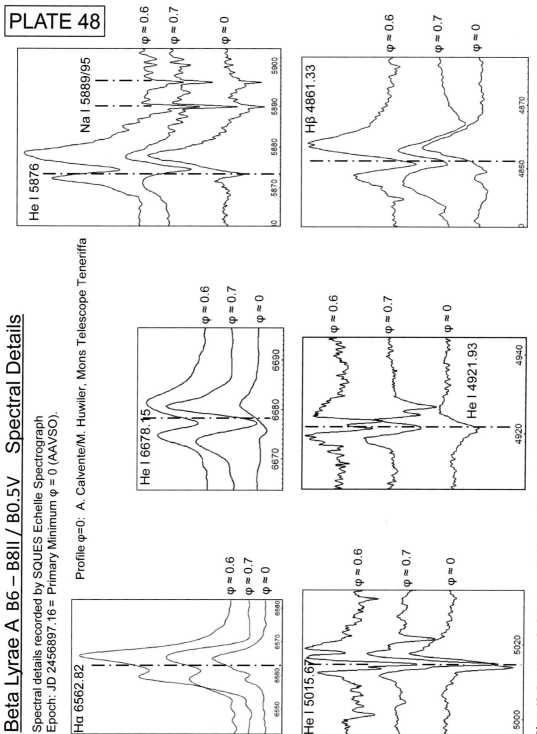

Beta Lyrae A B6 – B8II / B0.5V   Spectral Details

Spectral details recorded by SQUES Echelle Spectrograph
Epoch: JD 2456897.16 = Primary Minimum φ = 0 (AAVSO).

Profile φ=0: A. Calvente/M. Huwiler, Mons Telescope Teneriffa

PLATE 48

Plate 48 Semi-Detached System and Eclipsing Binary, Spectral Details

CHAPTER

24 Novae

24.1 The Phenomenon of Nova Outbursts

In this chapter mainly the spectroscopy-related aspects of nova outbursts are discussed. For a scientific investigation of such an event, additional detailed photometric and many other types of measurements are required.

Except for SN and gamma ray bursts (Chapter 25) nova outbursts are among the most spectacular events the Universe has to offer. The designation "nova" was coined in the sixteenth century by Tycho Brahe with "nova stella," meaning "new star." According to current knowledge this is of course a misnomer. The common reason for most of the nova outbursts is a nuclear runaway reaction in a semi-detached binary system (Chapter 23). The "primary" – here always an extremely dense white dwarf – is accumulating mass which is transferred from the "secondary" companion star, which has an overflowing Roche lobe – a process called tidal stripping (Chapter 23). The secondary is generally a late main sequence star or an expanded red giant.

At the surface of the white dwarf the temperature in the accumulated, extremely compressed hydrogen layer may exceed 10^6 K and so runaway nuclear fusion in the shell is triggered. The resultant explosion ejects a shell of matter with a terminal velocity of some 2000–4000 km s^{-1} [114]. At the very beginning, in the "fireball stage," the fast expanding shell is optically thick, forming a kind of pseudo-photosphere [115]. After the outburst nuclear reactions may continue on the surface at a lower rate, generating a mass loss by a wind with a terminal velocity $v_\infty \approx 1000$ km s^{-1} [114]. The expanding shell typically reaches a mass of ~1–10 $\times$ 10^{-5} $M_\odot$, and a temperature ~10,000 K. In the "post-fireball stage," due to the high expansion velocity, the density of the shell decreases rapidly. Novae show a bimodal distribution of their maximal absolute magnitudes at $M_{V\,max}$ −7.5 and −8.8 [Della Valle/Gilmozzi].

Figure 24.1 shows the remnants of the recurrent Nova T Pyxidis, in constellation Pyxis (3260 ly). The shell consists here of more than 2000 gaseous blobs within an area, measuring approximately one light year across [www.hubblesite.org]. In contrast to supernovae the star is not destroyed by the blast. Having shed its outer shell, the star can begin to accumulate mass again via tidal stripping and thus this

Figure 24.1 HST composite image: Nova T Pyxidis
(Credit: Mike Shara *et al.* and NASA)

event may recur repeatedly. Photometrically, novae belong to the cataclysmic variables (CV).

Whether a classical nova, a dwarf nova or merely a symbiotic star is observed, depends mainly on the involved masses and the distance between the binary components (see Table 23.1).

24.2 Classical and Recurrent Novae

These nova types occur in semi-detached binary systems with a mass accumulating white dwarf. A classical nova is a cataclysmic variable whose outburst has only been observed once. A recurrent nova has repeated outbursts with return periods observed to range from a few years up to a few decades. This is supposed to be the essential difference, with the exception that outbursts of the recurrent type may also be triggered by episodic mass transfer events [2]. For both types the origin of the outbursts is believed to be the nuclear runaway reaction on the surface of the white dwarf. The typical light curve may rise within a few days by ~7–19 mag [2]. Depending on the photometric subtype, it follows either a rather fast or a slow decline within several weeks or months, which may be superposed by some oscillations.

Spectra of classical novae are highly complex, exhibiting a more or less dense "forest" of emission lines. Depending on the phase of the outburst, both the species of the emissions as well as their intensities may change very quickly.

Recurrent novae we observe normally in the quiescent phase, showing the spectrum of the brighter component, frequently a red giant of the K or M class. Due to an ongoing mass transfer the TiO absorptions may sometimes appear to be superposed by emissions of the H-Balmer series. Typical examples are T Coronae Borealis and RS Ophiuchi.

24.3 Dwarf Novae

This nova type occurs in very close semi-detached or maybe even in contact binary systems with extremely small distances between the components. Accordingly, the orbital periods of the binary systems are extremely short – mostly just a few hours! Typical examples are SS Cygni with 6.5 h and U Geminorum 4.2 h [38]. Extreme cases are the hydrogen-poor, AM CVn stars with orbital periods < 1 h. The outbursts last from just a few days up to more than a month and the recurrence periods reach from a few days up to >30 years [116]. The intensity of the outbursts is much weaker here with a typical increase in brightness of just ~4–6 mag. Photometrically, dwarf novae

are subdivided into many subclasses, depending on the specific course of the light curve and return periods of outbursts.

In contrast to the classical and recurrent novae some of the proposed models postulate that thermal instabilities within the accretion disk momentarily increase the flow of matter onto the white dwarf and thus trigger the outburst [38]. However, many details are not yet understood and this is the focus of intense research.

Compared with classical novae, spectra of dwarf novae look rather simple. In the quiescent phase they usually display quite impressive emission lines of the H-Balmer series and He II, appearing superposed on a quite strong continuum. Remarkable here is the Balmer decrement, which often appears far too flat or even inverse (see Section 28.7) [3]. For this still enigmatic issue various hypotheses have been proposed. Most of them postulate processes near the surface of the accretion disk, enabling a significant modification of the generated decrement values.

Immediately after the outburst novae radiate most of their energy by a strong continuum, which is generated by the fast expanding pseudo-photosphere. Accordingly, the spectrum is dominated by an impressive continuum, which is just superposed by weak absorptions, mainly of hydrogen lines. Sometimes within the flat absorption sinks of the H-Balmer series, weak emission cores may show up [117]. In the phase of maximum brightness hydrogen (mostly Hα) and singly ionized helium He II may appear as weak emission lines.

24.4 Symbiotic Stars

Generally the term "symbiotic stars" is applied to very narrow but still detached binaries, exhibiting just a weak mass transfer. This causes emission lines, appearing superposed to the absorption spectra of cool M- or K-type giants. Compared with novae, the distances between the components are here much larger. Correspondingly, the orbital periods, i.e. from ~one year up to several decades, are significantly longer and thus the transfer rate of matter is correspondingly smaller. Typical examples are EG Andromedae and CH Cygni.

In such binary systems mostly a white dwarf is here in orbit with a cool red giant, which, however, in a detached binary system is not able to completely fill its much larger "Roche volume" (Chapter 23), [118]. Thus, no matter can flow beyond the Roche limit and both components are embedded in a common ionized nebula, which is responsible for the generation of emission lines. The weak mass transfer to the white dwarf takes place here by wind accretion, since such systems show very rarely any evidence of accretion disks [119]. Considering the separation between the components it exists a transition zone to the recurrent novae where sometimes weak outbursts may be observed. Symbiotic stars therefore often show similar spectra to recurrent novae in the quiescent phase.

Approximately 80% of the symbiotic stars belong to the subclass S ("stellar"), i.e. in such systems a normal red giant is involved. The remaining approximately 20% of the subclass D ("dusty") containing Mira variables, which are typically embedded in a dust shell [119]. The symbiotic S-class star BD Camelopardalis, documented on Plate 28, is an example of a D-type symbiotic binary – on this recording exhibiting no emission lines.

24.5 Nova-like Outbursts in LBV Stars

Probably triggered by internal instabilities, very massive single stars such as luminous blue variables (LBV) and the eponymous S Doradus-type may also occasionally exhibit impressive nova-like outbursts. Typical examples are P Cygni, sometimes called "Nova Cygni 1600" (Section 18.1), as well as Eta Carinae.

24.6 Evolution of the Outbursts with Classical Novae

The spectroscopic analysis and interpretation of classical novae is highly complex. Nova outbursts pass through different phases with extremely different excitation conditions, generating thereby very different kinds of spectra. Here, just a very rough overview can be provided. The articles "The Evolution and Classification of Postoutburst Novae Spectra" [120] and "The Tololo Nova Survey" [121], both by R. E. Williams *et al.* provide, even for amateurs, very helpful information in respect of line identification, classification of classical novae, as well as a rough understanding of their evolution.

24.7 Spectral Characteristics after Maximum Light

R. E. Williams has recognized that the later spectral evolution of a nova is strongly related to features in the profiles, obtained just after the maximum light. The criteria allow us to distinguish between two unequal basic types – Fe II and He/N. The physical reasons for this difference are not yet understood and are topics for hypotheses. The H-Balmer lines are almost the strongest, appearing in spectra just after maximum light. However, as classification lines the strongest non-H-Balmer emissions are applied. Since the gas density

at this time is still high, emission lines of shock-insensitive, permitted electron transitions dominate the spectra. For the estimation of the terminal velocity v_∞ in nova eruptions the FWHM value of the emission lines is replaced in Equation {17.1} by the value at half width at zero intensity (HWZI, see Appendix G) [121].

Fe II-class novae have:

- rather slim permitted emission lines, for example Fe II, He I or N I
- other low excitation transitions, for example Na I, O I, Mg I, Ca II
- frequent P Cygni profiles in H-Balmer and Fe II lines
- slow spectral evolution (several weeks)
- rather "moderate" terminal velocities with HWZI < 3500 km s^{-1}

In rare cases so called "hybrid novae" [121] may quickly evolve from an Fe II-class spectrum into an He/N profile, before developing any forbidden lines.

He/N-class novae have:

- strong permitted emission lines with higher excitation transitions (compared to Fe II class): He I λ5876, He II λ4686, N II λλ5001/5679 and/or N III λ4640
- emissions that are sometimes very broad and flat-topped
- usually, no P Cygni profiles
- faster spectral evolution (several days)
- higher terminal velocities, HWZI mostly 2500–5000 km s^{-1}

24.8 Evolution from the Permitted to the Nebular Phase

After the "fireball stage" the density of the expanding shell decreases rapidly, allowing the formation of forbidden lines which are sensitive to impacts.

In the Fe II class this means:

- First, higher excitation transitions of forbidden lines appear, the "auroral lines": [N II] λ5755, [O III] λ4363 [O II] λλ7319, 7330, [O I] λ5577 [2].
- Later, even more sensitive, lower excitation "nebular lines" follow: [O III] λ5007, [N II] λ6584, [Ne III] λ3869, [Fe VII] λ6087 [2].
- For HWZI >2500 km s^{-1} strong forbidden emissions of [Ne III] λ3869, and [Ne V] λ3426, may appear, typical features of "neon novae" [114].

For the He/N class, characteristic features in spectra are mostly higher ionized forbidden lines. The following three cases have been observed so far:

Table 24.1 Tololo classification system [121]

	Phase	Subclass	Criteria, based on emission lines
C	CORONAL	Designates the strongest non-Balmer line: a, he, he⁺, c, n, o, ne, s, fe (a = auroral transition)	Spectrum shows "coronal phase" whenever coronal line [Fe X] $\lambda6375$ is stronger than nebular [Fe VII] $\lambda6087$
P	PERMITTED	Designates the strongest non-Balmer line: he, he⁺, c, n, o, na, ca, fe	Spectrum shows "permitted phase" if not in the "coronal phase" and the strongest non-Balmer line is a permitted transition
A	AURORAL	Designates here the strongest auroral line: n, o, ne, s	Spectrum shows "auroral phase" if not in the "coronal phase" and the strongest non-Balmer line is an auroral transition
N	NEBULAR	Designates here the strongest nebular line: c, n, o, ne, s, fe	Spectrum shows "nebular phase" if not in the "coronal phase" and the strongest non-Balmer line is a nebular transition

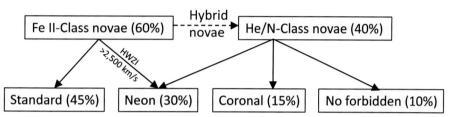

Figure 24.2 Possible evolutionary paths from the permitted to the nebular spectrum

- "coronal" – [Fe X] $\lambda6375$ and "nebular" emissions [Fe VII] $\lambda6087$
- strong forbidden Ne III and Ne V emissions, forming so called "neon novae"
- no forbidden lines at all in some cases

The schematic diagram in Figure 24.2 is adapted according to [114]. It shows the possible evolutionary paths from the permitted to the nebular spectrum supplemented with the observed approximate percentage of occurrence. "Auroral" and "nebular" emissions are collectively referred here by the term "standard."

24.9 The Spectroscopic Tololo Classification System

The Tololo system is presented here slightly reduced and adapted to the average equipment of an amateur astronomer. For a more detailed specification with an extension of the considered wavelength range and an enhanced notation system, refer to [2], [120], [121].

According to specific spectral features in a recorded profile, the Tololo system allows us to assign it to a certain evolutionary "phase." It can be determined stepwise, applying the instructions in Table 24.1 from top to bottom [121]. The range for the classification lines is limited here to the wavelength region $\lambda3300$–7600.

To specify the spectra, the Tololo system uses the following notation. For example: In P_{he} the "P" stands for a spectrum, classified in the "permitted phase" and the subscript "he" stands for neutral helium as the strongest non-H-Balmer line in the profile.

If the entire nova event is documented with several spectra, its evolutionary sequence can easily be specified. For example, nova V2214 Oph/88 was classified as $P_{fe,he}N_{ne}$ which means the initial Fe II class with strong He I lines developed in the final "nebular phase" into a "neon nova" [121].

24.10 Comments on Observed Spectra

Plate 49 Classical Nova: Development of the Spectrum During Outburst

Nova Delphini V339 Del, Fe II Class

Koichi Itagaki discovered the classical nova on August 14, 2013 at a brightness of $m_v \approx 6.8$, immediately followed by the first recorded spectra. The estimated peak brightness of $m_{v\,max} \approx 4.3$ was reached on August 16, 2013, some 1.85 days after the discovery. Later, the precursor star was determined as having a former apparent magnitude of $m_v \approx 17.6$ [38].

The two profiles on **Plate 49** have been recorded by Erik Wischnewski with the "Star Analyser," a slitless applied transmission grating (100 L mm⁻¹), achieving a remarkably good

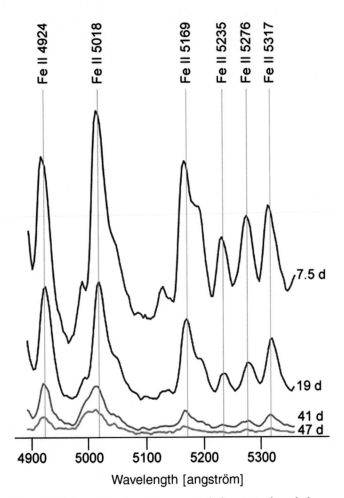

Figure 24.3 Transition from the permitted phase P to the nebular phase N
(Credit: Erik Wischnewski)

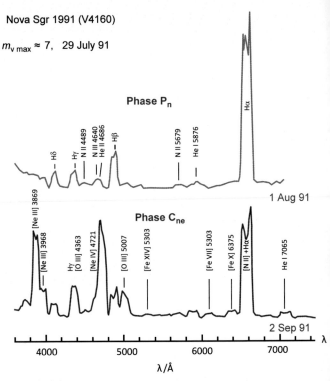

Figure 24.4 He/N-class Nova Sgr 1991, schematically after R. E. Williams *et al.* [121]

resolution! The first spectrum was obtained +2.5 d (days post maximum). The numerous Fe II emissions clearly indicate the classification as Fe II-class nova in the permitted phase P_{fe}. The second spectrum, obtained +47 d, shows the fading nova, reaching the final nebular phase N_o, exhibiting many forbidden emissions.

As a supplement Figure 24.3 by Erik Wischnewski, impressively shows the weakening of Fe II emissions during the transition from the permitted phase P to the nebular phase N.

Nova Sgr 1991 (V4160), He/N class

Due to lack of self recorded spectra, the spectral features of a classical He/N-class nova are demonstrated here by the example of V4160. The profiles in the montage of Figure 24.4 are schematically adapted after R. E. Williams *et al.* [121]. The nova was discovered on July 29, 1991 around the maximum brightness $m_{v\ max} \approx 7$.

The upper, blue profile was recorded three days later and shows a typical He/N nova in the permitted phase P with characteristic, very broad, boxy emissions with flattened tops

(HWZI ~4300 km s^{-1}). Sometimes the peaks appear split, probably an indication for the structure and transparency of the expanding shell. The strongest non-H-Balmer lines are N III λ4640 and He II λ4686. Similarly typical for the He/N class was the record-breaking fast brightness evolution.

The lower, red profile was recorded a month later and shows the coronal phase C with Fe VII, Fe X and Fe XIV whereby Fe X is stronger than Fe VII. The extremely intense and broadened Ne emissions classify this phase as neon nova.

Plate 50 Recurrent Nova at Quiescence

T Coronae Borealis (HD 143454)

Plate 50: T Coronae Borealis sometimes called the "Blaze Star" (~2600 ly) is a semi-detached binary system with a relatively long orbital period of ~228 days [109]. It consists of a red giant, classified as M3 IIIp [11] and a white dwarf. The brightness in the quiescent phase is slightly variable with $m_v \approx 10$. The first recorded outburst occurred 1866, the second in February 1946, while the brightness peaked at $m_v \approx 2.0$ [122]. Typically for quiescent novae, the profile displays here the spectrum of the M giant with a well developed system of TiO absorption bands. The white dwarf would be by far too faint to generate any spectral features in the profile directly. At the time of recording just Hα and relatively weakly Hγ and Hδ appeared in emission. Other published quiescent spectra of T CrB sometimes show the whole H-Balmer series in emission [2]. To optimize the clarity, the labeling of the lines is limited to some prominent lines.

Plate 51 Dwarf Nova

SS Cygni

Plate 51: The brightest of all dwarf novae, SS Cyg (~370 ly) is one of the most frequently observed variable stars. It was discovered 1896 by Louisa D. Wells, a member of Pickering's legendary female staff. According to AAVSO the white dwarf orbits here with an orange dwarf, classified on the main sequence as K5 V p [11]. Due to the extremely small surface distance of only ~100,000 km, even a K dwarf is capable here to fill its Roche volume, causing thereby an overspill of matter. A complete orbit takes just about 6.5 hours!

SS Cyg stays about 75% of the time in quiescence with an apparent brightness of $m_v \approx 12.2$. The outburst occurs very abruptly, reaching maximum light of $m_{v\,max} \approx 8.5$ normally within ~1 day. The return period is ~4–6 weeks and the outbursts last ~1–2 weeks. To schedule the observation and to obtain information about the current phase, the consultation of the AAVSO website is recommended [38].

In the quiescent phase, with an apparent magnitude of $m_v \approx 12.2$, the spectrum is dominated by the emissions of the H-Balmer series, as well as by neutral helium. The Balmer decrement here typically runs inversely, i.e. $I_{H\gamma} > I_{H\beta}$ (Section 28.7, [3]). These lines appear superimposed on an intense continuum, generated by the K5 V dwarf, whose spectral type, however, is barely recognizable in this profile.

In the outburst phase, the spectrum changes radically. At maximum brightness with $m_{v\,max} \approx 8.5$, as expected an intense continuum appears, overprinted with weak absorptions of the H-Balmer series. He II at $\lambda 4686$ and Hα appear in emission, while Hβ and Hγ can just be seen as small emission cores within the absorption sinks. Figure 24.5 shows these spectral details enlarged for the H-Balmer series.

The intensities of the two profiles in Plate 51 are comparable (see scale). The entire continuum intensity above the wavelength axis is not shown to scale here due to lack of space. Recording information for the spectrum in the quiescent phase: C8/DADOS, 200 L mm^{-1}, 50 µm slit, Atik 314L+, 2 × 1200 s, 2 × 2 binning.

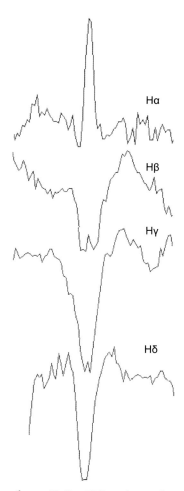

Figure 24.5 Dwarf nova SS Cyg, H-lines during the outburst phase

Plate 52 Symbiotic Star

EG Andromedae, Subclass S "Stellar"

Plate 52: Similar to the recurrent nova T CrB, a white dwarf is here in orbit with a red giant of spectral type M3 III [119]. However, the orbital period of 482.6 days is more than twice as long, because the masses of the components involved are correspondingly lower. In this detached binary system, the red giant fills its Roche volume to about 80%. Thus, a mass transfer takes place here not by an accretion disk, but rather by a faint wind accretion. The latter is responsible for the weak emission lines of the H-Balmer series, appearing superimposed on the impressive titanium monoxide bands. Otherwise, this profile looks very similar to that of T CrB, the recurrent nova in the quiescent phase (Plate 50). To optimize the clarity, the labeling of the lines is limited here to some prominent lines.

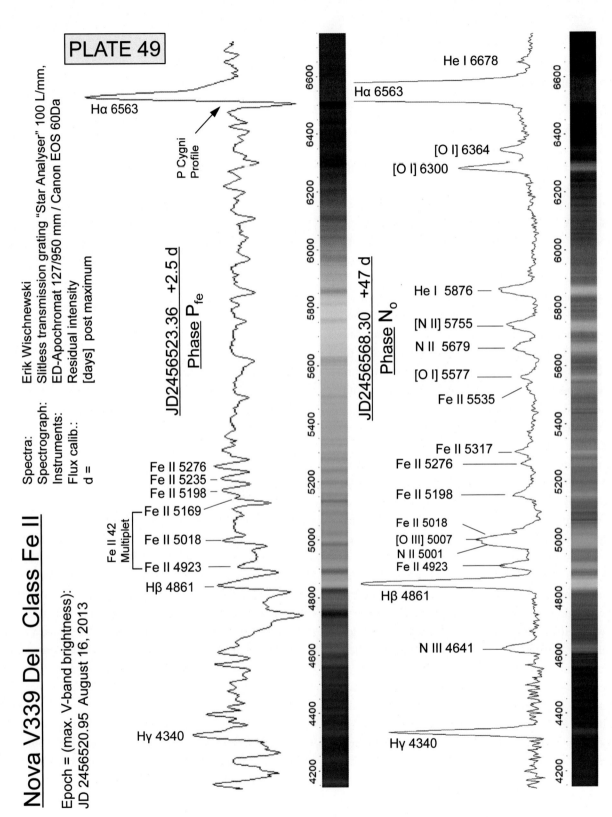

Nova V339 Del Class Fe II

PLATE 49

Epoch = (max. V-band brightness):
JD 2456520.95 August 16, 2013

Spectra: Erik Wischnewski
Spectrograph: Slitless transmission grating "Star Analyser" 100 L/mm,
Instruments: ED-Apochromat 127/950 mm / Canon EOS 60Da
Flux calib.:   Residual intensity
d =   [days] post maximum

JD2456523.36 +2.5 d
Phase P_{fe}

Hα 6563

P Cygni
Profile

Fe II 5276
Fe II 5235
Fe II 5198
Fe II 5169
Fe II 42
Multiplet
Fe II 5018
Fe II 4923
Hβ 4861

Hγ 4340

JD2456568.30 +47 d
Phase N_o

He I 6678
Hα 6563

[O I] 6364
[O I] 6300

He I 5876
[N II] 5755
N II 5679
[O I] 5577
Fe II 5535

Fe II 5317
Fe II 5276
Fe II 5198
Fe II 5018
[O III] 5007
N II 5001
Fe II 4923
Hβ 4861

N III 4641

Hγ 4340

6600 6400 6200 6000 5800 5600 5400 5200 5000 4800 4600 4400 4200

Plate 49 Classical Nova: Development of the Spectrum During Outburst

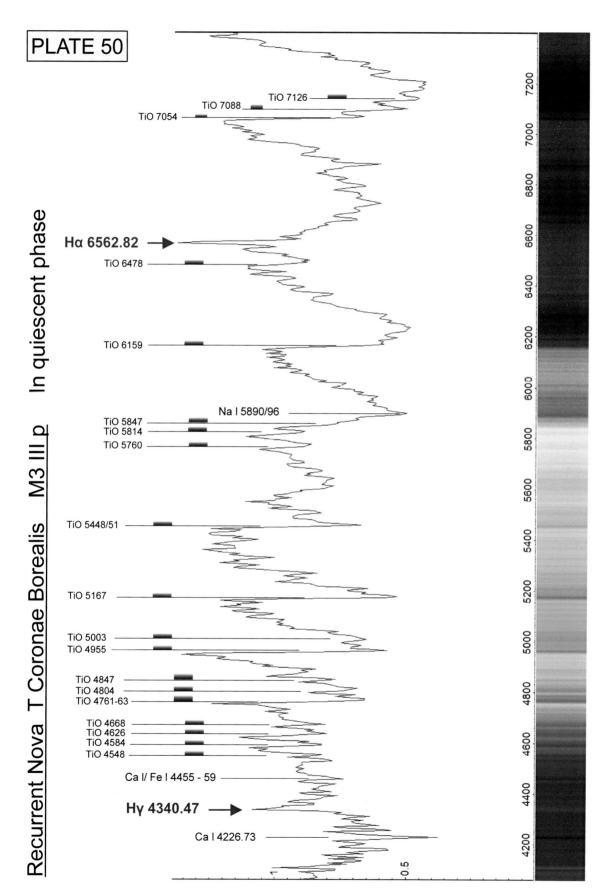

PLATE 50

Recurrent Nova T Coronae Borealis M3 III p In quiescent phase

Hα 6562.82 →

TiO 7054
TiO 7088
TiO 7126
TiO 6478
TiO 6159
Na I 5890/96
TiO 5847
TiO 5814
TiO 5760
TiO 5448/51
TiO 5167
TiO 5003
TiO 4955
TiO 4847
TiO 4804
TiO 4761-63
TiO 4668
TiO 4626
TiO 4584
TiO 4548
Ca I/ Fe I 4455 - 59
Hγ 4340.47 →
Ca I 4226.73

Plate 50 Recurrent Nova at Quiescence

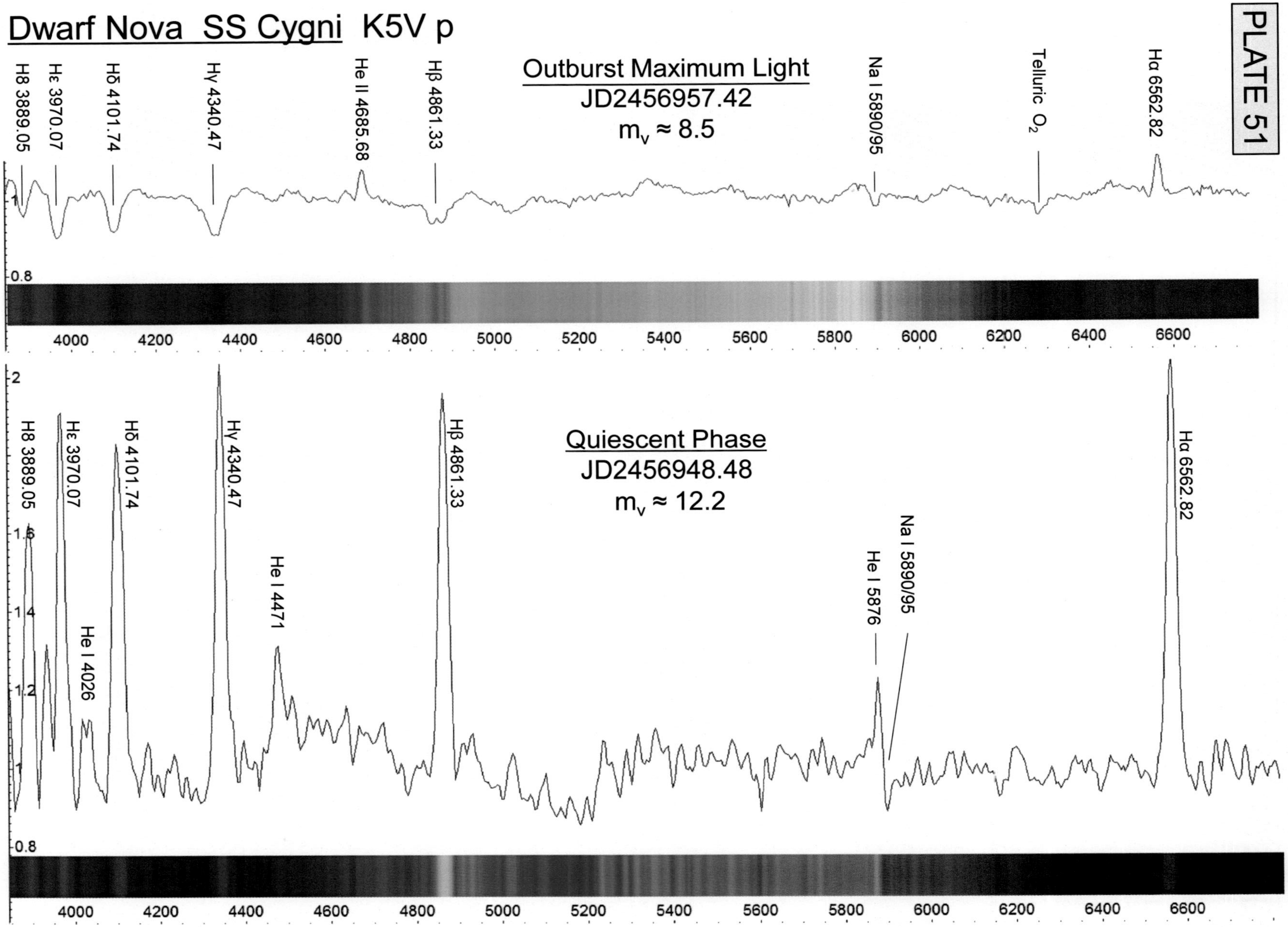

PLATE 51

Dwarf Nova SS Cygni K5V p

Outburst Maximum Light
JD2456957.42
$m_v \approx 8.5$

Hα 6562.82

Telluric O₂

Na I 5890/95

Hβ 4861.33

He II 4685.68

Hγ 4340.47

Hδ 4101.74

Hε 3970.07

H8 3889.05

Quiescent Phase
JD2456948.48
$m_v \approx 12.2$

Hα 6562.82

Na I 5890/95
He I 5876

Hβ 4861.33

He I 4471

Hγ 4340.47

Hδ 4101.74

He I 4026

Hε 3970.07

H8 3889.05

Plate 51 Dwarf Nova

PLATE 52

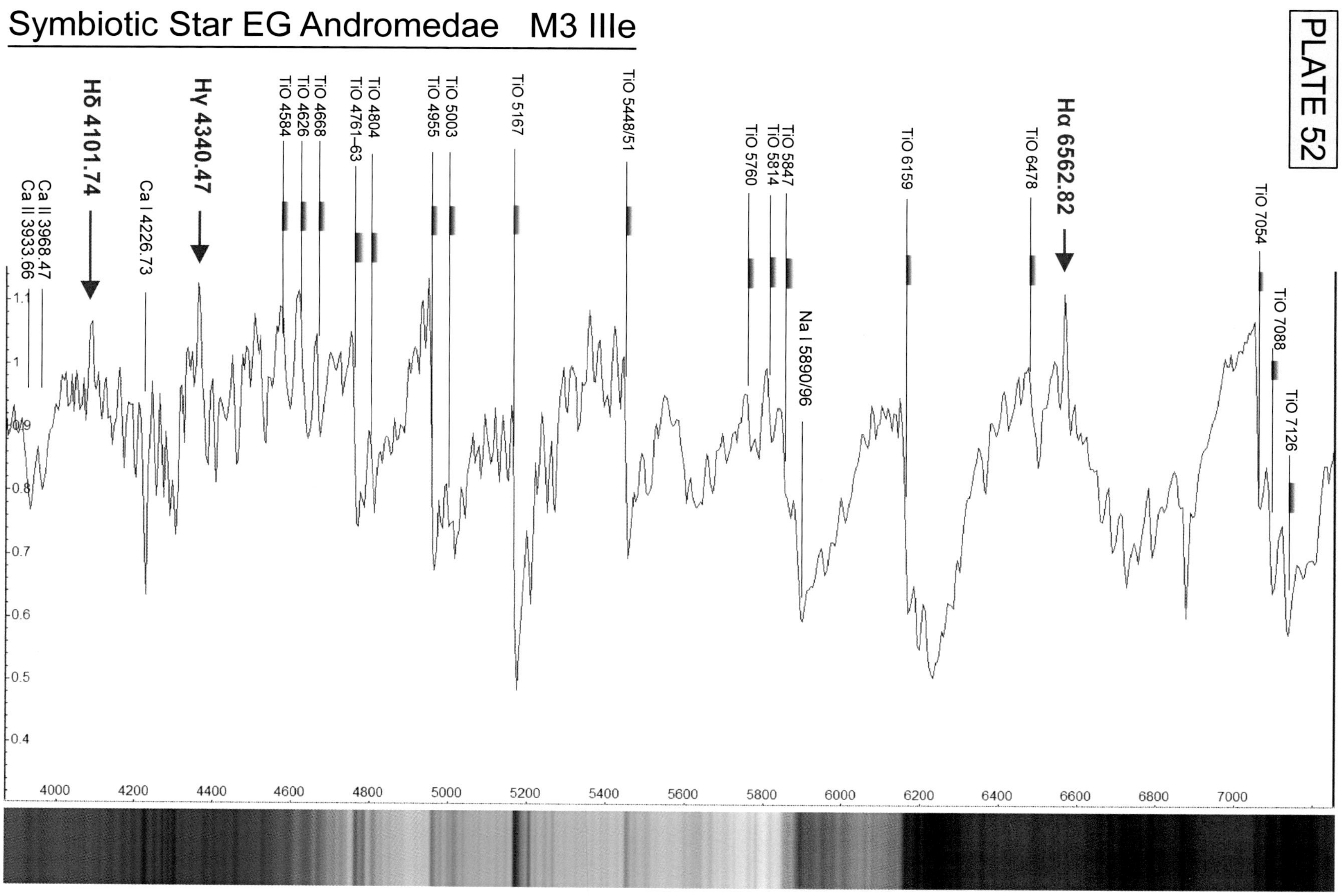

Symbiotic Star EG Andromedae   M3 IIIe

TiO 7126
TiO 7088
TiO 7054

7000

6800

6600
Hα 6562.82 →
TiO 6478

6400

6200
TiO 6159

6000

Na I 5890/96
TiO 5847
TiO 5814
TiO 5760

5800

5600

5400
TiO 5448/51

5200
TiO 5167

5000
TiO 5003
TiO 4955

4800
TiO 4804
TiO 4761–63

4600
TiO 4668
TiO 4626
TiO 4584

4400
Hγ 4340.47 →

4200
Ca I 4226.73

Hδ 4101.74 →

4000
Ca II 3968.47
Ca II 3933.66

1.1   1   0.9   0.8   0.7   0.6   0.5   0.4

Plate 52 Symbiotic Star

CHAPTER

25 Supernovae

25.1 The Phenomenon of Supernova Explosions

In this chapter the spectroscopy-related aspects are discussed. For a scientific investigation of such an event, detailed photometric and many other types of measurements are required.

In contrast to a nova, a supernova (SN) explosion destroys the star totally and forms the definitive end of its life. The cataclysmic, nuclear runaway reaction generates an unimaginable amount of energy and almost the entire stellar mass, is blasted in to the surrounding space, initially with speeds of $>10,000$ km s^{-1}. For comparison, the coronal mass ejections (CME) of our Sun may reach $>3,000$ km s^{-1} and the detonation velocity of our most rapid explosives just ~8 km s^{-1}

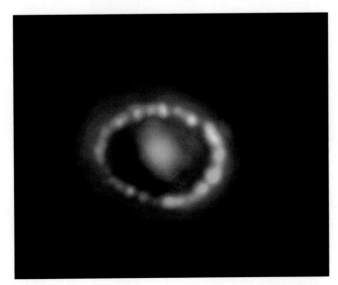

Figure 25.1 SN 1987 A (Type II). Composite Image: ALMA/HST/Chandra
(Credit: ALMA (ESO/NAOJ/NRAO)/A. Angelich)

(Nitropenta). As a result of such an explosion, the interstellar matter (ISM) becomes enriched with elements even heavier than iron, which decisively influences the later formation of stars, planets and also possible life. The diameter of old supernova remnants (SNR) may finally reach >100 ly, for example the famous, some 7000 year old Cygnus Loop (see Figure 28.4). In contrast, the diameter of the relatively young Crab Nebula M1 has grown up to just ~11 ly (Chapter 28).

Figure 25.1 shows a composite image of SN 1987A (SN type II) in the Large Magellanic Cloud (~168,000 ly), recorded in 2014 at very different wavelengths by HST, ALMA and Chandra. The propagating equatorial shockwave with the bright hot spots, visible here in green and blue, was recorded by HST and Chandra, the reddish dust cloud in the center of the SN remnant, by ALMA.

25.2 Designation of Supernovae

Supernovae are labeled with the letters SN, followed by the year of discovery and an ongoing assigned letter, such as SN2014J. Since several hundred SN are discovered each year with today's telescopes and automatic monitoring systems, after the first 26 events, the assigning of double letters becomes necessary.

25.3 Classification of SN Types

Supernovae are divided into two main types, labeled with the Roman numerals I and II. This rough subdivision is very easy even for amateurs, since the spectra of SN type II display absorptions or emissions of the H-Balmer series and those of SN type I show none. This quite simple relation was discovered as early as 1941 by Rudolph Minkowski. This classification, also named the Minkowski–Zwicky scheme, is based on the examination of absorption lines within these extremely complex spectra.

Immediately after the explosion SN radiate most of their energy by a strong continuum, which is generated by the fast expanding pseudo-photosphere, detectable up to far after maximum light. Similar to nova outbursts, some weeks to months later, towards the nebular phase, the profile turns increasingly into an emission spectrum, displaying even shock-sensitive forbidden lines. This appears particularly pronounced at the subclasses of the hydrogen-poor type I.

Using the example of SN2014J, type Ia, D. Van Rossum impressively shows which highly complex blends are generally formed by SN explosions [123]. Both the distinction between absorptions and emissions, as well as the determination of the real continuum-course, is very difficult. So the relative flux-calibration of the spectrum by a standard star really makes sense here. This process is outlined in [3].

Accordingly, in the context of SN spectra absorptions are often called "troughs" and their equivalent widths denoted as "pseudo-EW" (pEW) [123]. The elements and ions labeled in such profiles, are mostly the detected, and in some cases merely suspected, main causes of these impressive spectral features. A highly recommendable and relatively easily understandable introduction to this topic is provided in the publication by A. Filippenko "Optical Spectra of Supernovae" [124].

25.4 SN Type I

According to the current state of knowledge the characteristic lack of hydrogen at SN type I is mainly caused by two different scenarios. This requires the division into the subclasses Ia and Ib/Ic.

Type Ia

For stars with $\lesssim 8\ M_\odot$ the hydrogen envelope is repelled as a planetary nebula. This includes all former main-sequence stars classified later than about B4–B6. Finally, a white dwarf, mostly consisting of carbon and oxygen is all that remains. The spectra of type Ia supernovae are typically metal-rich, displaying mainly Fe, Si and S lines.

Type Ib/Ic

For stars with $\gtrsim 25\ M_\odot$ the hydrogen envelope may be repelled either as a Wolf–Rayet nebula [66] or by the expanding shell of an LBV star. Thus, roughly all former main-sequence stars of spectral class O are concerned. The subclass Ib shows helium lines whereas Ic displays none. Maybe at type Ic the WR star has already repelled its helium envelope before the onset of the explosion. Therefore, all these classes are also referred to as "stripped core-collapse supernovae." In contrast to the type Ia they barely show any silicon (Si) in their spectra.

25.5 SN Type II

The SN type II, which displays the characteristic hydrogen lines in the spectrum, covers the middle mass range of ~ 8–$25\ M_\odot$. Otherwise, considering the spectral class, this range is very small and concerns roughly all former main-sequence stars in the area of just the early B to the very late O class! A substantial contribution to this theory stems from the Swiss astronomer Fritz Zwicky in the 1960s. Spectroscopically a few of the supernovae are outliers, which are not yet fully understood. So, in rare cases a SN may start as type II but end up as type Ic [2]. Generally speaking, not all relationships are fully understood here.

25.6 Explosion Scenario: Core Collapse

The core-collapse SN forms the very end of all stars with an initial mass $M_i \gtrsim 8\ M_\odot$. This theory was proposed in 1938 by Walter Baade and Fritz Zwicky. At the end of the giant stage, successively heavier elements are generated in the core, resulting in a shell-like structure within the star. Table 25.1 shows this process schematically after Woosley and Janka [125], for a star with $\sim 15\ M_\odot$. Each subsequent fusion process lasts significantly shorter at massively increased density and temperature.

It will be fatal for the star as soon as it starts to produce iron. For the formation of this, and all the following even heavier elements, the fusion processes consume energy, thus allowing just the formation of a degenerate iron core approximately the size of the Earth [125]. As soon as the core reaches the Chandrasekhar mass limit (about $1.4\ M_\odot$) it collapses within a very short time to a diameter of about 30 km. This initial implosion then quickly turns in to a cataclysmic explosion whereby neutrinos seem to play a decisive role. Exclusively in this phase, all the elements heavier than iron are formed. Finally, depending on the initial stellar mass either a neutron star or a black hole is generated. During the core collapse the extremely large stars probably of the type Wolf–Rayet, LBV or generally hypergiants can further eject a highly intense gamma ray jet, headed parallel to the stellar

Table 25.1 Formation of heavy elements in stars with $M_i \approx 15\ M_\odot$

Stage	Fuel or Product	Ash or Product	Temperature [K]	Density [g cm^{-3}]	Time Scale [Years]	Shell Structure in the Core Area (schematic)
Hydrogen	H	He	3.5×10^7	5.8	11×10^9	
Helium	He	C, O	1.8×10^8	1390	2×10^9	
Carbon	C	Ne, Mg	8×10^8	2.8×10^5	2000	
Neon	Ne	O, Mg	1.6×10^9	1.2×10^7	0.7	
Oxygen	O, Mg	Si, S, Ar, Ca	1.9×10^9	8.8×10^6	2.6	
Silicon	Si, S, Ar, Ca	Fe, Ni, Cr, Ti	3.3×10^9	4.8×10^7	18 days	
Iron core collapse	Fe, Ni, Cr, Ti	Neutron star	$>7 \times 10^9$	$>7 \times 10^9$	~1 s	

rotation axis (gamma ray burst). In extreme cases a hypothetical hypernova may even occur, such is expected for Eta Carinae. For a rough overview on the physical effects of the shockwave within an expanding SNR, refer to Chapter 28.

25.7 Explosion Scenario: Thermonuclear Carbon Fusion

This scenario is also called thermonuclear explosion and is exclusively limited to the SN type Ia with an initial stellar mass of $M_i \lesssim 8\ M_\odot$. As a member of a semi-detached binary system, a white dwarf (Section 16.4) by accretion of matter and finally exceeding the critical Chandrasekhar mass limit of ~1.4 $M_\odot$, can explode as a SN type Ia. However, an additional criterion may possibly be a minimum annual rate for the accretion (ΔM year^{-1} > $1 \times 10^{-8}\ M$) [126]. Otherwise just small nova outbursts occur, limited to the currently accreted material on the stellar surface (Chapter 24). But if all conditions are met, the degenerate electron gas of the iron core can no longer withstand the gravitational pressure. In contrast to the core-collapse scenario the stellar core consists here mostly of reactive carbon and oxygen, which is why the object immediately explodes. Therefore the SN type Ia is sometimes also referred as carbon detonation supernova and leaves, in contrast to the core-collapse scenario, no residual object.

25.8 SN Type Ia: An Important Cosmological Standard Candle

As already mentioned the SN type Ia exclusively occurs with white dwarfs if they exceed by accretion the "quasi standardized" critical Chandrasekhar mass limit of about 1.4 $M_\odot$. Thus, this way a more or less uniform amount of energy of about 10^{45} J is released [127]. Further, both the photometric and the spectral profiles of such events are very similar. With SN type Ia the luminosity reaches the maximum after about 20–30 days, with an absolute magnitude in the blue range of $M_B \approx -19.5$ [128]. Compared with other types of SN explosions these values remain within a rather small stray area. The maximum brightness clearly reaches the top rankings, i.e. they are bright enough to outshine an entire galaxy for several months. Due to this uniform appearance the SN Ia explosions serve as indispensable "standard candles" for measuring a large part of the observable Universe. For this purpose the core-collapse SN types are less useful, because the intensity of such explosions depends strongly on the initial stellar mass M_i. The absolute maximum brightness occurs here within a very large stray area of $M_B \approx -15$ to -21 [128]. The extreme values are denoted here as "subluminous" and "over luminous."

25.9 Diagram for the Spectral Determination of the SN Type

The following diagram can be used for the spectroscopic identification of the SN type. In addition, the relevant explosion scenario, the type of the progenitor star, and finally the rough order of magnitude for the initial stellar mass M_i are assigned here. Not shown are the subcategories of type II, which are determined photometrically by the course of the light curve. The light curve of type II-P shows a plateau phase after the maximum light while the brightness of the type II-L ~ decreases rather linearly, and type II-n shows narrow lines [2].

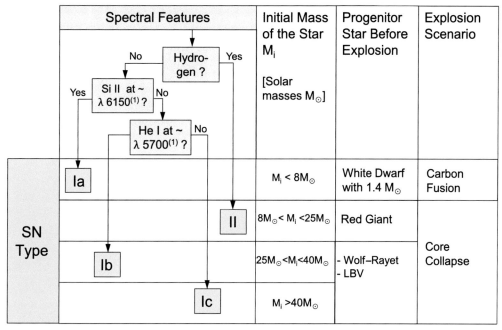

Figure 25.2 Spectral determination of the SN type
(adapted and supplemented after various templates)

25.10 SN Type Ia: Spectral Features in the Optical Range

Plate 53 demonstrates The spectral features of SN type Ia by the example of SN2014J.

Plate 53 Supernova Type Ia, Characteristic Spectrum and Chronology

Supernova SN2014J

Supernova SN2014J, located in the galaxy M82, was the brightest for decades and on January 30, 2014, reached the maximum apparent brightness of $m_v \approx 10.5$. It was discovered by chance and surprisingly late on January 21, on the occasion of a student exercise at the University of London. The apparent explosion date was subsequently determined from pictures, taken by automatic surveillance systems, to January 14, 2014 [129].

Figure 25.3 shows in the crosshairs SN2014J within the host galaxy M82. The upper red profile of Plate 53 shows the usual SN type Ia spectrum near the maximum light with the typical two sulfur sinks (S II) at $\lambda 5400$, also called "W-absorption," and the prominent silicon trough "silicon valley" at $\lambda 6150$. The latter forms the key feature for the identification of the SN type Ia which also allows the rough estimation of the detonation velocity.

Figure 25.3 SN2014J (Type Ia) in M82
(Credit: R. Stalder, Hubelmatt Observatory, Lucerne)

This huge Si II silicon absorption with the rest wavelengths of $\lambda\lambda 6347$ and 6371 appears here typically to be blueshifted impressively by $\sim\lambda 200$ Å into the range of $\sim\lambda 6150$. Processed by the spectroscopic Doppler law according to Equation {18.1} it results in a radial velocity of 9800 km s^{-1}. A similar result is obtained according to Equation {17.1} by the FWHM of the Si II trough at $\lambda 6150$. D. Van Rossum calls this the "Si II-velocity" [123].

Approximately 20 days later, i.e. ~37 days post-explosion (pe), in the blue bottom profile, as expected, several Fe II emissions are prevailing, suppressing for example the S II W-absorption at λ5400. However, the impressive silicon trough is still present here, showing a comparable intensity. A striking feature of both profiles is the intense Na I line, which is correspondingly interpreted by several publications as interstellar absorption within the extremely dusty host galaxy M82. A. Filippenko analyzes this line even as saturated [129]!

The two relatively flux calibrated profiles show a slight intensity increase towards the long-wave (red) direction. This is more strong evidence for the huge interstellar reddening because the "unreddened" model spectra show the intensity peak rather in the blue range [123]. Recording information for the red profile: January 31, 2014: Hubelmatt Observatory, Lucerne, 40 cm MFT Cassegrain, focal length 5550 mm, DADOS 200 L mm^{-1}, 50 µm slit width, Atik 314L+, 2×900 s, 2×2 binning. Recording information for the blue profile: February 20, 2014: Celestron C8, DADOS 200 L mm^{-1}, 50 µm slit width, Atik 314L+, 1×1800 s, 2×2 binning.

25.11 SN Type II: Spectral Features in the Optical Range

Due to the lack of self recorded spectra, the spectral features of SN type II are demonstrated by the famous example of SN1987A, which exploded at a distance of ~167,000 ly in the Large Magellanic Cloud (LMC). Supernova SN1987A was the closest of all observed since "Kepler's Supernova 1604" (type Ia), which occurred in the Milky Way (20,000 ly). The brightness reached a maximum of $m_v \approx 3$ and thus the object was easily visible by naked eye in constellation Dorado. The progenitor star could be identified here as a blue giant of

spectral class B3 [130]. The intense search for the collapsed core remains unsuccessful to date and the expected neutron star has not yet been detected. SN1987A was by the way the first object, except for the Sun, from which neutrinos have been detected!

This event still remains enigmatic. So, for example, the light curve was somewhat unusual for a core collapse SN and the brightness maximum was significantly lower than expected. The corresponding absolute magnitude of $M_B \approx -15$ is considered as clearly subluminous for SN type II! As a possible reason an asymmetrically running explosion is discussed. Further debated is the progenitor star, a blue B3 giant. According to the present understanding of stellar evolution, a red giant of a later spectral class would be expected.

The profile in Figure 25.4 is schematically adapted after J. Danziger and R. Fosbury [131]. It was obtained on February 27 in La Silla, four days after the explosion. Here the characteristic, spectroscopic features of SN type II appear very striking: the impressive P Cygni profile of the Hα line and the strong absorption troughs of Hβ and Hγ. Compared with the rest wavelength the trough of the strongly blueshifted Hα absorption indicates a radial velocity of $cz \approx 17,000$ km s^{-1} [131]. The weak Na I absorption remains unshifted and is probably of interstellar origin. Some five months later the H-Balmer lines still appeared strong but were supplemented by other blueshifted absorptions, mainly Fe II, O I and Na I [124].

25.12 SN Type Ib and Ic: Spectral Features in the Optical Range

Due to lack of self-recorded spectra, the spectral features of SN types Ib and Ic are demonstrated by the SN1984L (type Ib) and SN1987M (type Ic). The profiles in the montage

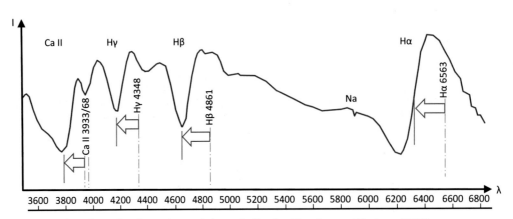

Figure 25.4 SN1987A, some four days after the explosion (schematically after Danziger and Fosbury [131])

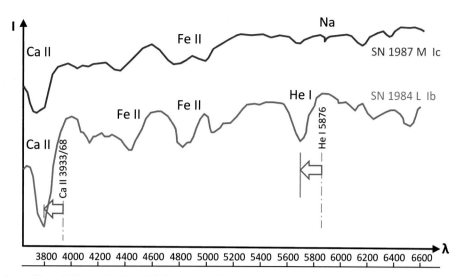

Figure 25.5 SN1984L (type Ib) and SN1987M (type Ic) (schematically after Filippenko [124])

of Figure 25.5, schematically adapted after A. Filippenko [124], were recorded ~1 week after the observed B-band maximum. The intensity of the continuum is here not to scale in respect of the intensity axis I.

At a first glance both profiles of these early recorded spectra look very similar. The lower profile of type Ib displays a strong He I absorption trough, which appears to be blue-shifted to ~5700 Å. Textbook-like, in the upper profile of type Ic, this line is missing. Both SN types showed in the following "nebular phase" some two months past maximum,

impressive emissions of Na I, Mg I and even of forbidden [O] and [Ca II] [124].

SN1984L was discovered on August 28, 1984 in NGC 991 by an Australian amateur. The date of maximum light was estimated afterwards to be August 20 ±4 d, with an apparent magnitude in blue of $m_b \approx 14.2$ and an absolute of $M_B \approx -17.4$ [132].

SN1987M was discovered in NGC 2715, on September 21, 1987. Apparent maximum light was $m_b \approx 15.3$ and the absolute magnitude was estimated to be $M_B \approx -18.4$ [133].

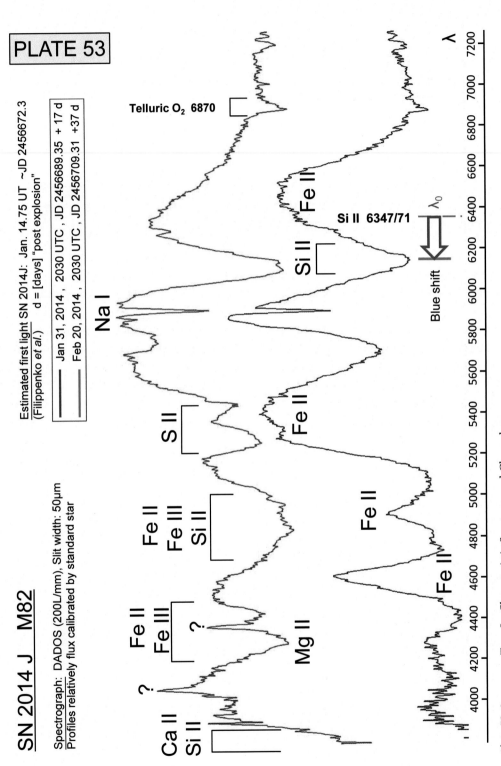

Plate 53 Supernova Type Ia, Characteristic Spectrum and Chronology

PLATE 53

SN 2014 J   M82

Spectrograph: DADOS (200L/mm), Slit width: 50μm
Profiles relatively flux calibrated by standard star

Estimated first light SN 2014J: Jan. 14.75 UT ~JD 2456672.3
(Filippenko et al.)   d = [days] "post explosion"

Jan 31, 2014 , 2030 UTC , JD 2456689.35 + 17 d
Feb 20, 2014 , 2030 UTC , JD 2456709.31 + 37 d

Telluric O$_2$ 6870

Na I

Si II

Fe II

Si II 6347/71

Blue shift

λ$_0$

S II

Fe II

Fe II
Fe III
Si II

Fe II

Fe II

Fe II
Fe III
?

Mg II

Fe II

?

Ca II
Si II

λ

7200 7000 6800 6600 6400 6200 6000 5800 5600 5400 5200 5000 4800 4600 4400 4200 4000

CHAPTER

26 Extragalactic Objects

26.1 Introduction

It is impossible with amateur equipment to record spectra of single stars within external galaxies. However, it is feasible to record composite spectra of galaxies and quasars! In contrast to profiles of individual stars the composite (or integrated) spectra show the superposed characteristics of hundreds of billions individual star spectra. Using Doppler spectroscopy, the radial velocities, specifically the z-values of such objects can also be measured. Further, except for pure "face on" galaxies, the rough distribution of the rotational velocity within the galactic disk can be estimated [3].

On a professional level, this has been practiced successfully since the beginning of the twentieth century and has contributed substantially to our present understanding of the Universe. The first who tried this with M31 was Vesto Slipher in 1912 at the Lowell Observatory in Flagstaff Arizona. He was able to measure the blueshift of the spectrum and derived a radial velocity of -300 km s^{-1}. Further, he detected the rotation of this "nebula." The fact that M31 is a galaxy outside the Milky Way was proved only later in the early 1920s. He also noted that most of the other galaxies appear redshifted and thus are moving away from us. Lemaître and Hubble used these shift measurements later for the correlation with the distance (Hubble constant).

26.2 Morphological Classification of Galaxies

The characteristics of the spectra are partly correlated with the morphological types of galaxies. Figure 26.1 shows the so-called Hubble–Sandage tuning-fork diagram.

It is based on a former faulty hypothesis that this sequence should represent the evolution of galaxies, starting from the elliptical shape of E0 and ending with the spiral types Sc, or SBc. Unfortunately today, similarly to the stellar spectral

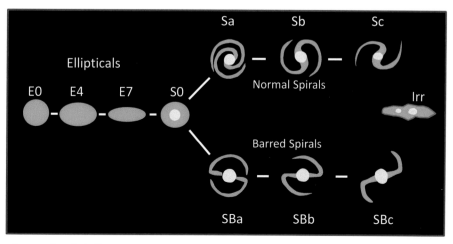

Figure 26.1 Hubble–Sandage tuning-fork diagram

classes, the elliptical are sometimes called "early" and the spirals "late" types. However, according to current knowledge, the elliptical galaxies represent the final stage, merged by a number of smaller spiral type galaxies. Extreme cases are the huge elliptical cD galaxies mainly located in the center of superclusters. During this merger process the irregular stage Irr is also passed through, represented by famous examples like the Antenna Galaxy NGC 4038/4039.

26.3 Spectroscopic Classification of Galaxies

By the spectroscopic classification these objects are distinguished according to their spectral features. In the 1920s it was recognized that such composite or integrated spectra contain similar information to those of individual stars. They also show profiles, for example, with or without emission lines, chiefly depending on the kind and activity of the core region. Such spectral profiles are powerful means to determine, for example, the content of stars and the development state of the galaxy [134]. The relative similarity of the extra galactic composite spectra to those of individual stars, was also a convincing argument in the historical dispute (mainly between H. Shapley and H. Curtis), that galaxies are not just dust or gaseous nebulae like M42, but extremely distant, huge clusters of individual stars!

In the second half of the twentieth century, classification systems were developed, which were based on the similarity with stellar spectral types. The "integrated spectral type" of galaxies was determined in this way by W. Morgan, beginning with A, AF, F and ending with the late K systems. At that time the following correlations were noticed [2]:

- The later the stellar spectral type with which the profile shows similarities, the stronger the galaxy is centrally concentrated.
- The elliptical classes E1–E4 remain dominated by absorption features of later spectral types. From here on, however, the characteristic changes until the profiles of the types Sc/SBc and Irr look similar to those of early spectral classes. Furthermore emission lines show up here with increasing intensity.

26.4 Rough Scheme for Spectroscopic Classification of Galaxies

Table 26.1 shows an attempt for a rather rough classification system, based on the presence, intensity and shape of emission lines. It also includes peculiar shapes such as Seyfert galaxies and quasars, which stand out by their extremely high core activity and therefore belong to the category of AGN (active galactic nuclei). The core activity within this table is increasing from top to down.

26.5 Absorption Line Galaxies

Plate 54 demonstrates the principle of the absorption line galaxy by the example of the Andromeda galaxy M31. With this type, star-like spectra appear, almost exclusively with absorption lines.

Plate 54 Absorption Line Galaxy: Comparison with Stellar G Class

Andromeda M31/NGC 224 and ε Vir G8 III

Table 26.2 shows the heliocentric parameters according to NED [12], and Evans *et al.* [135], (value includes suspected dark matter). The *z*-value of the redshift is defined by Equation {26.1}.

Plate 54 shows a composite spectrum of the core of M31 (Figure 26.2). According to [134] composite spectra of the galaxy types Sa–Sb are dominated mainly by developed giant stars. Indeed, here the pseudo-continuum of the recorded M31 spectrum fits best to that of a single star of the late G class that is called "integrated spectral type." For a comparison, the M31 profile is shown here in a montage with the pseudo-continuum of Vindemiatrix (ε Vir).

The galaxy M31 obviously belongs to the category of absorption line galaxies. The most dominant common features are the Fraunhofer G band, the magnesium triplet, and the sodium double line (Fraunhofer D line). The H-Balmer lines are just very faintly visible in absorption. According to [134], this finding fits rather to elliptical galaxies (type E). Also, for the type Sa–Sb a few emission lines caused by a younger stellar population may sporadically be recognizable. However, such are not detectable in this spectrum of M31.

The profile was recorded in the Nasmith focus of the 36 Inch CEDES Cassegrain Telescope in Falera – exposure time: 3 × 340 s. Clearly visible here is the expected blueshift of several ångstroms by the absolute wavelength-calibrated M31 profile, compared to the relative calibrated spectrum of ε Vir, based on known lines.

26.6 LINER Galaxies

Plate 55: The very numerous so-called LINER galaxies form a kind of transition between absorption lines and starburst galaxies. LINER stands for "low-ionization nuclear emission-line region." The spectral features of the galactic

Table 26.1 Rough scheme for spectroscopic classification of galaxies

Galaxy Type	Features within the Optical Spectral Range	Affected Objects and Responsible Effects	Examples
Absorption line galaxy	Almost exclusively absorption lines, sporadically weak emissions may show up.	Mainly elliptical galaxies with a rather old star inventory, low star formation rate and just weakly active nuclei.	M31, M33, M81
LINER galaxy	By the majority absorption lines. In the core region emission lines of weakly ionized or neutral atoms as O I, O II, N II and S II. Possibly weak emissions of ions with higher ionization energy as He II, Ne III and O III.	Mainly types E0–S0. Compared with absorption line galaxies LINER contain a younger star inventory. Slightly increased core activity.	M94, M104
Starburst galaxy	Intensive H-emission lines, possibly weak appearance of: [N II] ($\lambda6583$), [S II] ($\lambda\lambda6716/31$) and [O I] ($\lambda6300$), generally few or no absorptions.	Colliding, gravitationally interacting, or gas-rich young galaxies with giant H II regions and a correspondingly high rate of star birth.	M82, NGC 4038/4039
Seyfert galaxy	Intensive and Doppler-broadened H-emission lines, possible appearance of intensive forbidden emissions of [O III], [N II], [S II], and [O I]. Absorptions are generally faint or even absent.	Central, supermassive black hole with high accretion rate.	M77, M106, NGC 4151
Quasar	Mainly intensive and extremely Doppler-broadened H-emission lines	Central, supermassive black hole with extremely high accretion rate.	3C273
Blazar, BL Lacertae Object	Spectrum of the bright jet exhibits just a continuum mostly without any spectral lines and irregular, strong intensity fluctuations. Spectral lines measurable only in the very faint ambient galaxy, restricted to very large telescopes.	Blazars differ from quasars just by our perspective; here we look more or less directly into the synchrotron radiation of the jet.	Makarian 421

Table 26.2 M31 heliocentric parameters

Radial Velocity v_r	M_V	Redshift z	Distance	Morphology	Diameter	Mass [135]
$-300 \ \mathrm{km \, s^{-1}}$	-21.77	-0.001	2.6 M ly	Sb	141,000 ly	$1.23 \times 10^{12} \, M_\odot$

core region are emission lines of weakly ionized or neutral atoms such as O I, O II, N II and S II. Emissions of highly ionized atoms such as He II, Ne III and O III may appear just weakly. In contrast to the Seyfert galaxies (Section 26.9), the core of a LINER galaxy is still relatively faint. Well-known examples are M104 (Sombrero) and M94. Morphologically, the types E0–S0 are mainly concerned. This phenomenon, especially the origin of the involved ionization processes, is still subject of debate.

Plate 55 LINER Galaxy M94/NGC 4736

Table 26.3 shows the heliocentric parameters according to NED [12] and Trujillo *et al.* [136]. The *z*-value of the redshift is defined by Equation {26.1}.

In 2008, this typical "face on" galaxy (Figure 26.3) achieved general awareness in the media caused by discussions that its rotational behavior could be explained without an assumption of dark matter. Both profiles of Plate 55 have been recorded in the integrated light at a time interval of 48 hours and inevitably with different slit positions relative to the core region. The upper, red profile may rather represent the peripheral regions. Compared to M31, amazingly many absorption lines can be identified here with almost star-like shape. This composite spectrum shows strong similarities with a stellar profile of the late F class. Particularly striking here is one of their "brands" – the comparable intensity of the G band (CH molecular), and the directly adjacent Hγ line (Section 8.3). This striking "line-double" can exclusively be seen in the F class. This also fits to the intensity of the two

Table 26.3 M94 heliocentric parameters

Radial Velocity v_r	M_V	Redshift z	Distance	Morphology	Diameter	Mass [136]
+308 km s^{-1}	−19.37	+0.001	16.4 M ly	Sa	67,000 ly	$6.5 \times 10^{10} \, M_\odot$

Figure 26.2 M31
(Credit: Urs Fluekiger)

Figure 26.3 LINER galaxy M94
(Credit: ESA/Hubble and NASA)

Fraunhofer Ca II lines and the still rather modest absorption of the magnesium triplet ($\lambda\lambda$5167–5183). Compared to M31 the entire H-Balmer series appear here very prominently and also all other symptoms indicate a much younger star inventory in M94. The lower, blue profile may represent

rather the weakly active core of the galaxy. Some possible, unidentifiable emission peaks are marked here with red arrows. The most striking features are marked with red ellipses. The double peak emissions at about λ4507 and Hβ are possibly caused by Doppler effects due to the rotation of the core region (see also M77 and M82). Recording information: C8/DADOS/Atik 314L+: 1 × 1800 s, 2 × 2 binning.

26.7 Starburst Galaxies

Plate 56: Starburst galaxies are mostly colliding, gravitationally interacting, or young, gas-rich galaxies with giant H II regions and a correspondingly high star birth rate. Examples are M82 or the Antenna Galaxy NGC 4038/4039. In the profile of M82 the most striking spectral features are intense emissions of the H-Balmer series, a moderate intense sulfur [S II] doublet at $\lambda\lambda$6716/33, just weak emissions of [N II] (λ6583) and possibly also [OI] (λ6300) and finally the virtual absence of any absorptions.

Plate 56 Starburst Galaxy M82/NGC 3034

Table 26.4 shows the heliocentric parameters according to NED [12] J.P. Greco, P. Martini *et al.* and other sources. The z-value of the redshift is defined by Equation {26.1}.

The apparent chaotic structure of this galaxy can be attributed to gravitational interaction with the much larger neighboring galaxy M81, as well as effects of our specific perspective. The galaxy M82 (Figure 26.4) is a typical representative of the starburst galaxies with very few, weak absorption lines. Here Na I is most likely of interstellar origin, probably mainly absorbed within M82. Compared with galactic emission nebulae (Chapter 28) as well as the Seyfert Galaxy M77, the excitation class is very low here (E < 1). The forbidden [O III] line at λ5007 shows up here just very weakly but the probably shockwave-induced [S II] emission appears quite strongly. Further, compared to [N II] at λ6583, the Hα emission is much more intense. Striking here is the possibly rotationally induced double peak in the range of the Hγ emission (see also M77 and M94). M82 was also the host galaxy of the brightest supernova for decades, SN2014J, type Ia (Chapter 25, Plate 53). Recording information: C8/DADOS 200 L mm^{-1}, Atik 314L+: 1 × 1800 s, 2 × 2 binning.

Table 26.4 M82 heliocentric parameters

Radial Velocity v_r	M_V	Redshift z	Distance	Morphology	Diameter	Mass
+203 km s^{-1}	−19.42	+0.0007	12.6 M ly	Irr	47,000 ly	~10^{10} $M_\odot$

Figure 26.4 Starburst galaxy M82
(Credit: NASA, ESA and Hubble Heritage Team (STScI/AURA))

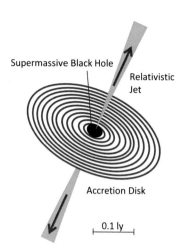

Figure 26.5 AGN with accretion disk and relativistic jets of matter

26.8 The Phenomenon of Active Galactic Nuclei (AGN)

The AGN phenomenon in the subsequently described Seyfert galaxies and quasars is caused by central supermassive black holes collecting vast amounts of matter from their surroundings, generating excessively high amounts of energy and intense X-ray radiation. This process is accompanied by an accretion disk in the equatorial plane and jets of matter which are ejected parallel to the rotation axis of this object with almost the speed of light (Figure 26.5). Seyfert galaxies form the largest group with AGN [138].

26.9 Seyfert Galaxies

Plate 57 demonstrates The principle of the Seyfert galaxy by the example of M77.

Plate 57 Seyfert Galaxy AGN

M77/NGC 1068

Table 26.5 shows the heliocentric parameters according to NED [12] and other sources. The z-value of the redshift is defined by Equation {26.1}.

Plate 57 shows a composite spectrum of the nucleus of M77. The galaxy is classified as Seyfert type 2 (Figure 26.6). In the

1940s Carl Kennan Seyfert (1911–1960) discovered in the core of some galaxies intensive emission lines of the H-Balmer series with Doppler broadenings of more than 1000 km s^{-1}. In addition, emissions of forbidden transitions, such as [O III] and [N II] can be detected. However, they cannot substantially be broadened, due to the shock sensitivity of the metastable initial states. In contrast to the emission of the H-Balmer series, they are probably generated far away from the turbulent core around the supermassive black hole, whose mass is estimated to be ~15 million $M_\odot$. In contrast to the permitted lines they show therefore virtually no intensity fluctuations.

Also remarkable are the double peaks in the emissions of Hγ and Ne III at λ3967. B. Garcia-Lorenzo *et al.* [139] and other authors suggest Doppler effects here are due to differently running streams of gas in the close vicinity around the black hole. Similar features are exhibited also in the spectra of M82 and M94. All emissions appear superimposed to a significant continuum.

Seyfert galaxies are divided into:

subclass 1 with strongly broadened lines, limited to the permitted transitions

subclass 2 with just slightly broadened lines, limited to the permitted transitions

Generally, it is now assumed, that this difference in classification is caused by effects of different perspectives, see "Seyfert unification theory" [140]. Thus, at subclass 2, the forming regions of the broad, permitted lines are possibly obscured by dust clouds and/or an unfavorable viewing angle.

Table 26.5 M77 heliocentric parameters

Radial Velocity v_r	M_V	Redshift z	Distance	Morphology	Diameter	Mass
+1137 km s^{-1}	−22.17	+0.0037	44 M ly	Sb/Sy2	115,000 ly	~1.0 10^{12} $M_\odot$

Figure 26.6 Seyfert galaxy M77
(Credit: NASA, ESA and A. van der Hoeven)

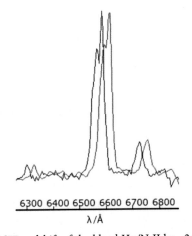

6300 6400 6500 6600 6700 6800
$\lambda / Å$

Figure 26.7 M77, redshift of the blend Hα/N II by ~24 Å

The galaxy M77 shows one of the larger redshifts among the Messier galaxies. In the range of the Hα line it yields about 24 Å ($z\lambda_0$). In Figure 26.7 the scale and the blue profile are based on the rest wavelength λ_0, calibrated with known lines. The red profile is scaled absolutely here with the calibration light; those on Plate 57 refer to the rest wavelength λ_0.

For obvious reasons the spectrum shows strong similarities to profiles of galactic emission nebulae. Accordingly, the excitation class E can also be determined here. The He II line at λ4686

shows E > 4. The criterion $\log\left(I_{N1+N2}/I_{He\ II\ (4686)}\right)$ yields here ~1.5, corresponding to an excitation class E 10 on the 12-step "Gurzadyan scale" (see Chapter 28). This high excitation level is also documented by the considerable intensity of the forbidden [O III] and [N II] lines, compared to the rather weak H-Balmer series. So here even [N II] (λ6583) clearly surpasses the Hα emission. Comparable to the supernova remnant M1, the sulfur [S II] doublet (λλ6717/6731) is here also strikingly intense – probably due to shockwave-induced excitation.

Veilleux and Osterbrock use in their classification scheme [138] among others, the decimal logarithms of the intensity ratios [O III]$_{(5007)}$/Hβ and [N II]$_{(6583)}$/Hα. According to Shuder and Osterbrock [141], the following intensity ratio applies as a rough criterion to distinguish Seyfert from H II or starburst galaxies (compare with M82, Plate 56): Criterion for a Seyfert galaxy: [O III]$_{(5007)}$/Hβ > 3.

The direct vicinity of the supermassive black holes, located at the center of Seyfert galaxies, appear almost star-like and very bright, while the rest of the galaxy remains relatively dark. This considerably simplifies the recording of the spectrum of M77 with an apparent surface brightness of $m_v \approx 8.9$ [11]. The spectrum from the core was recorded with the 25 μm slit and 200 L mm^{-1}. Exposure time: C8/DADOS 200 L mm^{-1}, Atik 314L+: 2 × 1200 s, 2 × 2 binning.

26.10 The Quasar Phenomenon

The term "quasar" is derived from quasi-stellar object (QSO), because these objects appear as point-shaped light sources. Maarten Schmidt discovered the first in 1963 at the object number 273 in Ryle's third Cambridge catalog of radio sources and was therefore called 3C273 (Figure 26.8). It quickly became clear that this object showed the largest redshift known at that time, and therefore could not possibly be a star. In addition, the obtained spectra differed dramatically from stellar profiles and appeared more like those of Wolf–Rayet stars, nova outbursts, or even supernova explosions.

According to today's knowledge, quasars are considered as the most energetic and luminous version of galaxies with active nuclei (AGN). As with the Seyfert galaxies, the cores

Table 26.6 Quasar 3C273 heliocentric parameters

Radial Velocity v_r	m_V	M_V	Redshift z	Distance	Type
+47,469 km s^{-1}	12.8	−27.70	+ 0.1583	~2 bn. ly	Quasar

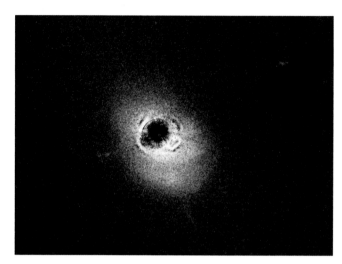

Figure 26.8 Quasar 3C273 Coronograph HST (Credit NASA/HST M. Clampin *et al.*)

of quasars host supermassive black holes, but much larger in size by several orders of magnitudes. By accretion disks, they accumulate huge amounts of matter from the surrounding galaxy, ejecting jets with relativistic velocity. Therefore, quasars are also strong sources of X-ray radiation and some of them also of radio emission, called "radio loud."

The transition between the fainter AGN of Seyfert galaxies to the brighter quasars is fluent. Sometimes as a very rough criterion for a quasar, a minimum absolute magnitude of $M_V \approx -23$ (Appendix H) is assumed. The mass estimation of the black hole is still difficult and uncertain. The literature, for example [142], shows strongly scattering values, which propose a mass of ~1 billion $M_\odot$. The point-shaped appearance of quasars can be explained by the enormous brightness of the nuclei, which in most cases totally outshine the rest of their host galaxies. Apart from the episodically occurring SN explosions, these are the most luminous, permanently radiating objects in the Universe and therefore of eminent importance for cosmological research.

Plate 58 Quasar: Emission Lines

Quasar 3C273: Details of the Spectral Profile

Plate 58: The apparently brightest quasar 3C273 ($m_V \approx 12.8$) in constellation Virgo is often called the most distant object which can still be seen with average amateur means, purely visually and without any use of astronomical cameras.

Heliocentric parameters according to NED [12] and other sources are shown in Table 26.6. The z-value of the redshift is defined by Equation {26.1}.

The spectrum is dominated by extremely broadened emissions of the H-Balmer series and by forbidden [O III] lines at $\lambda\lambda5007$ and 4959, fusing here to a blend. They appear superposed to a strong continuum. Due to quantum mechanical reasons, the forbidden [O III] lines cannot appear to be significantly broadened. It is therefore discussed whether the $\lambda5018$ emission of the Fe II (42) multiplet ($\lambda\lambda\lambda4923$, 5018 and 5169) supplies the major contribution to the intensity of this emission [143] [144]. This multiplet frequently appears in the spectra of active galactic nuclei (AGN), as well as in the profiles of protostars (Section 21.3). A striking feature is the lower intensity of the [O III] in relation to the Hβ emission. This observation is in contrast to the spectral profiles of the active nuclei in Seyfert galaxies (Plate 57), planetary nebulae and H II regions. This phenomenon was recognized by the discoverers of quasars in the 1960s.

The Ne III emission at $\lambda3869$ is undisputed. The other features are mostly broadband blends of different emissions, generated by various ions, significantly complicating the line identification [143]. Consequently, their exact composition is still unclear. Striking is a broad emission at $\lambda\lambda4500–4700$. The He II line at $\lambda4686$ and numerous emissions of C III and N III were suggested as the cause by J. B. Oke [145] – this in analogy with similar spectra of supernovae and Wolf–Rayet stars. A. Boksenberg *et al.* [143] assumes that the striking emission in the range of $\lambda5870$ is caused by He I at $\lambda5876$. The Na I doublet is also under discussion, and in certain phases it can be observed during nova outbursts. Due to the extreme shock sensitivity of the forbidden [O III] lines, and the very low ionization energy of Na I, these emissions must necessarily be generated at a considerable distance to the supermassive black hole. An indication for the contraction processes within the accretion disk are the inverse P Cygni profiles in the region around $\lambda\lambda6100–6400$, also observable in the spectra of the much smaller disks around the T Tauri and Ae/Be protostars (Chapter 21).

The Hα emission is redshifted here so far (1017 Å) that it coincides with the intense, telluric Fraunhofer A line. This is the cause why Hα appears split [145]. This circumstance complicates the determination of the redshift, applying this line, and seems further to contribute to the "flat" Balmer decrement $I_{Hα}/I_{Hβ} < 2.85$, (Section 28.7, [3]).

Because Hα appears split, the occurring radial expansion velocities are estimated according to Equation {17.1}, applying the FWHM of the Doppler-broadened and gauss-fitted Hβ line. It results in FWHM $_{Emission\ Hβ} \approx 88$ Å. This yields a terminal radial velocity of the matter of $v_r \approx 5400$ km s^{-1}. This is roughly within the strongly scattering values found in the literature. For the jet, however, based on X-ray analyses, speeds of up to 70% of light speed are postulated [146]. For amateurs this feature is in the optical spectral range neither detectable nor measurable.

Not only the brightness of the object may vary considerably (see AAVSO), but also the spectral characteristics can change significantly within a short time, such as the FWHM and EW values of the Hβ emission. It further shows that this central region cannot be indefinitely large, and therefore barely exceeds the approximate diameter of the solar system. Quasar 3C273 would certainly be a highly interesting candidate for a monitoring project. In addition to such considerations we should always be aware that these changes, observed within a very short time, took place about 2.4 billion years ago, when our planet Earth was still in the Precambrian geological age! Recording information: C8/DADOS 200 L mm^{-1}, Atik 314L+, 5 × 1200 s, 2 × 2 binning.

Plate 59 Quasar: Redshift

Quasar 3C273

On **Plate 59** redshifted wavelengths are indicated, measured at Gaussian fitted lines and related to the original wavelength scale, absolutely calibrated with the light source. The individual amounts of the shift $\Delta\lambda = \lambda - \lambda_0$ are rounded to 1 Å and vary in size, but behave strictly proportional to the rest wavelengths λ_0. Therefore, for all analyzed lines, the calculation of the z-values must theoretically yield the same amounts. This is also a good final check for the quality of the wavelength calibration.

With such strongly redshifted spectra the z-value is one of the most important measures of modern cosmology, remaining fully independent of the different debated models. It is a direct measure for both the "distance" as well as for the past [3]:

$$z = \frac{\Delta\lambda}{\lambda_0} \qquad \{26.1\}$$

Table 26.7

Line	λ Red shifted	λ_0	$\Delta\lambda$	z
Hβ	5632	4861	771	0,1586
Hγ	5023	4340	683	0,1574
Hδ	4748	4102	646	0,1574

The z-value, measured and calculated from this profile (Table 26.7), yields $z = 0.158$ and is consistent up to three decimal places with the accepted value of 0.1583 [11] [12]:

$$v_r = \frac{\Delta\lambda}{\lambda_0} c = cz \qquad \{26.2\}$$

The validity of this spectroscopic Doppler equation is strictly limited to kinematically induced motions. In the area of $z \gtrsim 0.1$ the expansion of the spacetime lattice is dominating and thus the estimation of the radial or "escape velocity" v_r, makes no more sense. Therefore cosmological models with appropriately chosen parameters should be applied instead. If we nevertheless follow the "classical approach" with $z = 0.158$ we obtain a "radial escape velocity" $v_r = 47{,}400$ km s^{-1}. Inserted into Equation {26.3} of the Hubble law, the "distance" D can at least very roughly be estimated to 650 Mpc or ~2.1 billion ly.

$$D = \frac{v_r}{H_{(0)}} \qquad \{26.3\}$$

Hubble constant:

$$H_{(0)} \approx 73 \text{ km s}^{-1} \text{ Mpc}^{-1}$$

In Appendix H a list of distant AGN and quasars brighter than magnitude $m_v \approx 16$ can be found, which should be at least visually achievable for ambitious amateurs with access to larger telescopes. The highest z-value in this table is 3.9.

26.11 Blazars and BL Lacertae Objects (BL Lacs)

The term "blazar" is composed of the terms "BL Lacertae" and "quasar," also abbreviated as "BL Lac." The term was coined in 1978 by E. Spiegel at a congress on BL Lacertae objects in Pittsburgh, on the occasion of a banquet speech. Basically, these objects differ from quasars just by our perspective, i.e. here we look directly into the synchrotron radiation of the jet matter, which is ejected from the central black hole, parallel to the rotation axis and almost with relativistic velocity of light (Figure 26.9).

Table 26.8 Blazar Makarian heliocentric parameters

Radial Velocity v_r	m_V	M_V	Redshift z	Distance	Type
+9000 km s^{-1}	12.8	−25.96	+ 0.030	~400 M ly	Blazar/BL Lac

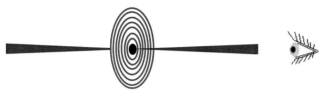

Figure 26.9 Blazar corresponds to a quasar with a direct line of sight into the jet

Therefore, we observe here a highly intense, aperiodically fluctuating radiation with a strong polarization across the whole electromagnetic spectrum. Similar to quasars, these objects were first interpreted as blue variable stars. Not until 1968 was their true nature discovered when BL Lacertae was identified as a radio source by Schmitt [147]. The spectra of these bright jets of matter typically show no spectral lines, neither in emission nor in absorption. Therefore, the redshift may here only be measured at the comparatively very faint and diffusely appearing galaxy, which is feasible just with large professional telescopes.

Figure 26.10 Blazar Makarian 421 and Galaxy 421–5 (Credit: NASA/STScI/Roberto Fanti)

Plate 60 Blazar, BL Lacertae Object

Makarian 421 (Mrk421)

Plate 60: Table 26.8 shows the heliocentric parameters according to NED [12]. The z-value of the redshift is defined by Equation {26.1}.

With a distance of 400 million ly, Makarian 421 (Figure 26.10) is one of the nearest objects of the category blazar/quasar. It is orbited by a small companion galaxy Mrk 421–5 [148]. Blazar Mrk 421 is one of the most intensive cosmic gamma-ray sources and therefore important for high-energy physics. Figure 26.10 shows an atlas picture from the 2Mass Survey.

The apparent brightness of the blazar is indicated by CDS [11] with $m_V \approx 12.9$, almost the same value as for the quasar 3C273 (Section 26.10). In most cases, the brightness is significantly weaker here, i.e. in the range $m_V \approx 13$–14 (AAVSO). Thus for amateurs this object is much more difficult to record, but otherwise very easy to find, because it is located in an eye-catching star pattern just nearby to 51 UMa ($m_V = 6.0$).

With amateur equipment only the continuum of the jet spectrum from Mrk 421 can be recorded, totally lacking any spectral lines, which of course does not allow any determination of the redshift. Plate 60 shows this profile, which just shows the telluric absorptions. Recording information: C8/DADOS, 200 L mm^{-1}, Atik 314L+, 1 × 1800 s, 2 × 2 binning.

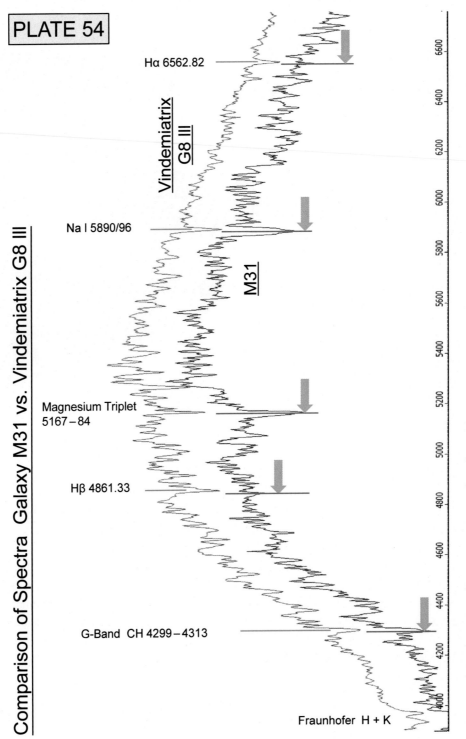

PLATE 54

Comparison of Spectra Galaxy M31 vs. Vindemiatrix G8 III

Vindemiatrix
G8 III

M31

Hα 6562.82

Na I 5890/96

Magnesium Triplet
5167 – 84

Hβ 4861.33

G-Band CH 4299 – 4313

Fraunhofer H + K

6600 6400 6200 6000 5800 5600 5400 5200 5000 4800 4600 4400 4200 4000

Plate 54 Absorption Line Galaxy: Comparison with Stellar G Class

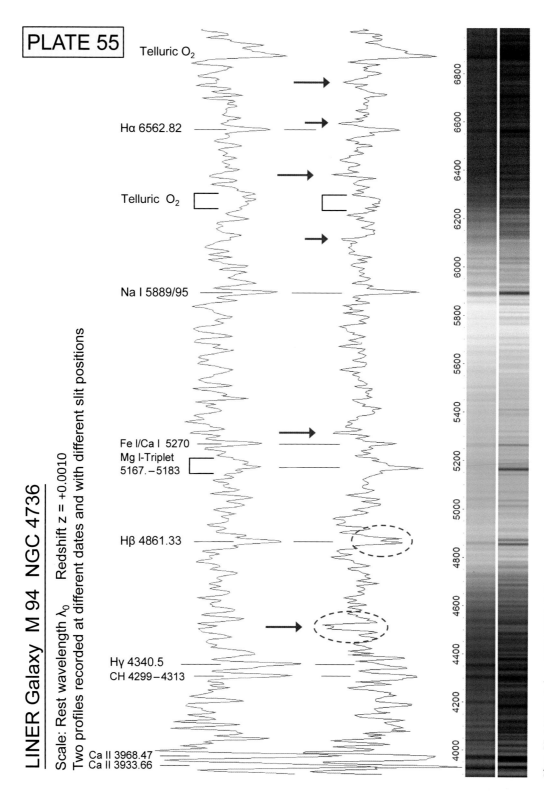

PLATE 55

LINER Galaxy M 94 NGC 4736

Scale: Rest wavelength λ_0 Redshift z = +0.0010

Two profiles recorded at different dates and with different slit positions

Telluric O$_2$

Hα 6562.82

Telluric O$_2$

Na I 5889/95

Fe I/Ca I 5270

Mg I-Triplet
5167. −5183

Hβ 4861.33

Hγ 4340.5

CH 4299−4313

Ca II 3968.47

Ca II 3933.66

Plate 55 LINER Galaxy M94/NGC 4736

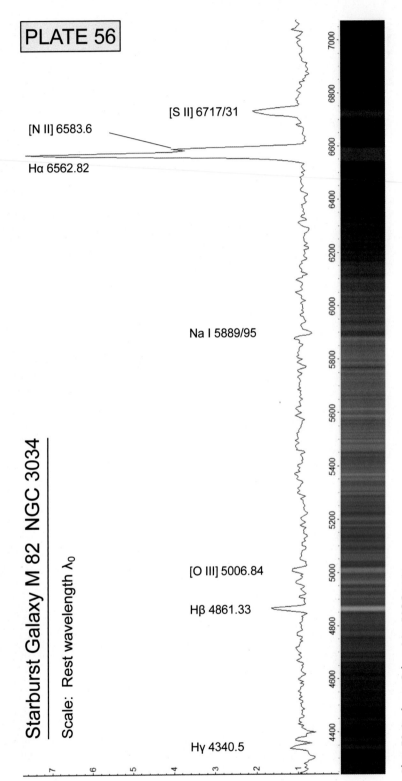

PLATE 56

Starburst Galaxy M 82   NGC 3034

Scale: Rest wavelength λ_0

[S II] 6717/31

[N II] 6583.6

Hα 6562.82

Na I 5889/95

[O III] 5006.84

Hβ 4861.33

Hγ 4340.5

Plate 56 Starburst Galaxy M82/NGC 3034

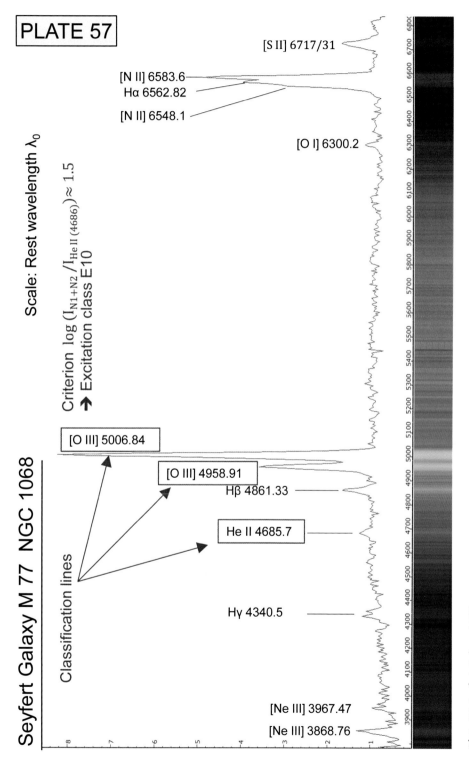

Plate 57 Seyfert Galaxy AGN

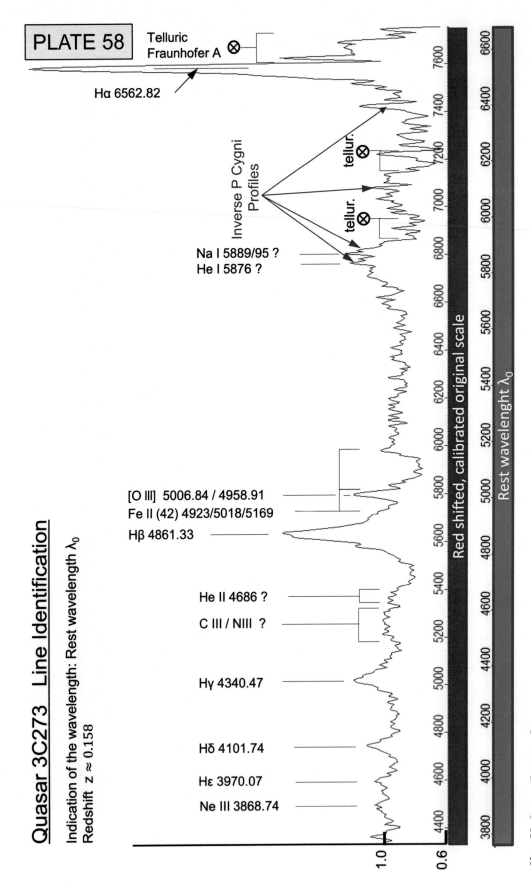

Plate 58 Quasar: Emission Lines

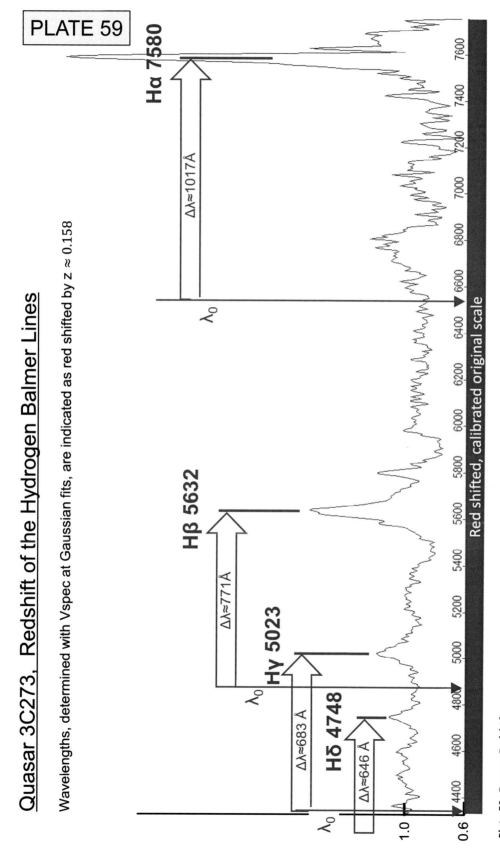

Plate 59 Quasar: Redshift

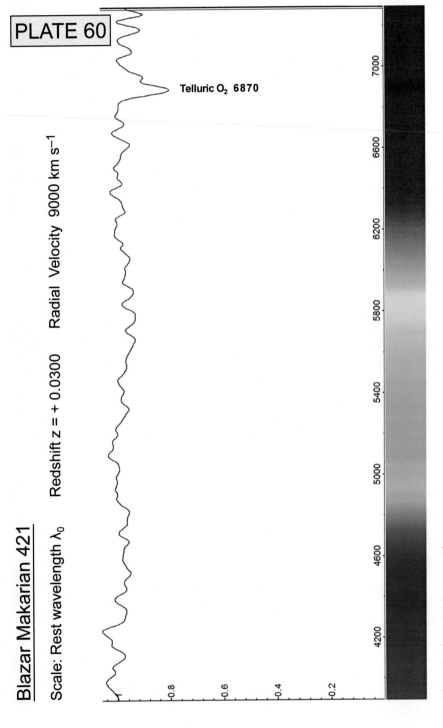

Plate 60 Blazar, BL Lacertae Object

CHAPTER

27 Star Clusters

27.1 Short Introduction and Overview

Star clusters are formed by compacted stellar accumulations within or in the vicinity of galaxies. Most of the members of such clusters have evolved from a common gas cloud, and therefore are about the same age. The individual clusters show a wide scattering range with respect to density, age and number of stars. Basically, star clusters can be divided into the two main categories with completely different properties. We distinguish open clusters and globular clusters.

27.2 Open Clusters (OCL)

Open clusters show an irregular shape and usually contain a few hundred (e.g. Pleiades) up to at most a few thousand stars (e.g. H+X Persei). The internal gravitational forces of such clusters are usually too weak to keep together their stars for more than at most a few hundred million years. Thus, they usually have a young star inventory, consisting of stars of population I (Section 27.4). One of the few exceptions is the very dense M67, whose age is estimated to be about 4 billion years. The typical diameters of open star clusters are relatively similar: for example Hyades: ~15 ly, M67: ~26 ly, Pleiades: ~14 ly. Very small accumulations, just consisting of a few stars and showing a common direction of movement, are referred to as "associations." Today, more than 1000 open clusters are registered, which are distributed over the entire visible region of the Milky Way. The observation of extragalactic open clusters, for example in M31, remains reserved to large professional telescopes.

27.3 Globular Clusters (GCL)

They differ from the open clusters in almost all aspects:

- they are strikingly spherical
- they are "packed" much denser
- they are all significantly older (population II stars, Section 27.4)
- their typical diameter is more than ~10 times larger
- they contain up to several hundred thousand stars

The typical distance between the stars in the outskirts of globular clusters is about 1 ly. In the core area these interspaces can even shrink roughly to the diameter of the solar system! Some typical diameters of the comparatively much larger globular clusters are: M2: ~182 ly, M3: ~180 ly, M5: ~178 ly, M13: ~145 ly, Omega Centauri: ~150 ly.

In the center of individual globular clusters (e.g. M15) intermediate-mass black holes of a few thousand solar masses have been detected. They show a certain similarity with dwarf galaxies. Currently in the Milky Way about 150 globular clusters are known. They all orbit our galactic center at a distance of about 130,000 light years, forming a halo this way. Thus, these objects could even be referred as "extragalactic" if they were not tightly gravitationally bound to the Milky Way, sometimes even dipping through the disk. The formation of globular clusters, but also details of their evolution, remains still largely unclear. However, it seems certain that their enormous age of about 12 billion years, is about the same as that of the entire Milky Way, i.e. slightly younger than the 13.7 billion-year-old Universe. In contrast to the open clusters the high density and internal gravitational forces, allow such structures virtually an arbitrarily long life. The observation of extragalactic globular clusters, for example in M31, remains limited to large professional telescopes.

27.4 Spectroscopic Analysis of Star Clusters

In contrast to the galaxies, where amateurs are limited to record integrated (composite) spectra, here both the open and, with significant restrictions, also the globular clusters, allow the analysis of brighter individual stars. In the professional field, multi object spectroscopy (MOS) allow the simultaneous recording of up to several 100 profiles. The main objective here is the determination of the metal abundance Z [Fe/H] [3]. This value allows direct conclusions about the age of a cluster. The first star generation, somewhat oddly called population II, was created with the birth of the Milky Way ~12 billion years ago, where the interstellar matter was still dominated by hydrogen, helium and lithium. The enrichment with heavier elements took place only later by ejecta from SN explosions or repelled planetary nebulae. This enriched material generated later on the metal-rich second star generation, similarly confusingly called Population I, to which our Sun belongs. The most efficient way for the determination of the Z-value, has proven to be the analysis of the Ca II calcium triplet in the near infrared at $\lambda\lambda 8542$, 8498 and 8662. This procedure is abbreviated as "CaT." The empirical relationship between the metallicity Z and the summed EW values of the Ca II absorptions has been refined over the last 25 years and is outlined in numerous publications, such as [149], [150], [151].

27.5 Spectroscopic Age Estimation of Star Clusters by Amateurs

The described CaT method is too demanding for most amateurs. So, for example, the Ca II calcium triplet is located in the infrared range, somewhat outside the reach of most amateur equipment and frequently contaminated by blends with other absorptions. Alternatively a very simple method is presented here, based on the correlation between the duration of stay on the main sequence of the HRD and the spectral class of a star. It relies on:

- The assumption that the individual stars of such clusters have been formed at about the same time from a common gas and dust cloud.
- The known relation that the earlier classified (or more massive) the star, the shorter the lifetime.
- The quintessence: within a stellar sample, the duration of the stay on MS for the earliest classified MS star determines very roughly the age of the cluster.

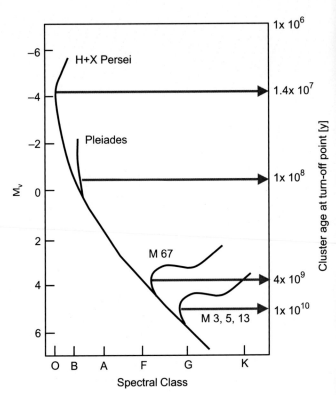

Figure 27.1 Age of the cluster depending on the location of the turn-off point in the HRD

- At the "turn-off point" the star reaches on the MS its earliest possible spectral class. Then it moves within the HRD to the top right on the giant branch, showing thereby significantly later classifications.
- The older the cluster, the more to the right in the HRD (i.e. "later") is the turn-off point. This means with increasing age of the cluster, the main sequence seems to "burn off" like a candle from top to bottom (Figure 27.1).

Table 27.1 shows a dramatic increase in the roughly estimated duration of stay on the MS towards later classified stars. Below the late G class it even exceeds significantly the present age of the Universe. Such data are a rough guide only. Based on computer simulations they often show (source dependent) a considerable spread.

As an example: If within a cluster a larger spectral sample yields an early A type on the MS as the earliest classification, its age can roughly be estimated to 350–500 million years. Any earlier classified MS stars of types B and O, are much more short-lived and thus either already exploded in a SN or migrated in the HRD to the top right of the giant branch.

In professional astronomy, such studies are often carried out photometrically in the B–V system. The major disadvantage is that these measured magnitude values always appear reddened by interstellar matter and need first of all to be adjusted with appropriate models. In Appendix C, the assignment of "de-reddened" B–V magnitudes to the corresponding spectral class can be found. The spectroscopic age estimation based on the spectral class remains clearly the first option for amateurs.

27.6 The Pleiades (M45): Analysis by Individual Spectra

The Pleiades (~390–460 ly), are so close, that this object appears more as a small constellation, rather than an open cluster. With the unaided eye, we see here just the brightest 6 up to maximum 10 stars which astonishingly all belong to the middle to late B class (Table 27.2). Their visibility depends mainly on the seeing conditions and the current brightness of the Be star Pleione.

Overall, the cluster contains about 500 stars. According to a study by Abt and Levato [152], the brightest 50 stars of M45 are spread over the various spectral classes as follows: B: 17, A: 24 and F: 9. The remaining stars are all classified as type F or later. It remains an enigma, why in this rather small cluster the number of massive stars, classified in the range of the middle to late B class, exceeds by orders of magnitude their statistically expected incidence of just some 0.12% (Section 4.12)!

Plates 61 and **62** impressively show the very similar spectra of the 10 brightest stars within M45. The middle to late B class can be recognized here mainly due to the relatively intense Balmer lines in combination with the very weak Ca II

Table 27.1 Duration of lifetime: spectral classes on main sequence

Spectral class	Lifetime on MS
O7	~6 My
O9	~8 My
B0	~12 My
B1	~16 My
B2	~26 My
B4	~43 My
B6	~95 My
A0	~350 My
A5	~1.1 Gy
F2	~2.7 Gy
G2	~10 Gy
K0	~20 Gy

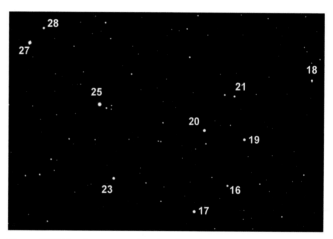

Figure 27.2 Bright stars in the Pleiades Cluster with Flamsteed numbers
(Credit: M. Huwiler)

Table 27.2 The brightest stars in the Pleiades cluster, spectral classes B6–B8

Star Name	Flamsteed No.	HD No.	m_v	M_V *	Spectral Class	Remarks [11]
Alcyone	25 Tau (η)	23630	~2.87	−2.74	B7 IIIe	Be and multiple star
Atlas	27 Tau	23850	~3.63	−2.00	B8 III	Binary star system
Electra	17 Tau	23302	~3.7	−1.84	B6 IIIe	Be star
Maia	20 Tau	23408	~3.87	−1.83	B8 III	Binary star system
Merope	23 Tau	23480	~4.18	−1.58	B6 IVe	Variable β Cep type
Taygeta	19 Tau	23338	~4.3	−1.31	B6 IV	Binary star system
Pleione	28 Tau	23862	~5.09	−0.58	B8 V ne	Be or Be shell star
Celaeno	16 Tau	23288	~5.46	−0.28	B7 IV	Variable
	18 Tau	23324	~5.64	+ 0.01	B8 V	
Asterope	21 Tau	23432	~5.76	+ 0.05	B8 V	Variable

* Note: By Abt and Levato [152], other data by CDS [11]

absorption at λ3933.66. Striking features of all profiles are also the unidentified small emission humps in the continuum at ~λλ5840 and 5350. Further visible here is the increase in intensity of the H-Balmer lines between the classes B6 to B8.

Plate 61 Stars of the Open Pleiades Cluster with Emission Line Classification

Pleione, Alcyone, Electra and Merope

Plate 61 shows a montage of those four spectra which are classified to show emission lines (index e). Three of them are even classified as Be stars [11]. However, in March 2014, just two of them showed intensive emission lines and Merope, 23 Tau, a very weak "filling in" within the core of its Hα absorption. As the only one the Be star Pleione shows in its profile coarser absorptions between the H-Balmer lines, possibly a hint to a pseudo-photosphere of a temporary Be-shell stage, [2]. The Be classified Alcyone cannot be classified according to Table 19.1, due to lack of a recognizable Hβ emission. Recording information: C8/DADOS, 200 L mm^{-1}, Atik 314L+, average 1 × 60 s, 2 × 2 binning.

Plate 62 Stars of the Open Pleiades Cluster with Absorption Lines

27.7 Age Estimation of M45

The earliest main sequence classification (V) here is B8 V. All earlier B types have already migrated to the giant branch. According to Table 27.1 this yields an age of some more than 100 million years. The accepted value is in the range of ~100–130 million years. An unsolved enigma here is the presence of white dwarfs, discovered by Cecilia Payne-Gaposchkin. Such objects do not match to the stellar evolution of such a young cluster.

27.8 Globular Clusters: Analysis by Integrated Spectra

Due to their enormous distances of about 25,000–40,000 ly the brightest individual stars of globular clusters reach at most an apparent magnitude of $m_v \approx 11$ (e.g. at M5 and M13). Thus, for an individual star analysis with an 8-inch telescope only the few brightest specimens could be recorded with a slit spectrograph, all belonging to the giants of the late F to M class. Further, the relatively low absolute magnitudes of the remaining main sequence stars are in a range of just M_v

~3–4.5. This makes it impossible to determine directly here the turn-off point of these extremely distant objects. The recommendation for amateurs is to record the composite spectrum in the integrated light, as it is applied also for the galaxies in Chapter 26. Due to the typically very old stellar inventory absorption lines can be seen here almost exclusively. With such spectra an "integrated spectral class" can be determined [153], mainly influenced by yellow and orange giants located on the RGB, HB and AGB. However, the late red giants of the M class, with spectra mainly consisting of broad TiO absorption bands, exert obviously hardly an impact (refer to Plate 63), which was recognized in the 1950s by W. Morgan [153].

Therefore the integrated spectral class provides just a very rough indication for the age of a globular cluster. In professional astronomy such profiles are also compared with synthetically generated spectra and the turn-off point is determined, by means of the H-Balmer lines among others. According to [154] the composite spectra are obtained here by sweeping over the cluster to prevent the profile being dominated by the stars of just one particular cluster zone. Due to the relatively short focal lengths, the frequently poor to mediocre seeing conditions and the usually rather inaccurate autoguiding, such an action is not really necessary at the amateur level. However, while recording the spectra the image was slightly defocused here in order to avoid any disproportionate influence of individual stars.

A still unsolved enigma is posed by single blue stars of earlier spectral type, the "blue stragglers." Such short-lived objects can be detected mainly in the central regions of all known globular clusters and do not fit at all into the picture of an extremely old cluster. There are several hypotheses under consideration – one of which is that due to the high star density within the central cluster zones, such "blue stragglers" could be generated by the fusion of two or more red giants.

Plate 63 Integrated Spectra of Globular Clusters

M3, M5 and M13

Plate 63 impressively shows the very similar looking integrated spectra of M3, M5 and M13. There are intensive absorptions in the profile of M13 at ~λ6600 (artifacts?) which are not identifiable here. Well defined and recognizable are the H-Balmer lines, the two Fraunhofer Ca II absorptions, as well as, surprisingly intense and well defined, the molecular CH band. The rough determination of the integrated spectral type with about F6–F7 is very easy here. The striking "brand" of the F class is the combined appearance of the CH absorption at λλ4299–4313 and the directly adjacent Hγ line (Section 8.3). The decimal subclass is derived here from the nearly equal intensity of these two absorptions. TiO absorption bands of late classified red giants are barely noticeable in these profiles. The applied exposure time with the C8 was here 1200 s in the 2 × 2 binning mode.

27.9 Age Estimation of M3, M5 and M13

A serious determination of the turn-off point with amateur means is not possible here. According to Table 27.1 the very rudimentary indication of the integrated spectral class yields for the middle to late F types a somewhat too young age for the clusters of just about 6–7 billion years. With 12 billion years the accepted value is significantly higher.

Figure 27.3 Globular cluster M3 (Credit Adam Block Mount Lemmon Sky Center University of Arizona) and central parts of M5 and M13 (Credit: NASA/ESA/HST)

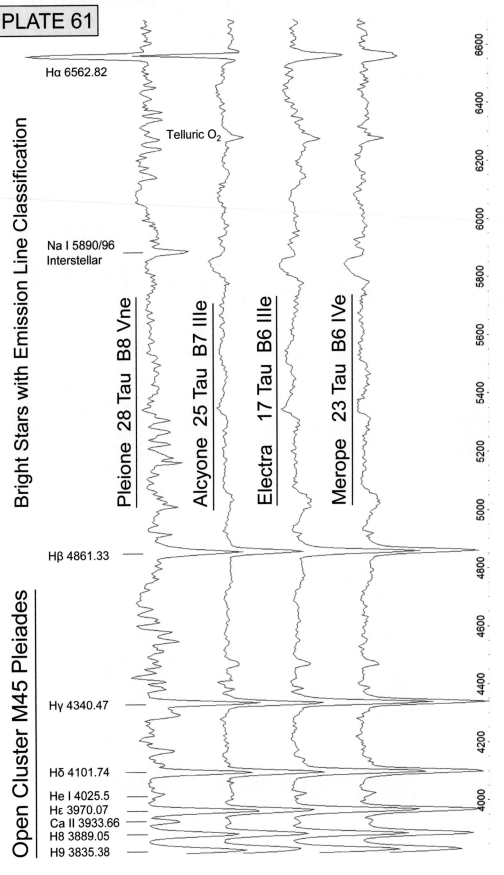

PLATE 61

Bright Stars with Emission Line Classification

Open Cluster M45 Pleiades

Hα 6562.82

Telluric O₂

Na I 5890/96
Interstellar

Pleione 28 Tau B8 Vne

Alcyone 25 Tau B7 IIIe

Electra 17 Tau B6 IIIe

Merope 23 Tau B6 IVe

Hβ 4861.33

Hγ 4340.47

Hδ 4101.74

He I 4025.5
Hε 3970.07
Ca II 3933.66
H8 3889.05
H9 3835.38

Plate 61 Stars of the Open Pleiades Cluster with Emission Line Classification

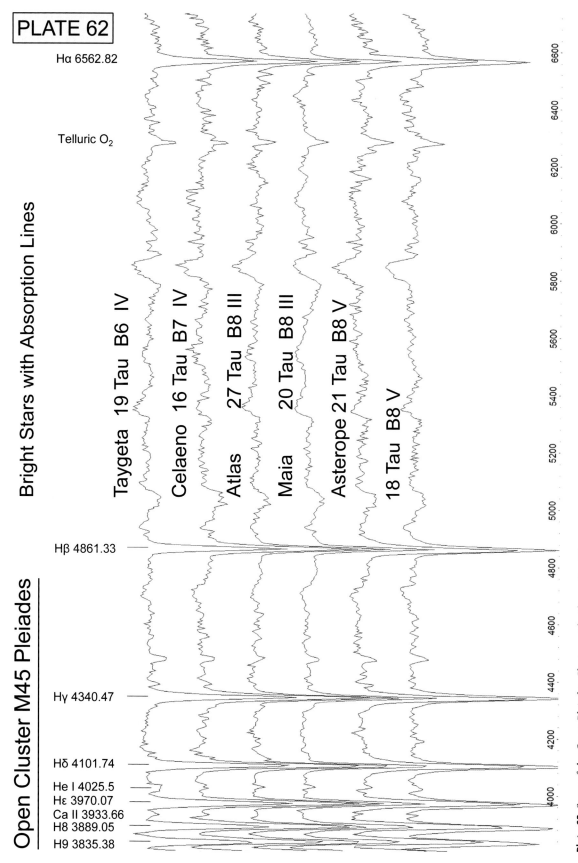

PLATE 62

Bright Stars with Absorption Lines

Open Cluster M45 Pleiades

Hα 6562.82

Telluric O₂

Taygeta 19 Tau B6 IV
Celaeno 16 Tau B7 IV
Atlas 27 Tau B8 III
Maia 20 Tau B8 III
Asterope 21 Tau B8 V
18 Tau B8 V

Hβ 4861.33

Hγ 4340.47

Hδ 4101.74
He I 4025.5
Hε 3970.07
Ca II 3933.66
H8 3889.05
H9 3835.38

6600
6400
6200
6000
5800
5600
5400
5200
5000
4800
4600
4400
4200
4000

Plate 62 Stars of the Open Pleiades Cluster with Absorption Lines

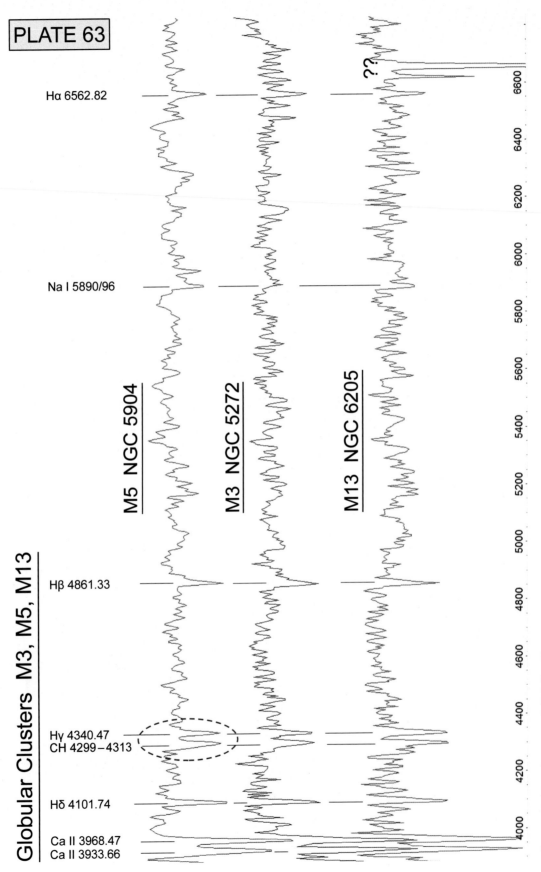

PLATE 63

Global Clusters M3, M5, M13

Hα 6562.82

Na I 5890/96

M5 NGC 5904

M3 NGC 5272

M13 NGC 6205

Hβ 4861.33

Hγ 4340.47
CH 4299–4313

Hδ 4101.74

Ca II 3968.47
Ca II 3933.66

??

4000   4200   4400   4600   4800   5000   5200   5400   5600   5800   6000   6200   6400   6600

Plate 63 Integrated Spectra of Globular Clusters

CHAPTER

28 Emission Nebulae

28.1 Overview and Short Introduction

Reflection nebulae are interstellar gas and dust clouds which passively reflect the light of the embedded stars. Emission nebulae, however, shine actively. This process requires that the atoms are first ionized by hot radiation sources with at least 25,000 K generating UV photons above the so-called Lyman limit of 912 Å. This corresponds to an energy of $>13.6\,\mathrm{eV}$ which can only be generated by very hot stars of the O and early B class. In this way the atoms of the surrounding nebula are partially ionized. By recombination the released electrons may be recaptured by the ions. Their subsequent cascade, down to the ground state, generates energy, which is compensated by emitted photons. Their frequency corresponds to the energy difference ΔE between the passed levels, corresponding to the Planck energy equation:

$$v = \Delta E / h \qquad \{28\}$$

Where ΔE is the energy difference between the passed levels [J], h is Planck's constant or quantum of action $(6.626 \times 10^{-34}\,\mathrm{J\,s})$ and v (Greek "nu") is the frequency of the photon $[\mathrm{s}^{-1}]$ of the spectral line.

Further interesting details of these processes are outlined in [3]. Thus, similar to some gas discharge lamps, these nebulae generate mainly quasi-monochromatic light, i.e. a limited number of discrete emission lines. Therefore specifically designed, narrow band nebula filters are effective here. Since the main part of the light is concentrated on a few more or less intense emission lines, these objects can still be detected even at extreme distances. For apparently bright and rather dot-shaped objects the brightest, mostly [O III] lines, become photographically detectible after very short exposure times.

These energetic requirements are mainly met by H II regions (e.g. M42), planetary nebulae (e.g. M57) and supernova remnants (e.g. M1). Further mentionable are the nuclei of active galaxies (AGN). The density of the nebulae is so extremely low that on Earth it can be generated only by the best ultra-high vacuum. The matter consists chiefly of hydrogen, helium, nitrogen, oxygen, carbon, sulfur, neon and dust (silicate, graphite etc.). Besides the chemical composition, the energy of the UV radiation, the temperature T_e as well as density N_e of the free electrons, determine the local state of the plasma. The intensity of the individual emission lines allows an easy first rough guess of important plasma parameters.

28.2 H II Regions

A famous textbook example is formed by parts of the Orion Nebula M42 (Figure 28.1). Here extremely hot stars of the O and early B class, all located in the famous Trapezium cluster, are primarily ionizing hydrogen atoms, besides some helium, oxygen and nitrogen. Such H II regions tend to have a clumpy and chaotic structure and may extend over dozens of light years. They show a high star formation rate and can still be detected even in distant galaxies. The reddish hue is caused here by the dominant Hα emission.

Further examples of bright H II regions are NGC 281 (Pac Man Nebula), NGC 604, NGC 2070 (Tarantula nebula), NGC 3372 (Carina nebula), NGC 6514 (Trifid nebula).

28.3 Planetary Nebulae

In the central part of these much higher energetic objects are mostly extremely hot, white dwarfs with effective

Figure 28.1 M42, Orion Nebula
(Credit: NASA, ESA, M. Robberto (STScI/ESA) and HST Orion Treasury Project Team)

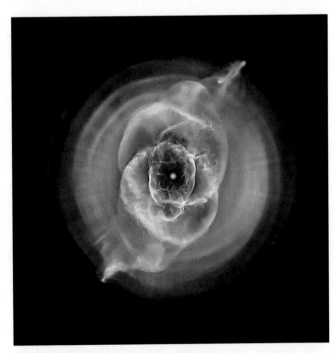

Figure 28.2 NGC 6543, Cat's Eye Nebula
(Credit: NASA/ESA/HEIC and the Hubble Heritage Team, R. Corradi *et al.*)

temperatures up to >200,000 K. They represent the final stage of stars at the end of the AGB with an initial stellar mass $M_i \lesssim 8\,M_\odot$, ionizing the atoms of their former

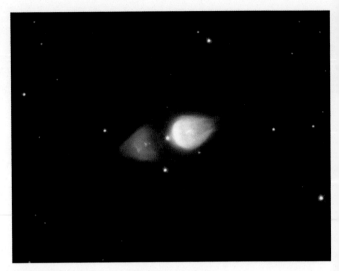

Figure 28.3 M1-92: Minkowski's Footprint
(Credit: NASA/ESA, HST)

stellar envelopes, expanding rather slowly, mostly with ~20–50 km s^{-1}. About 10% of central stars show similar spectra in the final stage like Wolf–Rayet stars, and thus are classified as WR types, denoted by "WRPN." Their absolute magnitude, however, is considerably lower. Planetary nebulae often show an ellipsoidal shape, in some cases with a regular fine structure. The reasons for the numerous existing forms are only partly understood, but are often influenced by orbiting companion stars. Figure 28.2 shows the Cat's Eye Nebula in the constellation Draco.

28.4 Protoplanetary Nebulae

These are the precursors of planetary nebulae. They are excited by the so-called post-AGB stars. These are former carbon stars, i.e. Mira variables at the upper end of the asymptotic giant branch (AGB), just beginning to repel their envelopes. They are still not hot enough to excite higher ionized emission lines such as [O III]. Figure 28.3 shows a famous example, the bipolar nebula "Minkowski's Footprint" M1-92, in the constellation Cygnus.

28.5 Supernova Remnants

Supernova remnants show a striking filamentary structure. The main part of the ionization energy is provided here by the collision of the rapidly expanding stellar envelope (a few 1000 km s^{-1}) with the interstellar matter. Figure 28.4 shows the Cygnus Loop (~1500 ly) recorded in UV light by GALEX.

In the center of SNR a remaining neutron star or pulsar emits a wind of relativistic electrons with nearly light speed.

Figure 28.4 Cygnus Loop in UV light
(Credit: NASA/JPL-Caltech)

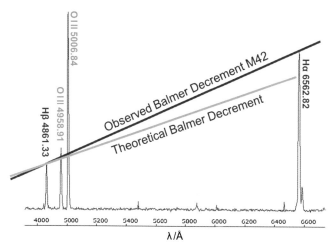

Figure 28.6 Typical features in the spectrum of the emission nebula M42

28.6 Wolf–Rayet Nebulae

The shells around the high energy Wolf–Rayet stars are excited in a similar way to SNR, but are much less intensive, such as the Crescent Nebula NGC 6888 **(Plate 74)** or Thor's Helmet NGC 2359 **(Plate 75)**. Figure 28.5 shows a part of the Carina Nebula NGC 3372 with the ionizing source WR22, recorded at the ESO Observatory in La Silla.

28.7 Common Spectral Characteristics of Emission Nebulae

In all types of emission nebulae, physically the same ionization processes are responsible for the generation of the spectral lines, albeit excited by very different objects and involved energies. This explains the very similar appearance of such nebular spectra. With exception of supernova remnants, emission nebulae generate only a very faint and diffuse continuum radiation. In Figure 28.6 the profile is slightly elevated in order to show the labeling of the wavelength axis. It displays an emission spectrum of M42 with mainly two noticeable features: the Balmer decrement and the intensity ratio between the [O III] lines.

Figure 28.5 WR22 and NGC 3372, Carina Nebula
(Credit: ESO)

It is deflected or slowed by magnetic fields within the plasma or electric fields around the ions. Such energy transformations are compensated by emitted photons, generating broadband synchrotron or bremsstrahlung, predominantly in the X-ray domain. Not until 2012 in the Cygnus Loop a possible candidate for a central neutron star could be found [155].

The Balmer Decrement

In nebular spectra the emissions of the H-Balmer series show a reproducible, quantum mechanically induced intensity loss towards decreasing wavelengths. This phenomenon is called the Balmer decrement D and is defined by the flux ratio of the Hα and Hβ emissions:

$$D = F(\text{H}\alpha)/F(\text{H}\beta) \tag{28.1}$$

The energy flux F of an emission line corresponds to its area, measured above the local continuum level (see Appendix G). For amateur purposes this ratio can be simplified and determined by the relative length or peak intensity of the line [3]. The theoretical value, calculated for nebular spectra is $D \approx 2.85$. Highly important for astrophysics, is the measured deviation from this theoretical decrement. It is a powerful indicator, for example, for the interstellar extinction (reddening) by dust particles. It is important to note that emission spectra generated by other astronomical sources may show different decrement values – such as Mira variables (Section 13.2) exhibiting even inverse decrements. Further details and analysis options are outlined in [3].

Intensity Ratio Between [O III] Lines

The intensity ratio of the brightest [O III] lines is always:
$I(5007)/I(4959) \approx 3$.

28.8 Plasma Diagnostics and Excitation Class E

Since the beginning of the twentieth century numerous methods have been proposed to determine the excitation classes of emission nebulae. This application is just one important part of the so called plasma diagnostics, explained in more detail in [3]. The 12-level "revised" system presented by Gurzadyan [156], which has also been developed by Acker, Aller, Webster and others, is one of the currently best accepted and appropriate also for amateurs [157], [158], [159]. It relies on the simple principle that with increasing excitation class, the intensity of the forbidden [O III] lines becomes stronger, compared with the H-Balmer series. Therefore as a classification criterion the intensity sum of the two brightest [O III] lines, in relation to the reference value of the Hβ emission, is applied. Within the range of the low excitation classes E1–E4, this value increases strikingly fast. In the following criterion terms, the [O III] lines at $\lambda\lambda 4959$ and 5007 are denoted as N_1 and N_2.

The criterion term for low excitation classes E1–E4 is:

$$I_{N1+N2}/I_{H\beta} \qquad \{28.2\}$$

Within the transition class E4 the He II line at $\lambda 4686$ appears for the first time. These ions require 24.6 eV for ionization, corresponding to about 50,000 K [160]. That is almost twice the energy as needed for H II with 13.6 eV. From here on, the intensity of He II increases continuously and replaces the now stagnant Hβ emission as a reference value in the criterion term. The ratio, for practical reasons, is expressed here

Table 28.1 Excitation classes for emission nebulae

Low E class $I_{N1+N2}/I_{H\beta}$		Middle E class $\log(I_{N1+N2}/I_{4686})$		High E class $\log(I_{N1+N2}/I_{4686})$	
E1	0–5	E4	2.6	E9	1.7
E2	5–10	E5	2.5	E10	1.5
E3	10–15	E6	2.3	E11	1.2
E4	>15	E7	2.1	E12	0.9
		E8	1.9	E12$^+$	0.6

logarithmically (decimal) in order to reduce the range of values of the classification system:

The criterion term for middle and high excitation classes E4–E12 is:

$$\log\left(I_{N1+N2}/I_{He\,II\,(4686)}\right) \qquad \{28.3\}$$

The 12 E classes are subdivided in to the groups: low (E = 1–4), middle (E = 4–8) and high (E = 9–12). In extreme cases 12^+ is assigned (Table 28.1).

28.9 Practical Aspects of the Determination of the E Class

The determination of the lower E classes 1–4 is very easy, since the Hβ line, compared to the [O III] emission, is relatively intense. At level E4 the He II line ($\lambda 4686$) begins to appear, at first very weak, requiring low-noise spectra, and a strong zoom into the intensity axis.

To compensate the attenuation effects, acting on the emissions by instrumental responses $D_{INST}(\lambda)$ and the influence of the Earth's atmosphere $D_{ATM}(\lambda)$, the measurement of the intensities should be carried out in a profile, relatively flux calibrated by a recorded standard star. The attenuating influences, due to interstellar matter $D_{ISM}(\lambda)$ which can be further corrected with help of the observed Balmer decrement, still remain. For the theoretical background and further information refer to [3].

The diagnostic lines are grouped quite closely together. Moreover, some of them are located in an area where the difference between the original and the pseudo-continuum, recorded with typical amateur equipment, is relatively low. This allows, even at raw profiles without any further corrections, a reasonable estimation of the excitation with accuracy of about ± 1 class. Due to the slightly greater separation of the He II diagnosis line ($\lambda 4686$), at middle and high excitation classes, the classification may result here in up to one step too low.

Table 28.2 Estimated temperatures for E classes

E Class	E1–2	E3	E4	E5	E7	E8–12
T_{eff} [K]	35,000	50,000	70,000	80,000	90,000	100,000–200,000

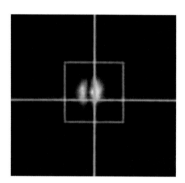

Figure 28.7 IC418, screenshot PHD Guiding

28.10 Practical Aspects of the Recording of Planetary Nebula

Spectra from the very small, disk-shaped and blue-greenish shining PN are quite easy to record. Thus, they are very quickly localized within a stellar group and the exposure time for the bright representatives takes only a few minutes. The brightest [O III] line often appears on the screen just after a few seconds (e.g. NGC 6210)! However, with rather small amateur telescopes, within these tiny apparent disks it is impossible to record selectively a specific area of the nebula. Moreover, the strongly varying line intensities are integrated here along the very short, exposed part of the slit. Furthermore, during long exposure times, small changes in the slit position relative to the nebula are observed as a result of inadequate seeing and/or guiding quality. In some cases even the central star itself may directly influence the recorded spectrum. Such effects may, for example, cause significantly different or even impossible results for the nebular Balmer decrement. Since the classification lines are quite closely grouped together, the influence on the excitation class was observed to be rather low!

Typical galactic decrement values are within the range of $D \approx 3.0\text{–}3.5$. However, there are stark outliers like NGC 7027 with $D \approx 7.4$ [156]! For extragalactic objects in most cases the decrement becomes $D > 4$ [159]. Figure 28.7 shows the small sliver of the Spirograph Nebula (IC418) on the 25 μm slit of the DADOS Spectrograph (PHD Guiding). Between the green autoguiding cross and the slit even the bright central star is visible.

By contrast, the apparently very large, famous planetary nebulae M27 and M57 require, with the C8 and the Atik

314L+, at least 20–30 minutes of exposure time (1 × 1 binning). Such objects allow a selective recording of spectra within specific areas of the nebula and even gain intensity curves for individual lines along the quite long exposed part of the slit (Plate 71). Furthermore the above mentioned problems are significantly reduced here.

28.11 The Excitation Class as an Indicator for Plasma Diagnostics

Gurzadyan (among others) has shown that the excitation classes are more or less closely linked to the evolution of the PN [156], [157]. The study with a sample of 142 PN showed that the E class is a rough indicator for the age of the PN, the temperature T_{eff} of the central star and the expansion of the nebula; however, in the reality the values may scatter considerably [161].

The Age of the PN

Typically PN start on the lowest E level and subsequently step up the entire scale with increasing age. The four lowest classes are usually passed very quickly. Later on this pace decreases dramatically. The entire stage finally takes just about 10,000 to >20,000 years, an extremely short period, compared with the total lifetime of a star!

The Temperature T_{eff} of the Central Star

The temperature of the central star rises also with the increasing E class. By repelling the shell, increasingly deeper and thus hotter layers of the star become "exposed." At about E7 in most cases an extremely hot white dwarf remains, generating a WR-like spectrum. This is demonstrated impressively in the table with PN in Appendix E. Hence, for T_{eff} the very rough estimates in Table 28.2 can be derived.

The Expansion of the Nebula

The visibility limit of expanding PN is restricted to a maximum radius of about 1.6 ly (0.5 parsec), otherwise at larger distances the density becomes too small [160]. With increasing E class,

Table 28.3 Mean values for E classes radii

E Class	E1	E3	E5	E7	E9	E11	E12+
R_n[ly]	0.5	0.65	0.72	1.0	1.2	1.4	1.6

also the radius R_n of the expanding nebula is growing. Table 28.3 shows Gurzadyan's mean values for R_n, which may scatter considerably for the individual nebulae [157].

28.12 Emission Lines Identified in the Spectra of Nebulae

Here follows a compilation of identified emission lines from different sources [160] [162]. "Forbidden lines" are written within brackets [].

Ne III 3869, [Ne III] 3967.5, He I 4026.2, [S II] 4068.6, Hδ 4101.7, C II 4267.3, Hγ 4340.5

[O III] 4363.2, He I 4387.9, He I 4471.5, He II 4541.6, [Mg I] 4571.1, [N III] 4641, He II 4685.7

[Ar IV] 4740.3, Hβ 4861.3, He I 4921.9, [O III] 4958.9, [O III] 5006.8, N l 5198.5, He II 5411.5

[Cl lII] 5517.2, [Cl III] 5537.7, [O I] 5577.4, [N II] 5754.8, He I 5875.6, [O I] 6300.2, [S III] 6310.2

[O I] 6363.9, [Ar V] 6434.9, [N II] 6548.1, Hα 6562.8, [N II] 6583.6, He I 6678.1, [S II] 6717.0

[S II] 6731.3, [He II] 6890.7, [Ar V] 7006.3, He I 7065.2, [Ar III] 7135.8, He II 7177.5, [Ar IV] 7236.0, [Ar IV] 7263.3, He l 7281.3

28.13 Comments on Observed Spectra

Because spectra of emission nebulae in most cases barely show any continuum, the profiles in the following plates are slightly shifted upwards in order to improve the readability of the scaled wavelength axis. The presentation of the following objects starts with the H II region M42. The following planetary nebulae are sorted according to ascending excitation classes. For most of the PN, the problem of distance estimation is still not really solved. The information may therefore vary, depending on the sources up to >100%! Correspondingly inaccurate are therefore also the estimated diameters of the nebulae!

By repelling the envelope and the progressive "exposure" of increasingly hotter, inner stellar layers, central stars of PN may generate spectra, simulating substantially more massive and luminous stars. This applies, for example, for all O, early B and WR classifications. However, the progenitor stars of planetary nebulae are limited to a maximum of ~8 $M_\odot$, corresponding to the middle B class.

Plate 64 Emission Nebula: H II Region

Orion Nebula: M42 NGC 1976

Plate 64 shows the emission spectrum of M42 (~1400 ly; excitation class: E1) taken in the immediate vicinity of the Trapezium cluster θ^1 Orionis. The main radiation source of this H II region is the C component, a blue giant of the very rare, early spectral type O6 (Plate 5) with a temperature of ~40,000 K [163]. Together with the other stars of the early B class, it is capable of exciting the surrounding nebula with the criterion value $I_{N1+N2}/I_{H\beta} \approx 5$ up to the border area between the classes E1 and E2. The Hβ line is here just slightly surpassed by the [O III] emission at λ4959. Due to the enormous apparent brightness the object is spectrographically easily accessible and requires just very modest exposure times (10 × 30 s, 1 × 1 binning).

Plate 65 Intensity Profiles of Hβ and [O III] (λ 5007) in the Central Area of M42 (image: Credit: NASA, C.R. O'Dell, S. K. Wong)

Orion Nebula: M42 NGC 1976

Plate 65 shows within the central region of the nebula for the [O III] (λ5007) and Hβ emission lines the intensity profiles along the entire length of the DADOS triple slit array, corresponding to ~2.5 ly (embedded image M42: Credit NASA, HST). The ratio of these intensities demonstrates indirectly the course of the excitation class. Thanks to the linear arrangement of the three slits, DADOS allows for apparent two-dimensional objects an improvised "long-slit" spectroscopy, which enables us with just one recording, to obtain rough spectral intensity information, combined with the spatial dimension for a certain ion. These Visual Spec profiles have been generated by 1D-intensity stripes, obtained along the whole slit array, processed and finally rotated by 90° with the IRIS software. On the narrow bridges between the individual slits the suspected course of the intensity curve is supplemented with dashed lines. After recording of the spectra in each case a screenshot by the slit camera was taken to document the exact location of the slit array relative to the nebula.

For this purpose the different width of the three slits (50, 25, 30 μm) plays a little role, since just the intensity course is analyzed here. Further, IRIS averaged here the intensity values within the whole slit width. The slit array was positioned on two places within the central part of M42 and approximately aligned in the north–south direction. The western section runs through the C component of the brightest Trapezium star θ^1 Orionis C (Plate 5) and ends after the so called Orion Bar. The eastern section runs along the "Schröter Bridge," then through the "Sinus Magnus" and finally ends after the impressive "Orion Bar," southeast of the Trapezium. The intensity scales of the profiles have been normalized within the Orion Bar at the peak value of [O III] = 1 and the peak of Hβ on the local ratio there of [O III]/Hβ ≈ 2.5.

Striking is the dramatic increase of intensity within the area of the Trapezium and in the huge ionization front of the Orion Bar. The latter marks the end of the ionized H II region or the so called Strömgren radius R_s. Its size is mainly influenced by the huge radiation power of the C component of Θ^1 Orionis (Section 5.1). Behind this transition zone the remaining H I nebula is heated up to just several 100 K, by the remaining UV photons with energies <13.6 eV, which could not be absorbed by the ionization processes within the H II area. Chemically increasing complex molecular clouds prevail in this peripheral area. Also remarkable is the sharp drop of the intensity within the dark cloud in the Sinus Magnus and the slight increase in the vicinity of the Schröter Bridge.

Plate 66 Post-AGB Star with Protoplanetary Nebula

Red Rectangle Nebula HD 44179

Plate 66: This highly interesting object (~2300 ly; excitation class: E ≪ 1), not discovered until 1974, is a protoplanetary nebula which is located in the constellation Monoceros. With an apparent variable magnitude of m_v ≈ 9.0 this post-AGB star has just left the last carbon star stage on the AGB, and is beginning to repel its envelope as a planetary nebula. In amateur telescopes the object is visible just as a star. Only large professional telescopes show the small bi-cone-shaped nebula (Figure 28.8). The object is also a spectroscopic binary. By repelling its envelope, the former carbon star has changed its spectrum radically and is now classified by CDS surprisingly as a main sequence star, B8V [11]. In contrast and much more plausible, in 1975 Cohen *et al.* [164] classified it as a giant (B9–A0 III), which fits significantly better to the strikingly slim absorption lines, suggesting a thin, expanding pseudo-photosphere. This spectral class would correspond to an effective temperature of >10,000 K. However, for the post-AGB component just 7500–8000 K is assumed [165]. The corresponding ionization energy is still far too low to significantly ionize the nebula, so here just the hydrogen Hα and the Na I lines appear in emission. This way an excitation class, according to Section 28.8, can of course not be determined. Highly resolved spectra show here the Na I emissions with double peaks, what is interpreted as a combination of

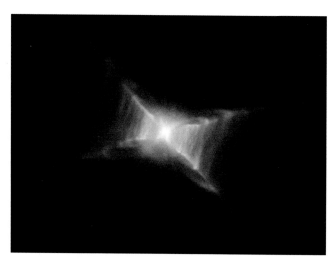

Figure 28.8 Red Rectangle Nebula (Credit: NASA, ESA, H. Van Winckel, M. Cohen)

perspective and Doppler effects, caused by the bipolar occurring mass loss [165]. Generally, several emissions in this domain are still enigmatic and subject of current studies, such as the Red Rectangle band (RRB) [166]. In such spectra sometimes C2 Swan band emissions testify of the former stage as a carbon star [167]. HD 44179 shows an iron abundance with a factor 1000 below solar [2]! Recording information: C8/DADOS 2 × 960 s, 2 × 2 binning.

Plate 67 Planetary Nebula: Excitation Class E1

Spirograph Nebula IC 418

Plate 67 shows the emission spectrum of a still very lowly excited PN (~2500 ly). The temperature of the highly variable, still unstable central star is here only 35,000 K [163] and therefore just able to excite this nebula with the criterion value $I_{N1+N2}/I_{H\beta}$ ≈ 2.8 to the lowest class E1 (Gurzadyan: E1 [157]). This is even lower than within the H II region M42. Further, Hβ here still outperforms the intensity of [O III] (λ4959). Accordingly, the PN stage must be very young and also the diameter of the ionized shell is estimated to be just 0.2 ly [163]. The complex spirograph pattern in the nebula is still not understood (Figure 28.9). Recording information: C8/DADOS 3 × 170 s, 1 × 1 binning.

Plate 68 Planetary Nebula: Excitation Class E3

Turtle Nebula NGC 6210

The excitation class in the spectrum of NGC 6210 (~6500 ly), presented in **Plate 68** is much higher than the previous example. The central star here has a temperature of ~58,000 K [163] and is classified as O7f (Figure 28.10). It excites this nebula with the

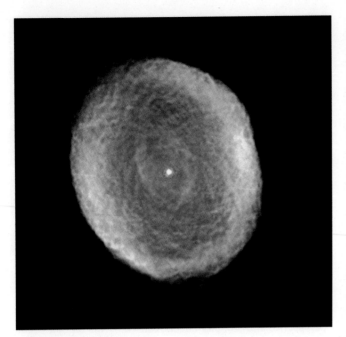

Figure 28.9 Spirograph Nebula
(Credit: NASA and the Hubble Heritage Team (STScI/AURA))

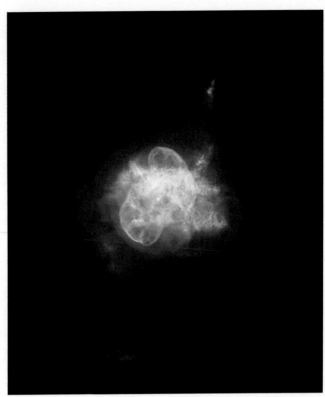

Figure 28.10 Turtle Nebula
(Credit: ESA/NASA/HST)

Plate 69 Planetary Nebula: Excitation Class E8

Saturn Nebula NGC 7009

Plate 69: The central white dwarf here has a temperature of ~90,000 K [163] and is some ~2000 ly distant. It excites this nebula with the criterion value $\log\left(I_{N1+N2}/I_{He\,II\,(4686)}\right) \approx 1.9$ up to the class E8. (Gurzadyan: gives a slightly different E7 [157].) The blue profile is strongly zoomed in the intensity to make the weaker lines visible. The Saturn Nebula was the very first whose expansion could photographically be detected (Figure 28.11) [160]. Recording info: C8/DADOS 7 × 240 s, 1 × 1 binning.

Plate 70 Planetary Nebula: Excitation Class E10

Ring Nebula M57, NGC 6720

Plate 70 shows the emission spectrum of M57 (~2300 ly), recorded approximately in the center of the Ring nebula. The central white dwarf with a temperature of ~150,000 K [163] excites this nebula with the criterion value of $\log\left(I_{N1+N2}/I_{He\,II\,(4686)}\right) \approx 1.4$ up to the class E10 (consistent with Gurzadyan: E10 [157]). The hydrogen lines are weak

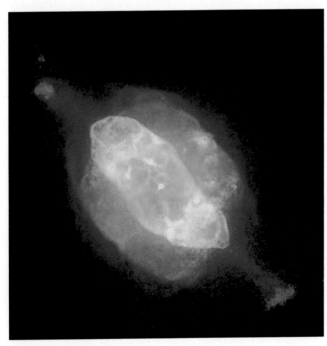

Figure 28.11 Saturn Nebula
(Credit: NASA, HST, B. Balick *et al.*)

criterion value $I_{N1+N2}/I_{H\beta} \approx 14$ slightly below the threshold of class E4 (Gurzadyan: E4 [157]). In the spectrum the He II line (λ4686) appears here just very weak. In some cases, this line can also be emitted directly by the central star [157]. Recording information: C8/DADOS 4 × 45 s, 1 × 1 binning.

here relative to [O III]. The Hα line just reaches parity with the weaker [N II] (λ6548) emission. The much stronger [N II] line at λ6584, however, is the second most intense emission behind [O III] (λ5007). Therefore the reddish color within the ring of M57 predominantly originates from [N II] instead of the by far weaker Hα line. Typical for this high excitation class is the old age of about 20,000 years, and the large extent of the nebula of about 1.4 ly. Recording information: 36 Inch CEDES Cassegrain, slit width 25 µm, 5 × 340 s, 1 × 1 binning.

Plate 71 Intensity Profiles of [O III] and [N II] in the Longitudinal Axis of M57 (Image Credit: H. Bond *et al.*, Hubble Heritage Team (STScI/AURA), NASA)

Ring Nebula M57, NGC 6720

Plate 71: In a similar way as presented on Plate 65 for M42, the intensity profiles for M57 (~2300 ly) of the strongest emission lines [O III] and [N II] are presented here, recorded along the entire 50 µm slit, positioned on the longitudinal axis of the Ring nebula.

Plate 72 Estimation of the Mean Expansion Velocity of a PN

Blue Snowball Nebula NGC 7662

Plate 72: The Blue Snowball Nebula (~1800 ly), is used here to demonstrate the estimation of the mean expansion velocity of a PN, based on the measured split up of emission lines. The theoretical background is explained in [3]. Due to the transparency of the PN, with the split up of $\Delta\lambda$, applied in the Doppler equation {18.1}, the total expansion velocity of the nebula can be estimated, related to the whole diameter of the shell, $2V_r \approx 48\,\mathrm{km\,s^{-1}}$. The radial velocity is finally obtained by halving this value. Due to the relatively low velocity a highly resolved spectral profile is required. It has been obtained here by the SQUES echelle spectrograph applying a slit width of 25 µm. Recording information: C8, 2 × 360 s, Atik 314+, 1 × 1 binning.

Plate 73 Emission Nebula: SNR Excitation Class E > 5

Crab Nebula M1 NGC 1952

Plate 73 shows the emission spectrum of the supernova remnant M1 (~6300 ly; excitation class: E > 5). The 50 µm slit runs in direction ~N–S through the central part of this

Figure 28.12 Crab Nebula
(Credit: NASA, ESA, J. Hester, A. Loll)

young SNR (embedded image M1: Credit NASA, HST). The causal SN type II was observed and documented in 1054 by Chinese astronomers. Today, the diameter of the expanding nebula reaches approximately 11 ly (Figure 28.12).

In 1913, Vesto Slipher recorded the first spectrum of M1. At that time he noticed a massive split up of the most intense emission lines. Unaware of the nebula expansion, he interpreted this spectral symptom wrongly as the newly discovered Stark effect (Section 7.3) caused by the interaction with electric fields. In 1919, R. F. Sanford exposed the M1 spectrum with the 2.5 m Hooker Telescope, and at that time the "fastest" film emulsions, took no fewer than 48 hours! The result he describes soberingly as "disappointingly weak," a clear indication that even today this cannot really be a beginner's object. The expansion of the nebula was not found until 1921 by C. O. Lampland, by comparing different photographic plates.

Applying a C8 telescope, with DADOS 200 L mm⁻¹ and a cooled Atik 314L+ camera in the 2 × 2 binning mode, 2 × 30 minutes were still needed to record a spectrum of passably acceptable quality. Due to the long exposure time, light pollution and airglow was recorded in a comparable intensity to the signal of M1. For details of the removal refer to [3].

This apparently almost 1000-year-old expanding shell is meanwhile diluted to such an extent, that, similar to planetary nebulae, it has become optically transparent. This effect also proves the redshifted peaks of the split and well shaped emission lines, often appearing of similar intensity like the blueshifted, but in some cases significantly weaker. For a detailed description of this spectral feature refer to [3].

At this low resolution the redshifted peak of the [O III] emission at λ4959, forms a blend with the blueshifted peak of the [O III] line at λ5007. Due to the transparency of the SNR, with the split up of $\Delta\lambda$ the total expansion velocity of the remnant can be estimated, related to the whole diameter of the shell, $2V_r \approx 1800$ km s^{-1}. The radial velocity is finally obtained by halving this value. At this young SNR it yields slightly below 1000 km s^{-1} but is ~50 times higher compared to PN. Thus, in contrast to the PN on Plate 72, a low resolved profile is sufficient here to demonstrate this effect.

The situation is totally different at the very beginning of new supernovae, novae or the very dense expanding shells of the LBV and Wolf–Rayet stars. Here, we see just one hemisphere of the expanding shell, heading towards us in our line of sight. In these cases, applying the Doppler equation, we can directly measure the radial velocity.

Even more spectral symptoms show that SNR are a special case within the family of the emission nebula. In the center of the nebula (profile B), due to synchrotron and bremsstrahlung processes (Section 28.5), a clear continuum is visible, which is very weak in the peripheral regions of the SNR (profile A). The latter profile was even slightly raised to enable the reading of the wavelength axis. In contrast to the SNR it is difficult to recognize a continuum in the spectra of PN and H II regions.

The line intensities I of the profiles $B_{1,2}$ were adapted relative to the continuum heights (I_P/I_C), to become roughly comparable with those of profile A. Apparently in profiles A and B the emissions in the range around λ6500 are of similar intensity. However, in the peripheral area of the nebula (profile A), the [O III] lines around λ5000 are several times stronger. Obviously the conditions for forbidden transitions are here much more favorable than in the vicinity of the high-energy pulsar, the stellar remnant of the SN explosion.

Due to shockwave-induced collisional excitation the strikingly intense sulfur doublet (λλ6717/31) becomes clearly visible. This feature is only weak in PN and almost completely absent in H II spectra. This also applies to the [O I] line at λ6300. At this resolution it can hardly be separated from the [O I] airglow line, located at the same wavelength (Chapter 31).

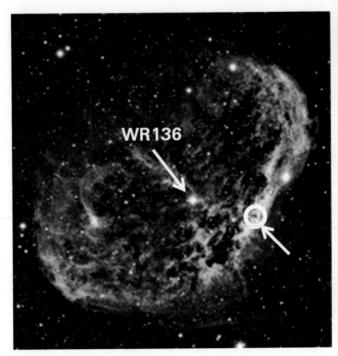

Figure 28.13 The Crescent Nebula
(Credit: Wikimedia Hewholooks)

Plate 74 Emission Nebula: Wolf–Rayet Excitation Class E1

Crescent Nebula NGC 6888

Plate 74 shows the emission spectrum of the Crescent Nebula NGC 6888, located in the constellation Cygnus (4700 ly; excitation class: E1). Figure 28.13 shows the nebula, the exciting source WR136, and the location of the recorded spectrum within the shock front.

The origin and ionizing source of NGC 6888 is the Wolf–Rayet star WR136, which is documented in Chapter 17, Plate 34. In the previous giant stage, the star repelled a part of its gas shell. After the transition to the WR stage, about 30,000 years ago [75], the mass loss intensified dramatically to about 10^{-5} to 10^{-4} $M_\odot$ per year [64] and the stellar wind accelerated to more than 1000 km s^{-1}. This violent stellar wind collides with interstellar matter and the gas layer, which was repelled earlier during the former giant stage of the star. This process apparently generates an elliptically shaped shock front, expanding with some 75 km s^{-1} [75] to a range of currently ~16 × 25 ly [74]. Similar to the SNR, this shock front is chiefly responsible for the ionization and also for the fluffy pattern of the Crescent Nebula. Within WR nebulae, these processes apparently run much less violently compared to SNR. For comparison, the shockwave of M1 expands with ~1000 km s^{-1}. This object is still some 30 times younger than NGC 6888. Further, in WR nebulae a central pulsar or neutron star is missing, which generates in SNRs a permanent, relativistic electron wind, combined with the effects described in Plate 73.

The repelling of the hydrogen shell happens at the very beginning of the Wolf–Rayet stage, then becoming very soon no longer visible. Later on, hydrogen can hardly be detected any more in the spectra of WR stars. After some 30,000 years, WR136 has just passed somewhat more than 10% of the entire estimated WR sequence of ~200,000 years.

The spectrum was recorded by DADOS 25 μm slit and 200 L mm^{-1}, just westerly of the star HD 192182 and within the outer shock front of the nebula. A continuum cannot be detected here. The profile in Plate 74 was shifted just slightly upwards, to make visible the labeling of the wavelength axis. In contrast to M1 the degree of excitation of the plasma is here very low with E1. The Hβ emission is even more intense than the [O III] line at λ4959. Certainly, this can also be attributed to the relatively advanced age of the nebula. In addition to the typical hydrogen and [O III] emissions, only neutral helium He I and forbidden lines of ionized nitrogen [N II] can be observed. The forbidden sulfur doublet [S II] at λλ6717/31, a characteristic feature for shockwaves, rises here barely above the noisy continuum level.

With the C8 and the cooled camera Atik 314L+, after an exposure time of 2 × 30 minutes in the 2 × 2 binning, resulted in a slightly noisy, but for this purpose anyway useful profile. On all shots, the separately recorded light pollution and the airglow (Plate 83) had to be subtracted (Fitswork). Under the prevailing conditions, the nebula remained invisible in the flip mirror, even with help of the [O III] filter. The slit of the spectrograph was positioned on the selected nebula filament with help of the field stars pattern.

Plate 75 Emission Nebula: Wolf–Rayet Star as its Ionizing Source

Thor's Helmet Nebula NGC 2359 and WR7 HD 56925

Plate 75: The Wolf–Rayet star WR7 (~15,000 ly;) in the constellation Canis Major is the ionizing source for the well known emission nebula NGC 2359, Thor's Helmet (excitation class: E3). Figure 28.14 shows the central part of the nebula. Plate 75 shows both the spectrum of the nebula as well as of the ionizing WR star with spectral type WN4. It is very similarly classified as WR133 (WN5) on Plate 33. In contrast to the latter, however, the profile of WR7 is here not imprinted

Figure 28.14 Thor's Helmet Nebula (Credit: ESO, B. Bailleul)

by the absorptions of a close companion star. It shows therefore the uncontaminated He II Pickering series, similar to WR136 (type WN6) on Plate 34. However, WR7 is classified earlier (type WN4) and thus somewhat older and already hotter than WR136. Whether this is the cause for the significantly higher excitation class of E3, measured at a comparable location within the outer shock front, is uncertain. The somewhat older NGC 2359 with an extension of about 30 light years is slightly larger than NGC 6888 with about 16 × 25 ly. The distance to NGC 2359 is about 15,000 ly and thus about three times as large as NGC 6888. This nebula appears therefore somewhat fainter. Recording information: C8, 200 L mm^{-1}, slit width 35 μm, Atik 314L+ 1 × 1800 s, 2 × 2 binning.

28.14 Distinguishing Characteristics in the Spectra of Emission Nebulae

Here, the main distinguishing features are summarized again. Due to the synchrotron and bremsstrahlung, especially in the X-ray part of the spectrum, SNR show a significant continuum. This appears especially pronounced in the X-ray domain, so X-ray telescopes are highly valuable to distinguish SNR from the other nebula types, particularly at very faint extragalactic objects. For all other types of emission nebulae the detection of a continuum radiation is difficult.

Table 28.4 Main features of the documented emission nebulae

Type of Nebula	Object	Excitation Class	[S II] Lines	Continuum
H II	M42	E1	none	weak
Proto PN	HD 44179	E $\ll$ 1	none	weak
PN	Various	E1–10	weak	weak
Starburst Glx	M82	E < 1	strong	significant
Seyfert Glx	M77	E10	strong	significant
SNR	M1	E5	strong	significant
WR Nebula	NGC 6888	E1	weak	weak
	NGC 2953	E3	significant	

In the optical part of SNR spectra, the [S II] and [O I] lines are, relative to Hα, more intense compared to PN, WR nebulae and AGN within starburst and Seyfert galaxies. This effect is caused here by shockwave-induced collisional ionization (Plate 73). At PN and WR nebulae the [S II] and [O I] emissions are very weak and almost totally absent in H II regions.

In H II regions, the excitation by O- and early B-class stars is relatively low and therefore the excitation class remains in the order of just E $\approx$ 1–2. Planetary nebulae usually pass through all 12 excitation classes, following the evolution of the central star. In this regard the SNR are also a highly complex special case. By the very young Crab Nebula (M1), dominate higher excitation classes whose levels are not homogeneously distributed within the nebula, according to the complex filament structure [69], [168]. The diagnostic line He II at λ4686 is therefore a striking feature in some spectra of M1.

In SNR the electron density N_e is very low, i.e. somewhat lower than in H II regions. In the highly expanded, old Cirrus Nebula it yields $\sim$300 cm^{-3}, however, in the still young and compact Crab Nebula it is $\sim$1000 cm^{-3}. In PN, N_e reaches highest usually in the range of $\sim$10^4 cm^{-3}. In the H II region of M42, N_e is within the range of 1000–2000 cm^{-3} [169]. The estimation of N_e and T_e, based on the line intensities, is presented in [3].

Table 28.4 summarizes the distinguishing main features of the nebular spectra, documented in this atlas. This sample of objects is of course too small to be statistically relevant, but provides anyway a rough impression about the involved spectral features and excitation classes.

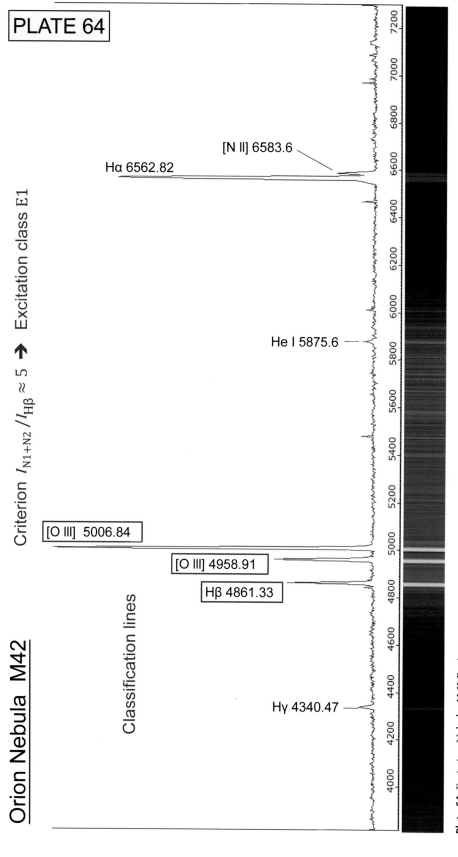

PLATE 64

Orion Nebula M42

Criterion $I_{N1+N2}/I_{H\beta} \approx 5$ → Excitation class E1

Classification lines

[O III] 5006.84

[O III] 4958.91

Hβ 4861.33

Hγ 4340.47

He I 5875.6

Hα 6562.82

[N II] 6583.6

Plate 64 Emission Nebula: H II Region

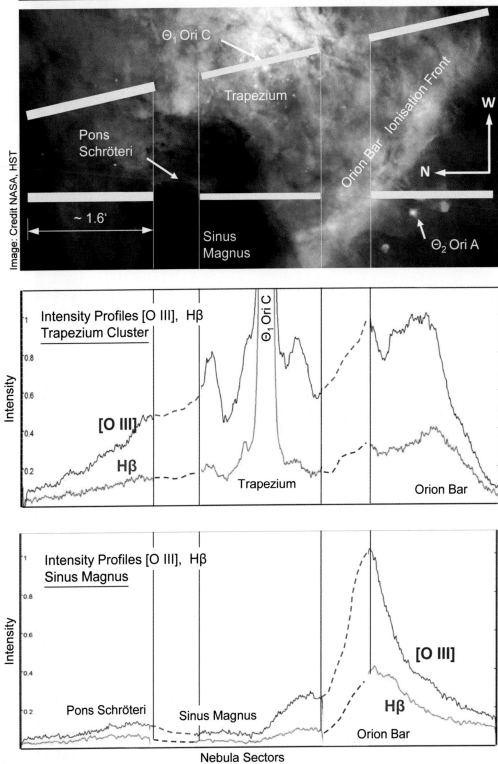

Plate 65 Intensity Profiles of Hβ and [O III] (λ 5007) in the Central Area of M42
(image: Credit: NASA, C.R. O'Dell, S. K. Wong)

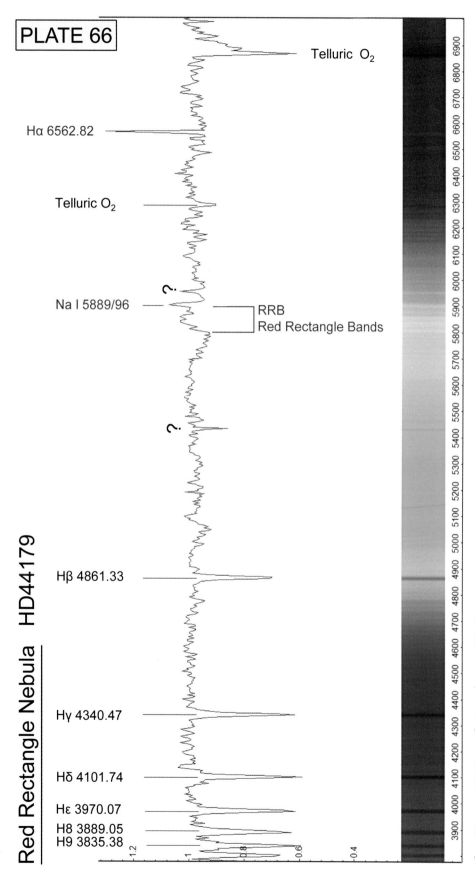

PLATE 66

Red Rectangle Nebula   HD44179

Telluric O$_2$

Hα 6562.82

Telluric O$_2$

Na I 5889/96

RRB
Red Rectangle Bands

Hβ 4861.33

Hγ 4340.47

Hδ 4101.74

Hε 3970.07

H8 3889.05

H9 3835.38

Plate 66 Post-AGB Star with Protoplanetary Nebula

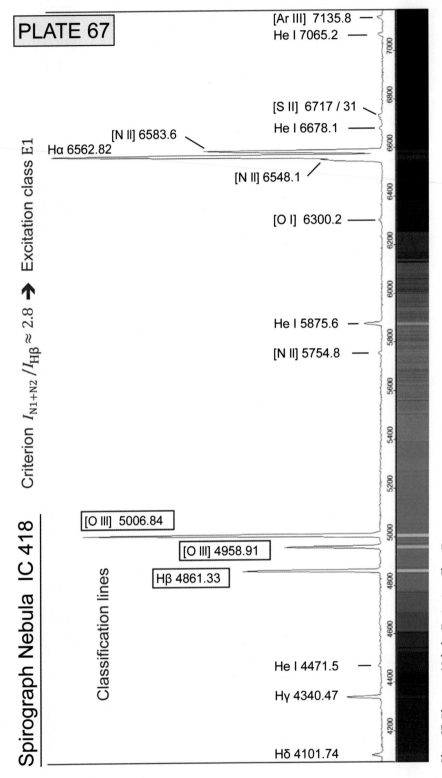

PLATE 67

[Ar III] 7135.8

He I 7065.2

[S II] 6717 / 31

He I 6678.1

[N II] 6583.6

Hα 6562.82

[N II] 6548.1

[O I] 6300.2

He I 5875.6

[N II] 5754.8

[O III] 5006.84

[O III] 4958.91

Hβ 4861.33

He I 4471.5

Hγ 4340.47

Hδ 4101.74

Excitation class E1

Criterion $I_{N1+N2}/I_{H\beta} \approx 2.8$

Spirograph Nebula IC 418

Classification lines

Plate 67 Planetary Nebula: Excitation Class E1

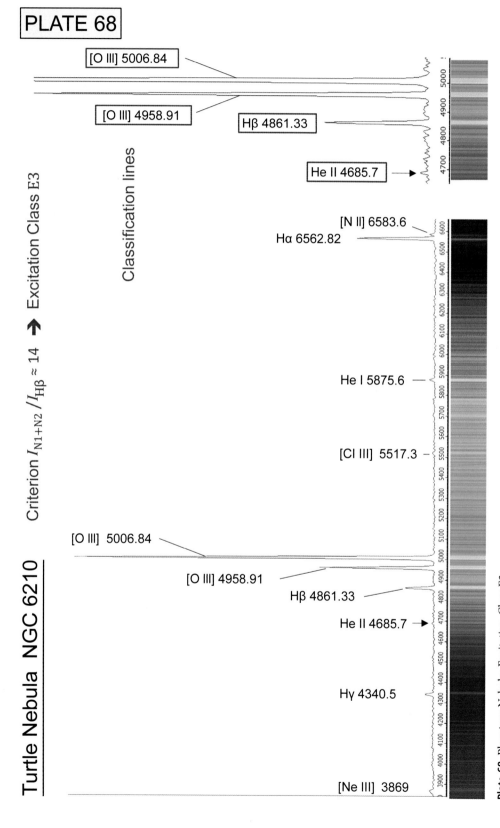

PLATE 68

[O III] 5006.84

[O III] 4958.91

Hβ 4861.33

He II 4685.7

Classification lines

Criterion $I_{N1+N2}/I_{Hβ} ≈ 14$ → Excitation Class E3

[N II] 6583.6

Hα 6562.82

He I 5875.6

[Cl III] 5517.3

[O III] 5006.84

[O III] 4958.91

Hβ 4861.33

He II 4685.7

Hγ 4340.5

[Ne III] 3869

Turtle Nebula NGC 6210

Plate 68 Planetary Nebula: Excitation Class E3

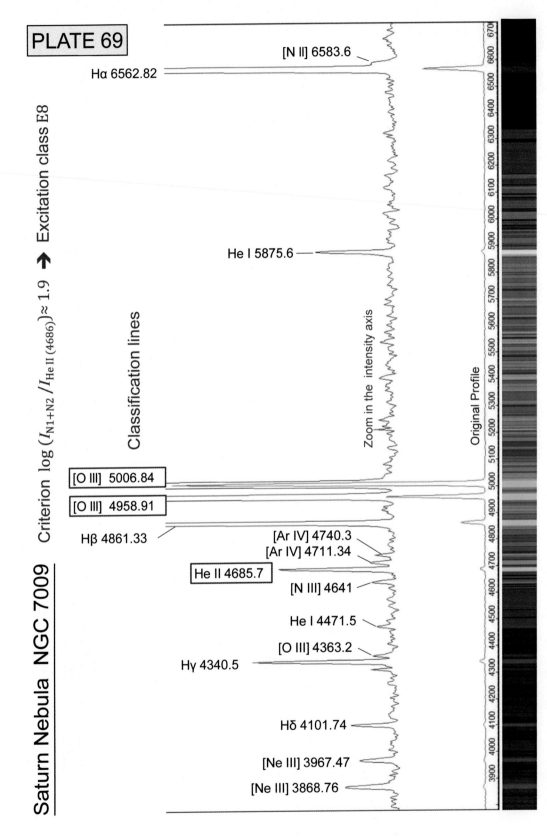

Plate 69 Planetary Nebula: Excitation Class E8

PLATE 69

Saturn Nebula NGC 7009

Excitation class E8

Criterion $\log \left(I_{\text{N1+N2}} / I_{\text{He II (4686)}} \right) \approx 1.9$

Classification lines

[N II] 6583.6

Hα 6562.82

He I 5875.6

Zoom in the intensity axis

Original Profile

[O III] 5006.84

[O III] 4958.91

Hβ 4861.33

[Ar IV] 4740.3

[Ar IV] 4711.34

He II 4685.7

[N III] 4641

He I 4471.5

[O III] 4363.2

Hγ 4340.5

Hδ 4101.74

[Ne III] 3967.47

[Ne III] 3868.76

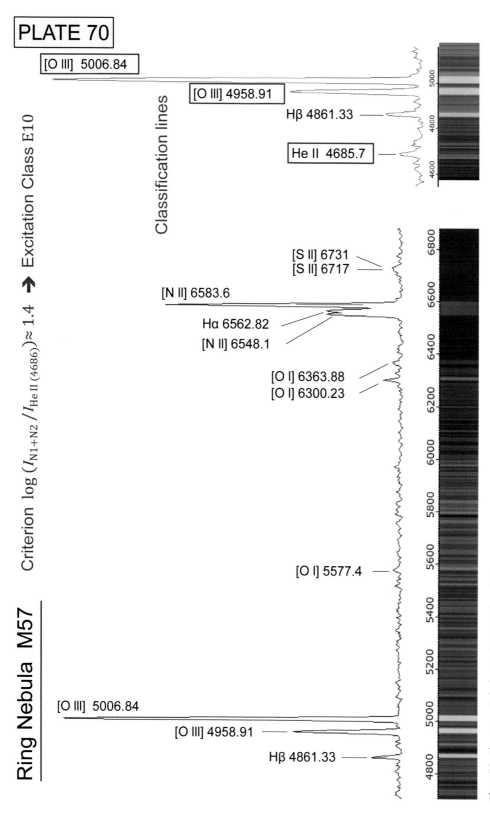

Plate 70 Planetary Nebula: Excitation Class E10

PLATE 71

Ring Nebula M57

Longitudinal section through M57. The DADOS 50 µm slit is approximately aligned with the longitudinal axis of the nebula. With Vspec the intensity profiles of the strongest emissions [O III] and [N II] have been generated along the entire length of the slit. The scales of the two profiles have been normalized on the peak value of [O III] = 1, and proportional to their line intensities in the spectrum.

[N II] chiefly generates the reddish color of the peripheral part, [O III] mainly produces the turquoise-blue color in the center of the nebula.

Image: Credit NASA, HST

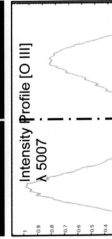

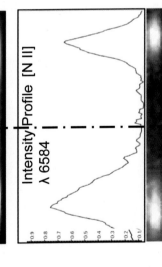

Plate 71 Intensity Profiles of [O III] and [N II] in the Longitudinal Axis of M57
(Image Credit: H. Bond *et al.*, Hubble Heritage Team (STScI/AURA), NASA)

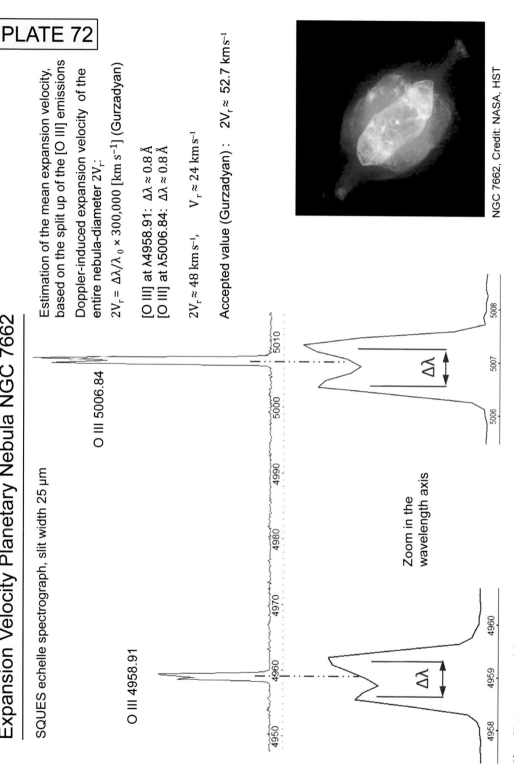

Expansion Velocity Planetary Nebula NGC 7662

SQUES echelle spectrograph, slit width 25 μm

O III 5006.84

O III 4958.91

Estimation of the mean expansion velocity, based on the split up of the [O III] emissions

Doppler-induced expansion velocity of the entire nebula-diameter $2V_r$:

$$2V_r = \Delta\lambda/\lambda_0 \times 300{,}000 \ [\text{km s}^{-1}] \ (\text{Gurzadyan})$$

[O III] at λ4958.91: $\Delta\lambda \approx 0.8 \ \text{Å}$
[O III] at λ5006.84: $\Delta\lambda \approx 0.8 \ \text{Å}$

$2V_r \approx 48 \ \text{km s}^{-1}, \quad V_r \approx 24 \ \text{km s}^{-1}$

Accepted value (Gurzadyan) : $2V_r \approx 52.7 \ \text{km s}^{-1}$

NGC 7662, Credit: NASA, HST

Zoom in the wavelength axis

$\Delta\lambda$

$\Delta\lambda$

Plate 72 Estimation of the Mean Expansion Velocity of a PN

PLATE 72

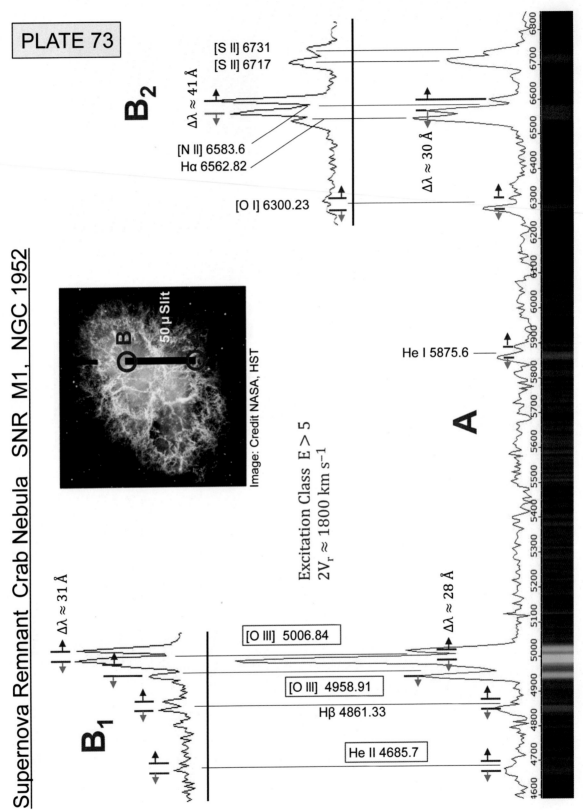

PLATE 73

Supernova Remnant Crab Nebula SNR M1, NGC 1952

Image: Credit NASA, HST

50 μ Slit

B₂

[S II] 6731
[S II] 6717

$\Delta\lambda \approx 41\,\text{Å}$

[N II] 6583.6
Hα 6562.82

[O I] 6300.23

$\Delta\lambda \approx 30\,\text{Å}$

He I 5875.6

A

Excitation Class E > 5
$2V_r \approx 1800\ \text{km s}^{-1}$

B₁

$\Delta\lambda \approx 31\,\text{Å}$

$\Delta\lambda \approx 28\,\text{Å}$

[O III] 5006.84

[O III] 4958.91

Hβ 4861.33

He II 4685.7

Plate 73 Emission Nebula: SNR Excitation Class E > 5

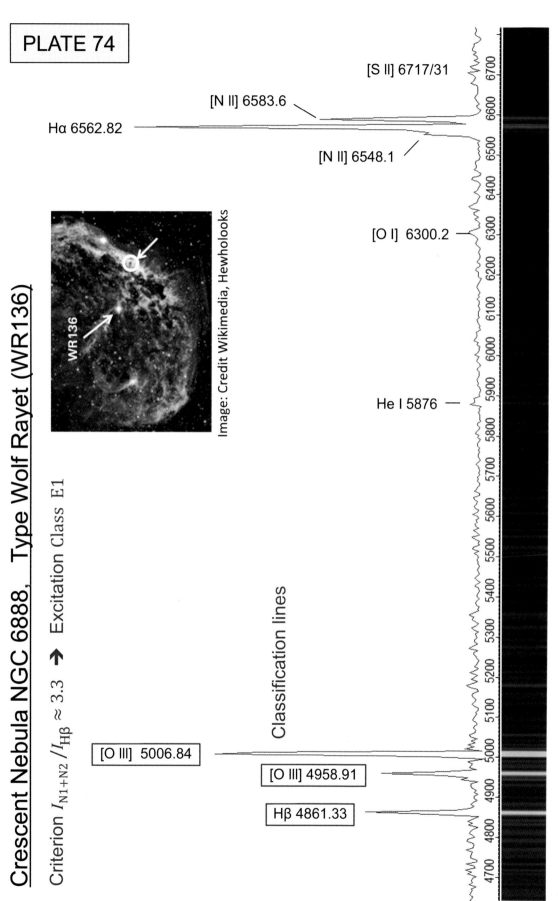

PLATE 74

Crescent Nebula NGC 6888, Type Wolf Rayet (WR136)

Criterion $I_{N1+N2}/I_{H\beta} \approx 3.3$ ↑ Excitation Class E1

Image: Credit Wikimedia, Hewholooks

WR 136

Classification lines

[S II] 6717/31

[N II] 6583.6

Hα 6562.82

[N II] 6548.1

[O I] 6300.2

He I 5876

[O III] 5006.84

[O III] 4958.91

Hβ 4861.33

Plate 74 Emission Nebula: Wolf–Rayet Excitation Class E1

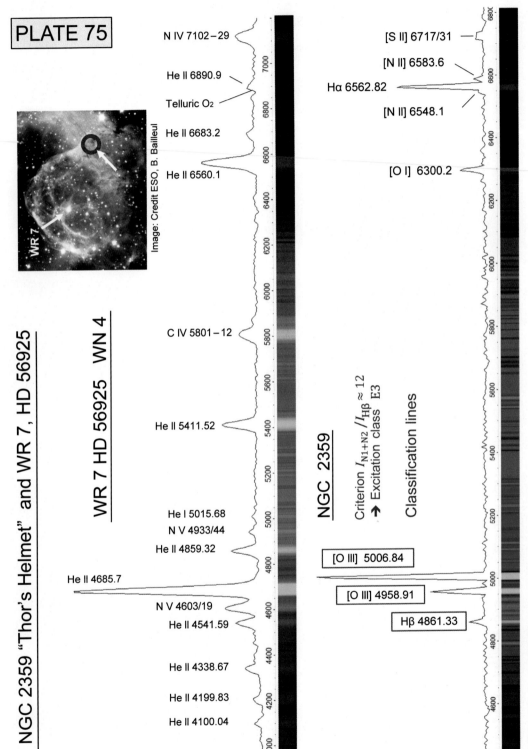

NGC 2359 "Thor's Helmet" and WR 7, HD 56925

PLATE 75

Image: Credit ESO, B. Bailleul

WR 7 HD 56925 WN 4

N IV 7102−29

He II 6890.9

Telluric O₂

He II 6683.2

He II 6560.1

C IV 5801−12

He II 5411.52

He I 5015.68
N V 4933/44
He II 4859.32

He II 4685.7

N V 4603/19
He II 4541.59

He II 4338.67

He II 4199.83

He II 4100.04

NGC 2359

[S II] 6717/31

[N II] 6583.6

Hα 6562.82

[N II] 6548.1

[O I] 6300.2

Criterion $I_{N1+N2}/I_{Hβ} \approx 12$
→ Excitation class E3

Classification lines

[O III] 5006.84

[O III] 4958.91

Hβ 4861.33

Plate 75 Emission Nebula: Wolf–Rayet Star as its Ionizing Source

CHAPTER

29 Reflectance Spectra of Solar System Bodies

29.1 Overview

The objects in our solar system are not self-luminous, and are visible only by reflected sunlight. Therefore, with exception of comets, these spectra always show not surprisingly the absorption lines of the Sun. On the other hand the spectral continua of the reflected profiles are overprinted, because certain molecules, for example CH_4 (methane), in the atmospheres of the large gas planets, are reflecting or absorbing the light differently within specific wavelength ranges (wavelength-dependent albedo). In 1871, Angelo Secchi discovered these dark, and at that time not yet identifiable, bands. The correct answer for this phenomenon was not found until 1930 by Rupert Wild and Vesto Slipher. This chapter presents a rough overview to this topic. At the professional level many more options exist to analyze such reflectance spectra.

29.2 Comments on Observed Spectra

According to their characteristics the planetary reflectance spectra are distributed here on three plates, and compared with the continuum of sunlight. All profiles have been recorded with DADOS $200\,L\,mm^{-1}$, at an elevation of some $30–40°$ above the horizon and are equally normalized to unity.

Plate 76 Reflectance Spectra of Mars and Venus

The extremely dense atmosphere of Venus generates on the surface a pressure of about 92 bar, i.e. approximately 90 times as high as on Earth. It consists of about 96% carbon dioxide (CO_2). The remaining shares are mainly nitrogen (N_2), water vapor (H_2O), and sulfur compounds in the form of sulfur dioxide (SO_2) and sulfuric acid (H_2SO_4) [170].

The extremely thin atmosphere of Mars, similar to Venus, consists of about 95% CO_2, but here under a surface pressure of only 0.006 bar, i.e. <1% of the value at the surface of the Earth. Here particularly the rocky surface of the planet might determine the reflectance properties.

In the displayed range on **Plate 76** the spectra of neither Venus nor Mars show significant deviations from the shape of the Sun's spectral continuum. In higher resolved spectra, of course, experts can recognize and analyze differences.

Plate 77 Reflectance Spectra of Jupiter and Saturn

The outer atmosphere of Jupiter consists of about 89% hydrogen and 10% helium. These gases have hardly any influence on the reflectance characteristics. The rest, however, mainly consisting of 0.3% methane (CH_4) and trace amounts of ammonia (NH_3), have a tremendous impact on the course of the continuum.

Saturn's outer atmosphere is composed slightly differently. It consists of about 96.3% hydrogen and only ~3.25% helium and some 0.45% of methane and trace amounts of ammonia and other gases. Impressive to see here, concentrated in the near-infrared range, are the very broad methane (CH_4) absorption gaps in the spectral continuum. Within this wavelength domain, these differences appear most pronounced in the ranges at $\lambda\lambda 6200$ and 7300.

Plate 78 Reflectance Spectra of Uranus, Neptune and Saturn's Moon Titan

The impressive reflectance spectra of the two outer gas giants look strikingly similar, whereby the absorptions are even more intense in the profile of Uranus. Due to the much greater distance from the Sun the temperatures are here so low that atmospheric components like ammonia are below their specific freezing point. Further, the share of methane is here somewhat higher. In contrast to Jupiter and Saturn, these spectral absorption features extend even to the short-wave spectral range. The outer atmosphere of Uranus consists of about 82.5% hydrogen, 15.2% helium and an increased share of 2.3% methane. For Neptune the composition is ~80.0% hydrogen, 19.0% helium and a significant 1.5% methane.

Titan, with a diameter of 5150 km is, after Ganymede, the second largest moon in the solar system, but the only one which has a dense atmosphere. The stratosphere chiefly consists of 98.4% nitrogen, 1.4% methane and the rest mainly hydrogen. The surface and outer mantle of the moon consist of ice and methane hydrate. Similar to the Earth, Titan has a liquid cycle, but working here with methane instead of water. Similar to Jupiter and Saturn the corresponding absorptions are limited here to the long-wavelength (red) section of the spectrum. In the reflectance spectra here, the absorptions of the solar spectrum are hardly recognizable.

Plate 79 Spectra of Comet C/2009 P1 Garradd

Comets, like all other objects in the solar system, reflect sunlight. However, on a comet's course into the inner solar system, core material increasingly evaporates, flowing out into the coma, and subsequently into the mostly separated plasma and dust tails. The increasing solar wind, containing highly ionized particles (mainly protons and helium cores), excites the molecules of the comet. Thus the reflected solar spectrum becomes more or less strongly overprinted with molecular emission bands – chiefly due to vaporized carbon compounds of the cometary material. The most striking features on **Plate 79** are the C_2 Swan bands, above all the band heads at $\lambda\lambda\lambda$5165, 5636 and 4737 (see also Section 33.4 and the comments to Plate 102). Frequently occurring emissions are CN (cyanide) at $\lambda\lambda$4380 and 3880, NH_2 (amidogen radicals) and C_3 at λ4056. In 1910 the discovery of cyan in the spectrum of comet Halley caused worldwide excitement, because by the traversing of the comet tail the formation of hydrocyanic acid in the Earth's atmosphere was feared. Sometimes sodium (Na I) lines can also be detected.

Overall, this results in very complex molecular composite spectra. Which of these numerous possible components and with which intensity they appear in the spectrum, is mainly dependent on the current intensity of the eruptions, the chemical composition of the core, as well as on our specific perspective on the coma, the plasma and dust tail. A spectrum of sunlight exclusively reflected by the dust tail becomes only slightly modified. A comprehensive catalog of possible cometary emission lines, can be found in [171] with additional information in [55].

On Plate 79, the coma profile of comet C/2009 P1 Garradd is presented, recorded on November 17, 2011 (C8, DADOS 200 L mm^{-1}, 3 × 900 s, 1 × 1 binning). A montage of the comet's profile is displayed, together with the C_2 Swan bands generated by a butane gas burner. This comparison clearly shows that in this spectrum of comet Garradd only two of four C_2 bandheads are visible at $\lambda\lambda$5165 and 4715. The missing two are overprinted by molecular CH, CN and NH_2 emissions. Absorptions of the solar spectrum are hardly recognizable here. This showed a test by a superposition of the comet profile with the solar spectrum. At the bottom of Plate 79 the zones of influence of the different molecules on the emissions of the spectrum are presented, according to tables in [170]. Those are based on spectral profiles, which were obtained with a high-resolution echelle spectrograph (R ~40,000). It is noticeable here that, apart from some isolated emissions, an overlapping of the influence zones can barely be found but clearly separated and strongly populated ranges dominate.

29.3 Reflectance Spectrum of a Total Lunar Eclipse

Marc Trypsteen recorded indirectly a transmission spectrum of the Earth's atmosphere by reflected sunlight during the total lunar eclipse, the "supermoon" of September 28, 2015 (Figure 29.1). In this way he obtained an exoplanet-like profile, showing also the "bio-signatures" and absorptions of

Figure 29.1 Lunar eclipse September 2015
(Credit: Marc Trypsteen)

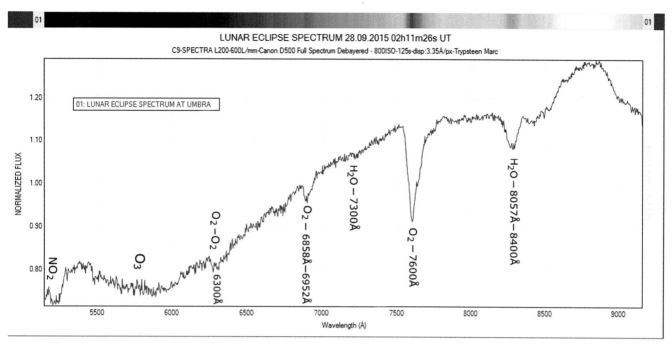

Figure 29.2 Reflectance spectrum of lunar eclipse, September 28, 2015 (Credit: Marc Trypsteen)

further molecules within the Earth's atmosphere as O_2, O_3, O_4, H_2O, NO_x (type NO_2/N_2O). Absorptions of the solar spectrum are probably extinguished by the refraction in the Earth's atmosphere and therefore barely visible in this profile. This spectrum (pseudo-continuum) was recorded just after the "Japanese lantern effect," at the very beginning of the entry into the umbra, to avoid influences of vertical absorption and lunar albedo effects (Figure 29.2). The slit was positioned in the region Mare Humorum east of crater Gassendi G, which provided an optimal albedo. In the region of $\lambda\lambda 5750$–6030 are the Chappuis ozone absorption bands and around $\lambda 6300$ those generated by O_4 or O_2–O_2 oxygen dimer. It is impressive to see here the increase of intensity towards the infrared domain, starting at $\sim\lambda 6000$ and creating in this way the characteristic impression of the "blood moon." The slightly increased intensity around $\sim\lambda 5600$ is caused by Rayleigh scattering. Recording information: C9, Spectra L200 with $600\,L\,mm^{-1}$ grating, debayered Canon d500 DSLR.

PLATE 76

Reflectance Spectra of Mars and Venus Compared to Sunlight

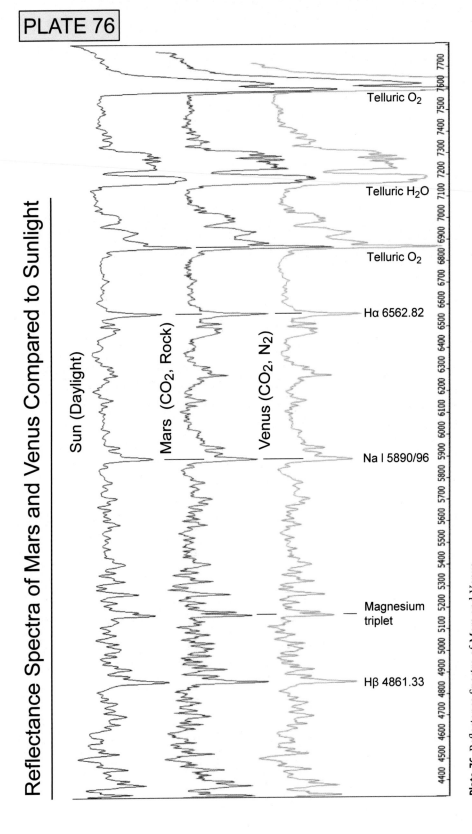

Plate 76 Reflectance Spectra of Mars and Venus

Reflectance Spectra of Jupiter and Saturn Compared to Sunlight

PLATE 77

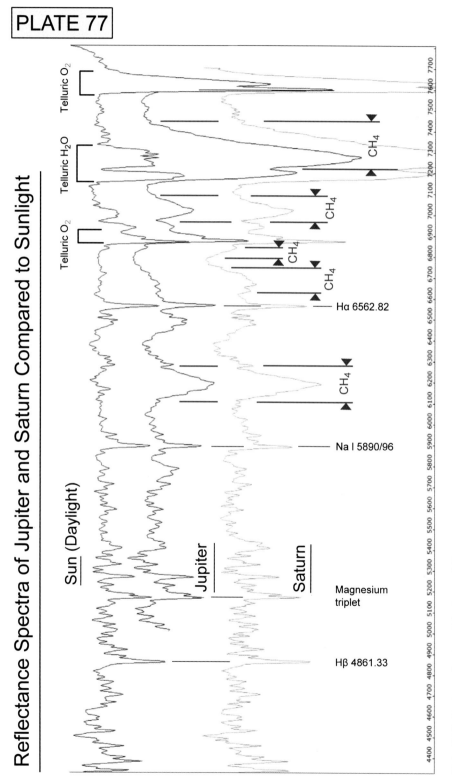

Plate 77 Reflectance Spectra of Jupiter and Saturn

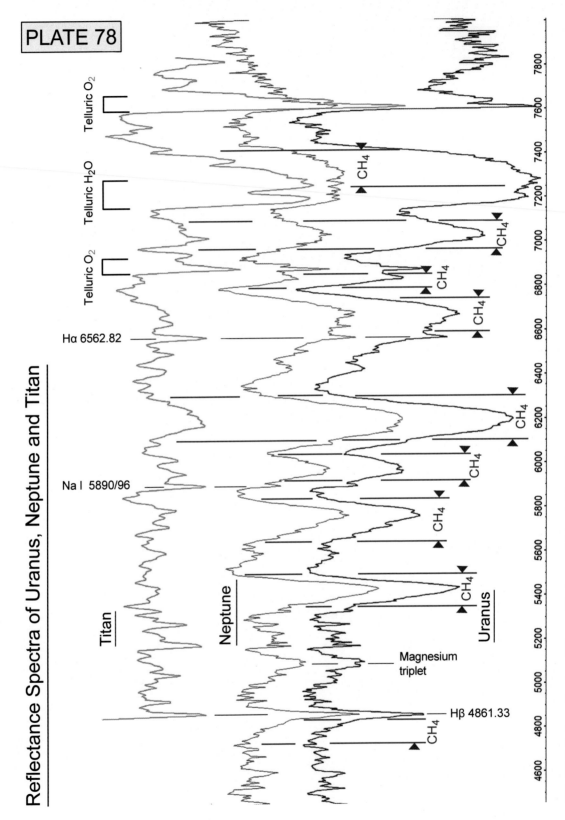

Plate 78 Reflectance Spectra of Uranus, Neptune and Saturn's Moon Titan

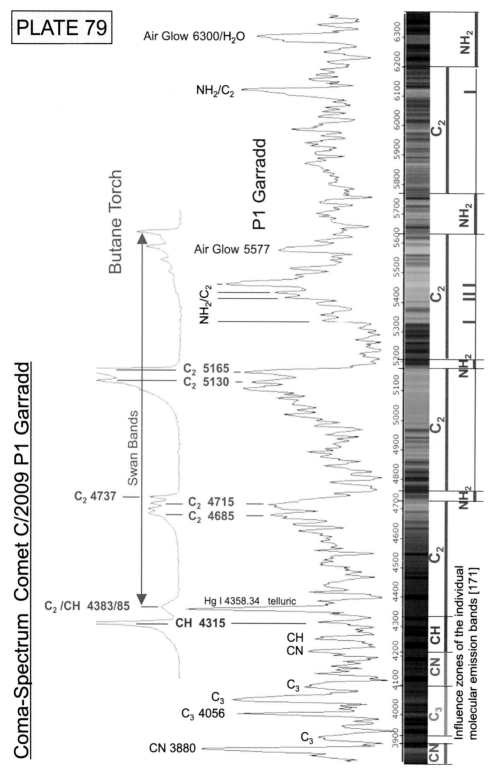

PLATE 79

Coma-Spectrum Comet C/2009 P1 Garradd

Air Glow 6300/H₂O

NH₂/C₂

P1 Garradd

Butane Torch

Air Glow 5577

NH₂/C₂
NH₂/C₂

C₂ 5165
C₂ 5130

Swan Bands

C₂ 4737

C₂ 4715
C₂ 4685

Hg I 4358.34   telluric

C₂/CH 4383/85

CH 4315

CH
CN

C₃

C₃
C₃ 4056

C₃

CN 3880

NH₂

C₂

NH₂

C₂

NH₂

C₂

NH₂

C₂

CH

CN

C₃

Influence zones of the individual
molecular emission bands [171]

CN

Plate 79 Spectra of Comet C/2009 P1 Garradd

30 Telluric Molecular Absorption

30.1 The Most Significant Molecular Absorptions by the Earth's Atmosphere

Plate 80 Telluric Absorptions in the Solar Spectrum

Plate 80: The region ~$\lambda\lambda6200$–7700 literally swarms with molecular H_2O and O_2 absorption bands caused by the Earth's atmosphere. Few of such "telluric" lines even appear beyond $\lambda5700$, unfortunately pretending to be stellar absorptions. "Tellus" is Latin and means "Earth." On Plate 80 the solar spectrum within the domain of $\lambda\lambda6800$–7800 is displayed ($900\,L\,mm^{-1}$). These features appear so impressively here that Fraunhofer labeled them with the letters A and B. At that time he could not know that these lines do not originate from the Sun, but arise due to absorption in the Earth's atmosphere.

For astronomers they are only a hindrance, unless they need fine water vapor lines of known wavelength to calibrate the spectra. These "calibration marks" are generated by complex molecular vibration processes, appearing as a broadly scattered swarm of absorptions. Atmospheric physicists deduce from the H_2O absorptions moisture profiles of the troposphere. The O_2 bands (mainly Fraunhofer A and B) allow conclusions about the layered temperatures of the atmosphere [172].

For most amateurs just the awareness is important, that particularly in low resolved spectra, the line identification within this area requires great caution. In most of the cases only the $H\alpha$ line unambiguously overtops the "jungle" of the telluric absorption lines and bands. This is particularly a problem for the middle spectral classes, whose spectra show here numerous metal lines. Exceptions are the Wolf–Rayet and Be stars and those exhibiting

a mass loss due to strong P Cygni profiles. In the latter two cases, at least the helium line He I at $\lambda6678$ can safely be identified.

Stars of the late K and all M classes, as well as the carbon stars, predominantly radiate in the infrared part of the spectrum. Therefore particularly intense titanium monoxide (TiO) absorption bands are capable of overprinting these telluric lines. Further, the reflectance spectra of the large gas planets show here mainly the impressive gaps in the continua of their spectral profiles. These telluric bands and lines can be reduced to a certain extent with a relatively large effort – for example, by subtraction of synthetically produced standard profiles of the telluric lines (see e.g. the Vspec manual) and further by comparison with profiles of standard stars. Very suitable for this purpose is the spectral class A0 which shows just few and very weak stellar lines within this range. Complicating facts are the influences of weather conditions, elevation angle of the object etc. For further information refer to [172], [173]. Moreover, there is a highly recommendable freeware program by P. Schlatter, which allows a large suppression of H_2O lines [174].

30.2 Telluric H_2O Absorptions around the $H\alpha$ line

Plate 81 Telluric H_2O Absorptions around $H\alpha$ in the Spectra of the Sun and δ Sco

Plate 81: Telluric water vapor lines are often used for the calibration of highly resolved spectra, particularly around the $H\alpha$ line. Unfortunately even in highly resolved profiles, most of the H_2O absorptions appear blended, mostly with other water or sometimes also stellar metal lines, which become strongly influenced by the specific spectral class of the object. Therefore just the few remaining, undisturbed H_2O absorptions are really useful for a high precision calibration. As an orientation aid these are shown here at two spectral profiles (R ~20,000) with different spectral classes – δ Scorpii, B0.5 IVe and the Sun, G2 V, recorded with the SQUES echelle spectrograph. The identification and the wavelengths of the lines are based here on SpectroWeb [9]. In case of a precision calibration for other spectral classes it is worth to review the location and blends of the water lines on a similar profile in SpectroWeb. The indication of the wavelength is limited here to unblended lines which are suitable for the calibration.

30.3 Telluric O_2 Absorptions within Fraunhofer A and B Bands

Plate 82 Telluric Fraunhofer A and B Absorptions in Spectra of the Sun and δ Sco

Highly resolved profiles, recorded by the SQUES Echelle spectrograph. The line identification and the according wavelengths are based on [175].

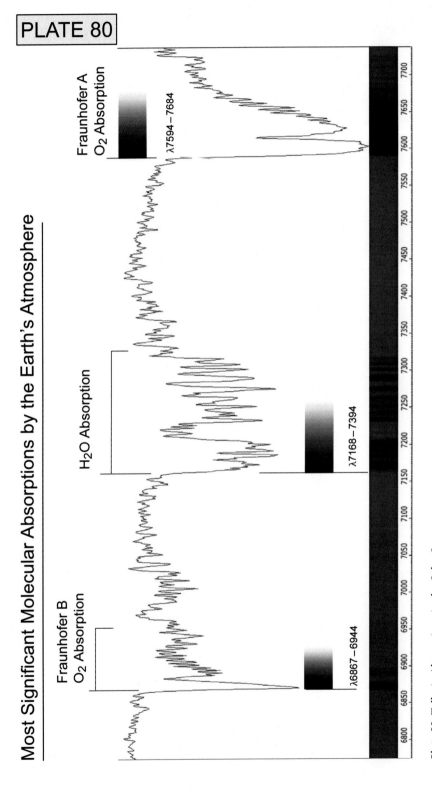

Plate 80 Telluric Absorptions in the Solar Spectrum

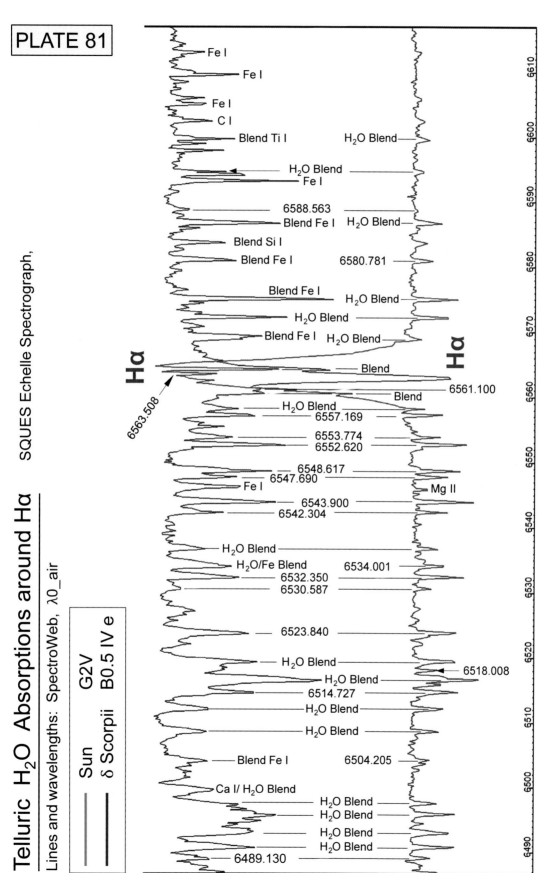

Telluric H₂O Absorptions around Hα

Lines and wavelengths: SpectroWeb, λ0_air

SQUES Echelle Spectrograph,

PLATE 81

Sun — G2V
δ Scorpii — B0.5 IV e

Fe I
Fe I
Fe I
C I
Blend Ti I   H₂O Blend
                    H₂O Blend
                    Fe I
              6588.563
              Blend Fe I   H₂O Blend
       Blend Si I
       Blend Fe I   6580.781
         Blend Fe I   H₂O Blend
         H₂O Blend
         Blend Fe I   H₂O Blend
Hα   Blend
                                        6561.100
6563.508   Blend
              H₂O Blend
              6557.169
              6553.774
              6552.620
            6548.617
         6547.690
    Fe I   Mg II
            6543.900
            6542.304
       H₂O Blend
       H₂O/Fe Blend   6534.001
         6532.350
         6530.587
         6523.840
       H₂O Blend
         H₂O Blend   6518.008
         6514.727
       H₂O Blend
       H₂O Blend
  Blend Fe I   6504.205
  Ca I/ H₂O Blend
         H₂O Blend
         H₂O Blend
         H₂O Blend
         H₂O Blend
  6489.130

Hα

6610 6600 6590 6580 6570 6560 6550 6540 6530 6520 6510 6500 6490

Plate 81 Telluric H₂O Absorptions around Hα in the Spectra of the Sun and δ Sco.

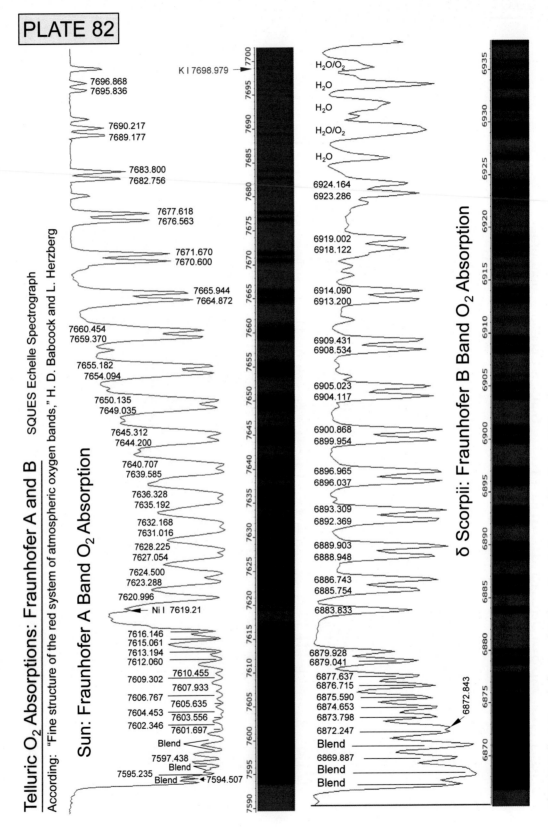

PLATE 82

Telluric O₂ Absorptions: Fraunhofer A and B — SQUES Echelle Spectrograph

According: "Fine structure of the red system of atmospheric oxygen bands," H. D. Babcock and L. Herzberg

Sun: Fraunhofer A Band O₂ Absorption

7696.868
7695.836

7690.217
7689.177

7683.800
7682.756

7677.618
7676.563

7671.670
7670.600

7665.944
7664.872

7660.454
7659.370

7655.182
7654.094

7650.135
7649.035

7645.312
7644.200

7640.707
7639.585

7636.328
7635.192

7632.168
7631.016

7628.225
7627.054

7624.500
7623.288

7620.996

Ni I 7619.21

7616.146
7615.061
7613.194
7612.060

7609.302
7610.455
7607.933
7606.767
7605.635
7604.453
7603.556
7602.346
7601.697

Blend
7597.438
Blend
7595.235 Blend 7594.507

K I 7698.979

H₂O/O₂
H₂O
H₂O
H₂O/O₂
H₂O

δ Scorpii: Fraunhofer B Band O₂ Absorption

6924.164
6923.286

6919.002
6918.122

6914.090
6913.200

6909.431
6908.534

6905.023
6904.117

6900.868
6899.954

6896.965
6896.037

6893.309
6892.369

6889.903
6888.948

6886.743
6885.754

6883.833

6879.928
6879.041

6877.637
6876.715
6875.590
6874.653
6873.798
6872.247

6872.843

Blend
6869.887
Blend
Blend

Plate 82 Telluric Fraunhofer A and B Absorptions in Spectra of the Sun and δ Sco

CHAPTER

31

The Night Sky Spectrum

31.1 Introduction

Mainly due to light pollution and airglow the night sky is significantly brightened and the astronomical observations thereby seriously hampered. The light pollution is mainly caused by street lamps and other terrestrial light sources. This light is chiefly scattered backwards by molecules and particles in various layers of the atmosphere.

The airglow is generated during the day in the atmosphere by photoionization of oxygen as a result of solar UV radiation and chemical reaction chains. At night recombination takes place, causing emission lines at discrete frequencies. Really striking here are only two of the [O I] lines at $\lambda\lambda 5577.35$ and 6300.23. The latter is visible only under a very good night sky, and therefore lacking in the spectrum of Plate 83. An additional contribution is supplied by the mesospheric sodium layer (Chapter 32). Airglow, however, also includes the rotational and vibrational bands of OH molecules in the near infrared range, detected in 1950 by A. B. Meinel. Further influences, particularly over the continuum of the night sky spectrum, are the diffuse galactic light (DGL), the integrated starlight (ISL) and the reflected zodiacal light (ZL). The latter may also contribute features of the solar spectrum.

31.2 Effects on the Spectrum

Depending on the quality of the night sky, at very long exposure times the emissions of airglow and light pollution may disturbingly superimpose the recorded signal of the investigated object. The effects of airglow to the spectrum are much less harmful than the light pollution, which may consist of dozens of emission lines, depending on the type of terrestrial light sources in the wider surroundings, as well as on meteorological factors. Under a perfect night sky, at least theoretically, only the airglow remains visible in the spectrum. The removal of light pollution and airglow is explained in [3].

Plate 83 The Night Sky Spectrum

Plate 83: This night sky spectrum was recorded to clean the disturbed spectrum of M1 (Plate 73). It was taken at about 610 m above sea level, under a rather moderately good rural sky with a limiting magnitude of about $m_v = 4$–5 and an elevation angle of ~50°. For such long exposures (30 minutes) the C8 was equipped with a long dew cap to reduce the influence of scattered light from the surrounding area. The lighting of the residential area here consists of gas discharge lamps, of which some but not all spectral lines appear in the spectrum. Major roads with sodium vapor lamps are some 100 m distant and located off a direct line of sight. The literature consistently refers to sodium high pressure lamps as the most harmful source of pollution, since they produce a bell-shaped emission around the Fraunhofer D lines $\lambda\lambda 5890/96$ (refer to Plate 86). This feature can hardly effectively be filtered out without impairing the desired signal. The night sky of the Canary Islands is protected by the famous sky contamination Act "La Ley del Cielo," which strongly limits the use of sodium high pressure lamps on the islands of Tenerife and La Palma [176].

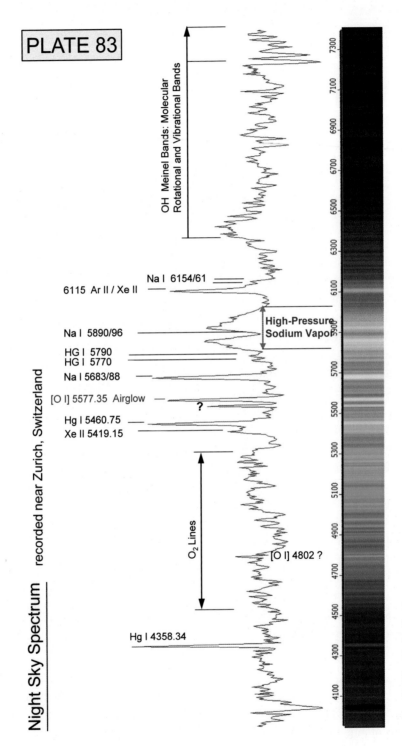

PLATE 83

Night Sky Spectrum — recorded near Zurich, Switzerland

OH Meinel Bands: Molecular Rotational and Vibrational Bands

Na I 6154/61

6115 Ar II / Xe II

High-Pressure Sodium Vapor

Na I 5890/96

HG I 5790
HG I 5770

Na I 5683/88

[O I] 5577.35 Airglow

?

Hg I 5460.75
Xe II 5419.15

O₂ Lines

[O I] 4802 ?

Hg I 4358.34

Plate 83 The Night Sky Spectrum

CHAPTER

32 The Mesospheric Sodium Layer

32.1 Overview

In the upper part of the mesosphere, at an altitude of ~85–100 km, a layer is formed which is enriched with sodium atoms and partially excited by the solar radiation. Due to fluorescence, the "sodium nightglow" is generated, a faint yellowish shine at the wavelengths of the Fraunhofer D_1 and D_2 lines. It contributes to the general airglow and is observable from space (Figure 32.1). Above this layer, the sodium is ionized, below it is bound in molecules, both conditions very effectively prevent the fluorescence effect.

This layer was discovered in 1929 by Vesto Slipher. The sodium there consists of accumulated debris from meteors. Today this layer is of great importance for ground-based optical astronomy. It helps to compensate the image degradation due to atmospheric turbulence (seeing). By laser excitation artificial "guide stars" are produced there, which allow control of the adaptive optics of large telescopes (Figure 32.2). In the 1970s many such layers were discovered at other heights, which, for example consist of iron, potassium and calcium.

32.2 Spectroscopic Detection of the Sodium Layer

To prove this sodium layer with amateur means, a simple test exists which has been described by P. Schlatter [177]. In order to do this, in the twilight after sunset one needs to aim the open spectrograph without any optics to the zenith. An optimum time to record this "sodium flash," a solar elevation angle of –8° below the astronomical horizon has been proven. Then the observation location is located

Figure 32.1 Sodium nightglow and airglow, observed from ISS (Credit: NASA)

Figure 32.2 Artificial Laser Guide Star, a part of the VLT's Adaptive Optics System
(Credit: ESO/G, Hüdepohl)

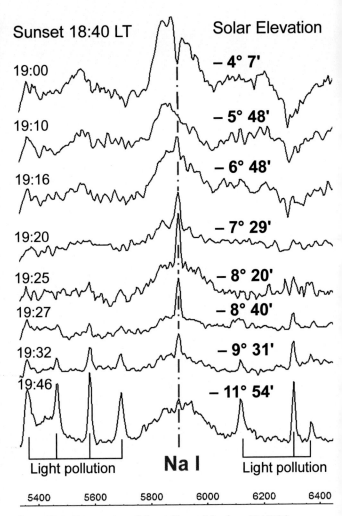

Figure 32.3 Time series of the sodium nightglow, DADOS 200 L mm^{-1}

already in the dark while the mesosphere is still excited by the solar radiation.

The time series in Figure 32.3 was recorded October 14, 2014 with DADOS 200 L mm^{-1}. It shows how shortly after sunset, the Na I absorption becomes slowly transformed into an emission. Subsequently this feature weakens and gives way to the ordinary light pollution spectrum (Chapter 31). The exposure times must be adjusted to the light conditions of the respective phase: Atik 314L+, ~10–300 s, 2 × 2 binning.

CHAPTER

33 Terrestrial and Calibration Light Sources

33.1 Spectra of Common Gas Discharge Lamps

Gas discharge lamps play a key role for astronomers. They can be useful, for example, to calibrate spectral profiles. The wavelengths of their emission lines are known with very high accuracy. They are also a source of interference from the light pollution of road and municipal lighting. For beginners they are also useful exercise objects, particularly during cloudy nights. Unfortunately all gas discharge lamps, even those for calibration purposes, must be operated with usually >100 V. If supplied with mains voltage, this may pose a serious safety risk. Particularly for outdoor operation, at least minimal electrical knowledge is required about relevant safety measures such as isolation transformers or GFCI devices. The safest and exclusively recommended solutions are autonomous low power DC/AC inverters 12 V DC/120 V AC or 12 V DC/230 V AC, which can be directly connected to the low voltage supply of today's telescopes.

Plate 84 Gas Discharge Lamp (Ne)

Orange neon glow lamps are used as indicator lights, for example for stoves, irons, etc. They produce a large number of emission lines, the intense ones of which unfortunately remain limited to the red range of the spectrum. In earlier times they were very popular among amateurs, applying these lamps as calibration light sources. Nowadays, for the calibration of broadband overview spectra and even higher-resolved profiles, low cost solutions with modified glow starters are preferred (Section 33.2). (Recorded using DADOS 200 L mm^{-1}.)

Plate 85 Gas Discharge Lamp (ESL)

In earlier times amateurs sometimes used energy saving lamps (ESL) for calibration purposes, mainly as a complement to the neon glow lamps, which are limited to the red part of the spectrum. Today their importance is reduced to experimental and educational purposes. Energy saving lamps contain several gases and substances performing different tasks – among others, fluorescent substances, usually rare earth metals. The mixture depends largely on the color the lamp produces. Only the intense lines of the auxiliary gases, for example argon (Ar), xenon (Xe), and mercury (Hg), are useful for calibration purposes. Unfortunately, some of these line positions are located very close together and are therefore difficult to distinguish, such as Ar II at λ6114.92 and Xe II at λ6115.08. (Recorded using DADOS 200 L mm^{-1}.)

Plate 86 Gas Discharge Lamp (Na)

This lamp is used worldwide for street lighting. The sodium generates light in the domain of the Fraunhofer D1 and D2 lines. Due to the high gas pressure it is properly speaking not a monochromatic light. The continuum, as a result of pressure and collision broadening, as well as self-absorption effects, shows a bell-like shape. Unfortunately for astronomers this spectral feature generates probably the worst part of light pollution because, unlike the discrete emission lines, it cannot simply be removed from the spectra. Depending on the specific product, added auxiliary gases, for example xenon, may produce some discrete emission lines in the spectrum. (Recorded using DADOS 200 L mm^{-1}.)

Plate 87 Gas Discharge Lamp (Xe)

Such high pressure gas discharge lamps are used for lighting of sport stadiums, position lamps on mountain tops etc. The various required auxiliary gases are interesting here. (Recorded using DADOS 200 L mm^{-1}.)

33.2 Spectra of Glow Starters Modified as Calibration Light Sources

In the professional sector or even at the senior amateur level, relatively expensive hollow cathode lamps are applied for the calibration of highly resolved spectra, producing a fine raster of, for example, iron–argon or thorium emission lines. Martin Huwiler, in collaboration with the author, has developed an effective low cost alternative by modified glow starters for fluorescent lamps. They consist of small bulbs of glass, filled with either just one or a mixture of several noble gases and contain a bimetal switch. To modify it to a calibration light source, with a supply voltage of 230 V, the current must be limited by a series resistor of some 24–30 kΩ, so a reasonable light intensity is achieved without causing a short cut by closing of the switch. For necessary safety measures see Section 33.1.

Table 33.1 shows the noble gases in some glow starters from different suppliers, which of course may change the gas mixture without any announcement. With some of the bulbs sometimes emissions of H I and O I may be observed, probably generated by dissociation of water vapor within the bulb. Numerous additional lines, which are not identified here, are generated by alloying constituents, coatings and dopants, such as tungsten (W), lanthanum (La), cerium (Ce), hafnium (Hf), as well as Fe, Ti, Cr, Sn, Ni.

Plates 88–90 were recorded using DADOS 200 L mm^{-1}.

Plate 88 Calibration Lamp with Ne and Xe

Glow starter Philips S10

Plate 89 Calibration Lamp with Ar

Glow starter OSRAM ST 111

Plate 90 Calibration Lamp with Ar, Ne and He

Glow starter RELCO SC480

In addition to argon, SC480 contains also neon and helium. Thus, in the optical part of the spectrum ~270

Table 33.1 Noble gases in the glow starters from various suppliers (2014)

Glow Starter	Osram ST 111	GE 155/500	Philips S10	Relco SC480	Relco SC422
Noble gases	Ar	Ar	Ne, Xe	Ne, Ar, He	Ar

lines are available, which is even sufficient for the calibration of highly resolved echelle spectra. Calibration lamps with this gas mixture are even applied in professional astronomy.

Plate 91 Calibration Lamp with Ar, Ne and He

Glow starter RELCO SC480, DADOS 900 L mm^{-1}, λλ3850–6300

Plate 92 Calibration Lamp with Ar, Ne and He

Glow starter RELCO SC480, DADOS 900 L mm^{-1}, λλ5900–8000

Plates 93–98 Calibration Lamp with Ar, Ne and He

For the modified glow starter RELCO SC480, the charts of Plates 93–98 show the individual emission lines, distributed to the orders of the SQUES echelle spectrograph (R ≈ 20,000). These are well suited for the calibration of individual, reasonably straight and horizontally running echelle orders. The appropriate wavelengths have been copied from the tool "elements" of the Vspec software. Line pairs which appear scarcely resolved here and therefore can hardly be selectively marked with the cursor, are referred with "blend." Lines which, due to overlapping orders, appear twice are labeled in color. As a supplement and with relatively high probability, in the rather sparsely populated orders 39, 40 and 43, Ti emissions could be identified, a "?" indicates these lines. Within the order 29 (λλ7600–7900), only two evaluable emission lines are generated by noble gases, which, however, are supplemented here by a striking oxygen triplet. To improve the visibility of the weaker calibration lines a stronger zoom into the intensity axis is required. In order to avoid transmission errors, the data can be copied by "ctrl c" from pdf or Word files and be directly transferred, for example, into the Vspec calibration popup field, by "shift insert."

33.3 Calibration and Processing of an Entire Echelle Spectrum

Plates 99–101 Calibration Lamp with Ar, Ne and He

Images with SQUES Echelle Orders

These plates show the numerous echelle orders as they are depicted by the SQUES spectrograph. They are applied for the relatively challenging calibration and processing of an entire echelle spectrum, for example by ISIS or MIDAS software. Per individual order in most of the cases significantly more than three emissions can be processed. The numbering of the lines corresponds to the wavelengths listed in Table 33.2. For the overall processing and calibration of the entire spectrum, for example with ISIS, it is recommended to select the more intense lines only. Applying different echelle spectrographs with comparable or even the same grating configuration, some of the emission lines, located here at the outermost edges of the orders, may appear to be moved into the neighboring orders. Further, the line intensity, not only of the hydrogen Balmer series, is noticeably subjected to the manufacturing tolerance of the starter production.

33.4 Spectra of Hydrocarbon Flames

Plate 102 Swan Bands/Hydrocarbon gas flames in Comparison with a Montage of Spectra

Butane gas torch, Comet Hyakutake and Carbon Star WZ Cassiopeiae

The Swan bands, described in Chapter 15 and Section 29.2, are of great importance for astrophysics. They are generated, for example, in the cool atmospheres of carbon stars as absorptions and in the comets of the solar system, appearing as emission bands. Molecular band spectra are generated by complex rotational and vibrational processes of heated molecules. The required excitation energy to generate Swan bands is relatively low. Therefore this spectral detail can be easily simulated by the intense combustion of hydrocarbons with do-it-yourself equipment from the hardware store! Plate 102 shows the Swan bands generated with a butane torch. In this plate, spectra of the butane gas flame, comet Hyakutake and the carbon star WZ Cassiopeiae (cut out from Plate 29) are superposed. The course of the Hyakutake profile (ESO/Caos) was schematically scaled and transferred into the drawing.

The amazingly similar emission spectra of the comet Hyakutake and the butane gas flame (C_4H_{10}) within the domain

Figure 33.1 Recording of an acetylene flame, workshop Urs Flükiger

of the C_2 Swan bands are striking! That is why for both cases the same physical process is taking effect. WZ Cassiopeiae shows the Swan bands in absorption instead of emission. Therefore the course of this profile runs inversely to the others. The wavelengths of the most intensive band heads are $\lambda\lambda6191, 5636, 5165, 4737$ and 4383. Further a number of fainter C_2 absorptions are still recognizable. Some of these lines are also visible in the profiles of the carbon stars in Chapter 15. The line identification is based amongst others on [55], [171]. With similar results spectra of acetylene flames (C_2H_2), have also been recorded (Figure 33.1). Figure 33.2 shows a highly-resolved echellogram of a butane gas flame, covering the orders 38–52.

33.5 Terrestrial Lightning Discharges

Since the beginnings of spectroscopy in the nineteenth century attempts have been made to obtain spectra of lightning discharges. Also, at the beginning of the twentieth century, well-known astronomers were involved, like Pickering and Slipher [178]. Figure 33.3 shows the spectrum of a lightning flash that has hit the ground at a distance of approximately 220 m from the observer. During the night and by chance Martin Huwiler filmed this thunderstorm through the closed window pane with a Canon G1X and a 300 L mm^{-1} transmission grating mounted in front of the camera lens.

Plate 103 Spectra of Terrestrial Lightning Discharges

Integrated Light of Several Lightning Discharges

Plate 103: For this purpose, the C8 telescope with the DADOS spectrograph and the Atik 314L+ was built up at night in the dark living room. This setup pointed through the closed window

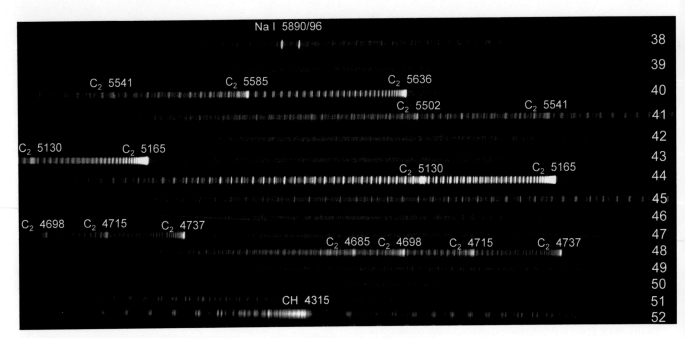

Figure 33.2 SQUES echellogram of a butane gas flame generating C_2 Swan and CH bands

Figure 33.3 Spectrum of a lightning stroke, recorded with a transmission grating from a distance of ~220 m
(Credit: M. Huwiler)

towards the approaching thunderstorm on the western horizon. Originating from Vesto M. Slipher, the idea was to gain a spectra of lightning, reaching temperatures of >30,000 K in the conducting channel. On the evening of July 24, 1917 with this intention, Slipher directed the spectrograph at the Lowell Observatory in Flagstaff to a thunderstorm which raged at a distance of about 10 km above the south slopes of the San Francisco Peaks [178].

Here three shots each of 180 s in the 2×2 binning have been processed. For each image the integrated light of some 5–10 lightning discharges was recorded. Since the cloud base was very low, on all shots the light pollution had to be subtracted. Striking here is the very intense CN emission at approximately $\lambda 3900$. According to [179] this is a characteristic feature of discharges with relatively long-lasting "continuing" currents, causing an increased fire risk. This phenomenon is generated mainly by the type of "cloud-to-ground lightning." In the 1980s, in the USA there was even discussion about detecting this spectral feature with satellites, as an early warning criterion for possible forest fires [179]. On all recorded shots, with the integrated light of several lightning strikes, this CN emission appeared in comparable intensity. Otherwise, most of the lines in the lightning spectrum are rather complex, broad blends of O II, N II, O I, N I as well as emissions of the H-Balmer series. The raw profile is here calibrated only in wavelength.

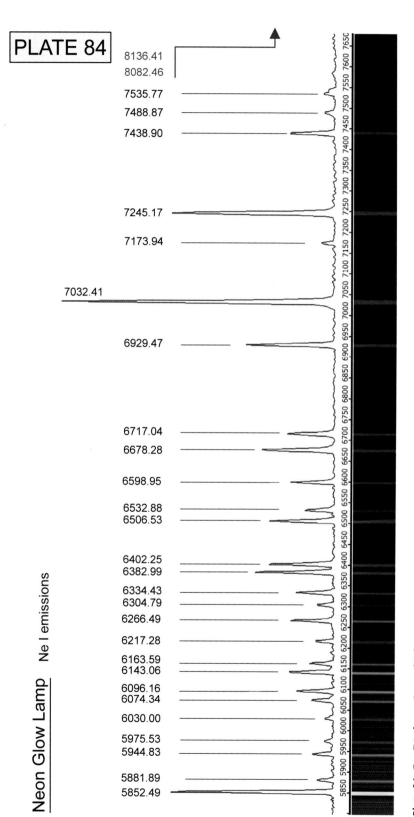

PLATE 84

Neon Glow Lamp   Ne I emissions

8136.41
8082.46
7535.77
7488.87
7438.90
7245.17
7173.94
7032.41
6929.47
6717.04
6678.28
6598.95
6532.88
6506.53
6402.25
6382.99
6334.43
6304.79
6266.49
6217.28
6163.59
6143.06
6096.16
6074.34
6030.00
5975.53
5944.83
5881.89
5852.49

Plate 84 Gas Discharge Lamp (Ne)

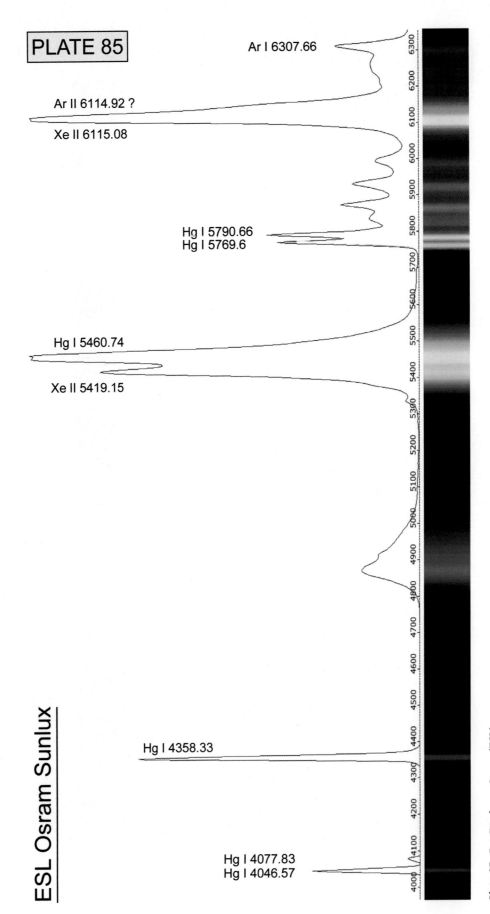

PLATE 85

Ar I 6307.66

Ar II 6114.92 ?

Xe II 6115.08

Hg I 5790.66
Hg I 5769.6

Hg I 5460.74

Xe II 5419.15

ESL Osram Sunlux

Hg I 4358.33

Hg I 4077.83
Hg I 4046.57

Plate 85 Gas Discharge Lamp (ESL)

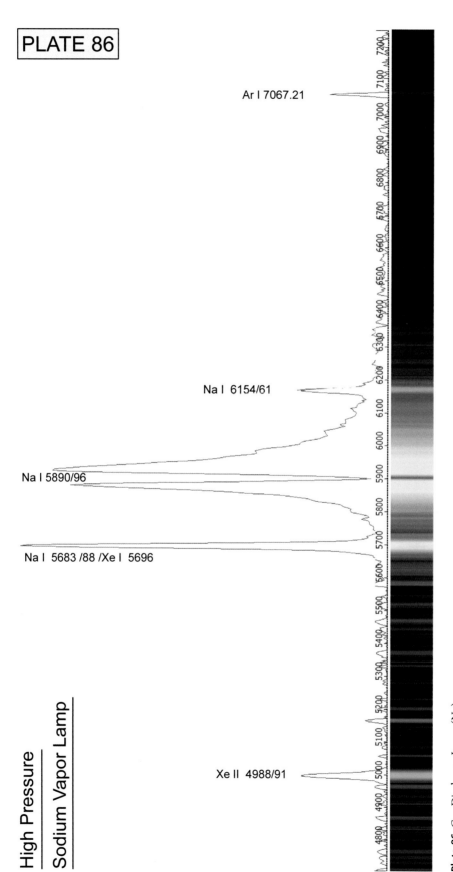

Plate 86 Gas Discharge Lamp (Na)

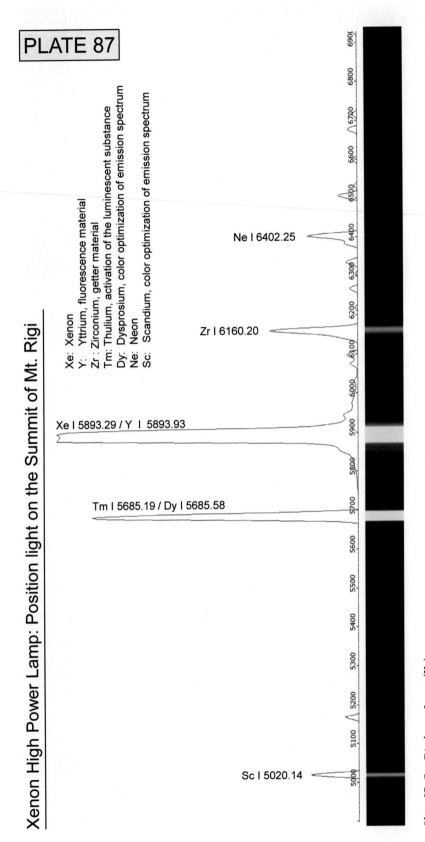

PLATE 87

Xenon High Power Lamp: Position light on the Summit of Mt. Rigi

Xe: Xenon
Y: Yttrium, fluorescence material
Zr : Zirconium, getter material
Tm: Thulium, activation of the luminescent substance
Dy: Dysprosium, color optimization of emission spectrum
Ne: Neon
Sc: Scandium, color optimization of emission spectrum

Ne I 6402.25

Zr I 6160.20

Xe I 5893.29 / Y I 5893.93

Tm I 5685.19 / Dy I 5685.58

Sc I 5020.14

Plate 87 Gas Discharge Lamp (Xe)

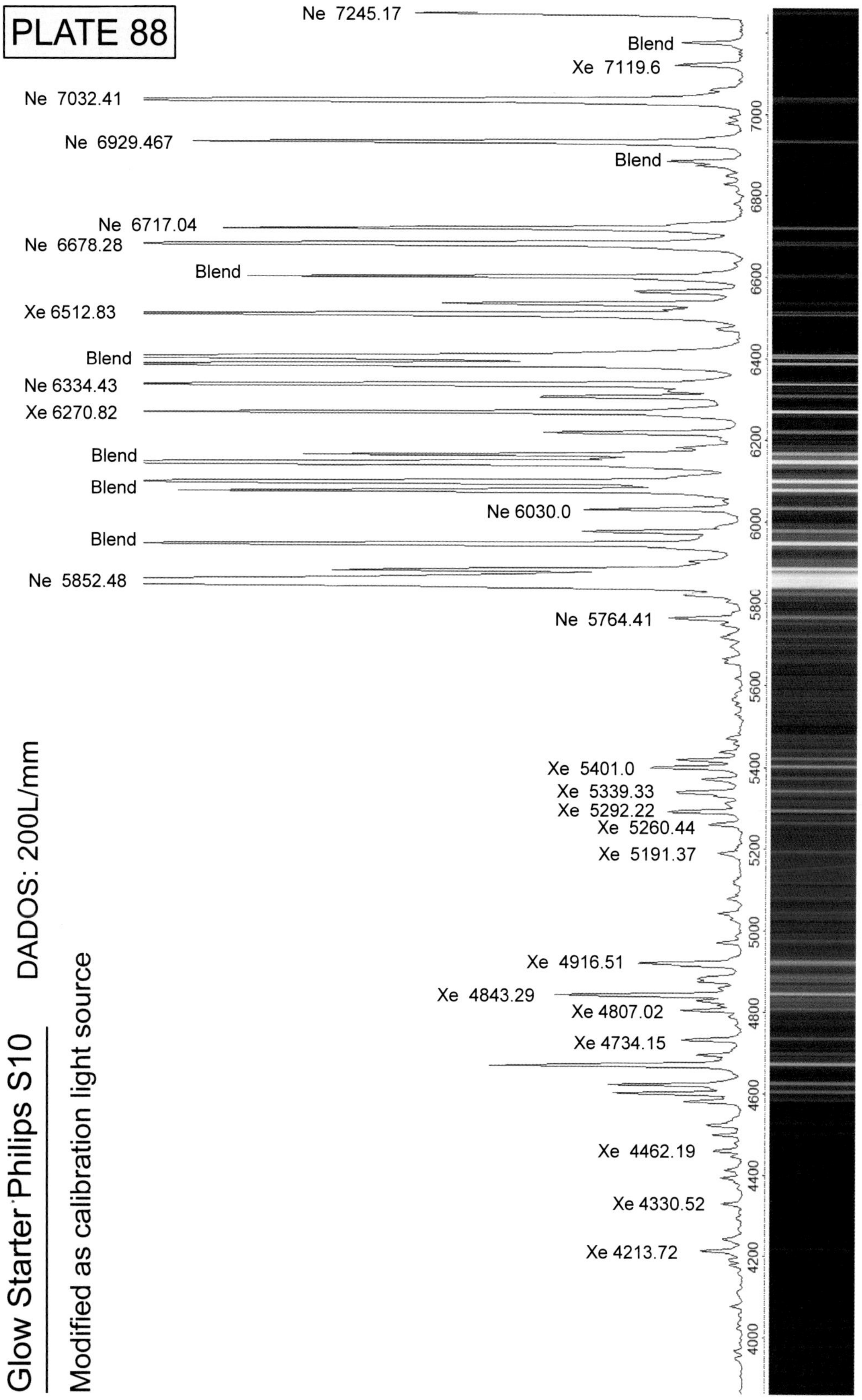

PLATE 88

Glow Starter·Philips S10   DADOS: 200L/mm

Modified as calibration light source

Ne 7245.17

Blend

Xe 7119.6

Ne 7032.41

Ne 6929.467

Blend

Ne 6717.04

Ne 6678.28

Blend

Xe 6512.83

Blend

Ne 6334.43

Xe 6270.82

Blend

Blend

Ne 6030.0

Blend

Ne 5852.48

Ne 5764.41

Xe 5401.0

Xe 5339.33

Xe 5292.22

Xe 5260.44

Xe 5191.37

Xe 4916.51

Xe 4843.29

Xe 4807.02

Xe 4734.15

Xe 4462.19

Xe 4330.52

Xe 4213.72

7000

6800

6600

6400

6200

6000

5800

5600

5400

5200

5000

4800

4600

4400

4200

4000

Plate 88 Calibration Lamp with Ne and Xe

PLATE 89

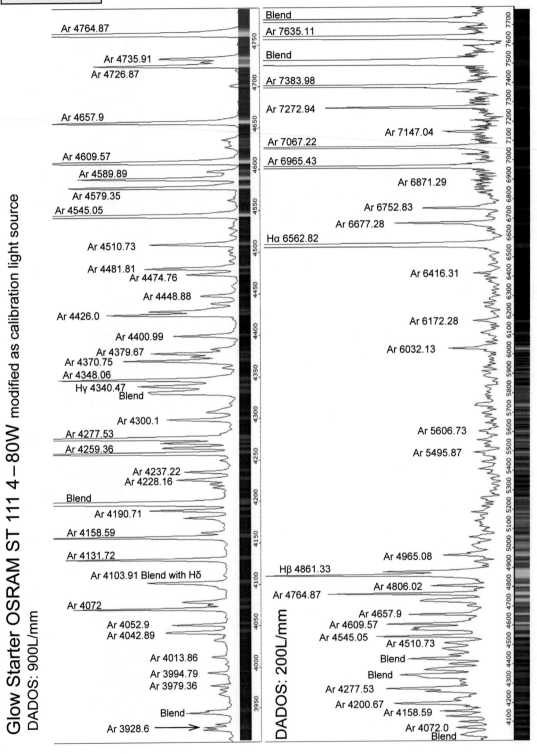

Glow Starter OSRAM ST 111 4 – 80W modified as calibration light source
DADOS: 900L/mm

Ar 4764.87
Ar 4735.91
Ar 4726.87
Ar 4657.9
Ar 4609.57
Ar 4589.89
Ar 4579.35
Ar 4545.05
Ar 4510.73
Ar 4481.81
Ar 4474.76
Ar 4448.88
Ar 4426.0
Ar 4400.99
Ar 4379.67
Ar 4370.75
Ar 4348.06
Hγ 4340.47
Blend
Ar 4300.1
Ar 4277.53
Ar 4259.36
Ar 4237.22
Ar 4228.16
Blend
Ar 4190.71
Ar 4158.59
Ar 4131.72
Ar 4103.91 Blend with Hδ
Ar 4072
Ar 4052.9
Ar 4042.89
Ar 4013.86
Ar 3994.79
Ar 3979.36
Blend
Ar 3928.6

DADOS: 200L/mm

Blend
Ar 7635.11
Blend
Ar 7383.98
Ar 7272.94
Ar 7147.04
Ar 7067.22
Ar 6965.43
Ar 6871.29
Ar 6752.83
Ar 6677.28
Hα 6562.82
Ar 6416.31
Ar 6172.28
Ar 6032.13
Ar 5606.73
Ar 5495.87
Ar 4965.08
Hβ 4861.33
Ar 4806.02
Ar 4764.87
Ar 4657.9
Ar 4609.57
Ar 4545.05
Ar 4510.73
Blend
Blend
Ar 4277.53
Ar 4200.67
Ar 4158.59
Ar 4072.0
Blend

Plate 89 Calibration Lamp with Ar

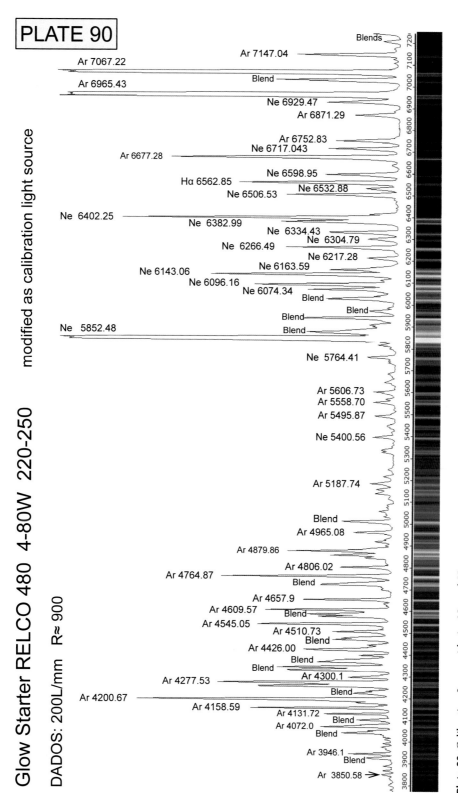

PLATE 90

modified as calibration light source

Glow Starter RELCO 480 4-80W 220-250

DADOS: 200L/mm R≈ 900

Blends

Ar 7147.04

Ar 7067.22

Blend

Ar 6965.43

Ne 6929.47

Ar 6871.29

Ar 6752.83
Ne 6717.043

Ar 6677.28

Ne 6598.95

Hα 6562.85

Ne 6532.88

Ne 6506.53

Ne 6402.25

Ne 6382.99

Ne 6334.43

Ne 6304.79

Ne 6266.49

Ne 6217.28

Ne 6163.59

Ne 6143.06

Ne 6096.16

Ne 6074.34

Blend

Blend

Blend

Blend

Ne 5852.48

Blend

Ne 5764.41

Ar 5606.73
Ar 5558.70
Ar 5495.87

Ne 5400.56

Ar 5187.74

Blend

Ar 4965.08

Ar 4879.86

Ar 4806.02

Ar 4764.87

Blend

Ar 4657.9

Ar 4609.57

Blend

Ar 4545.05

Ar 4510.73

Blend

Ar 4426.00

Blend

Blend

Ar 4277.53

Ar 4300.1

Blend

Ar 4200.67

Ar 4158.59

Ar 4131.72

Blend

Ar 4072.0

Blend

Ar 3946.1

Blend

Ar 3850.58

Plate 90 Calibration Lamp with Ar, Ne and He

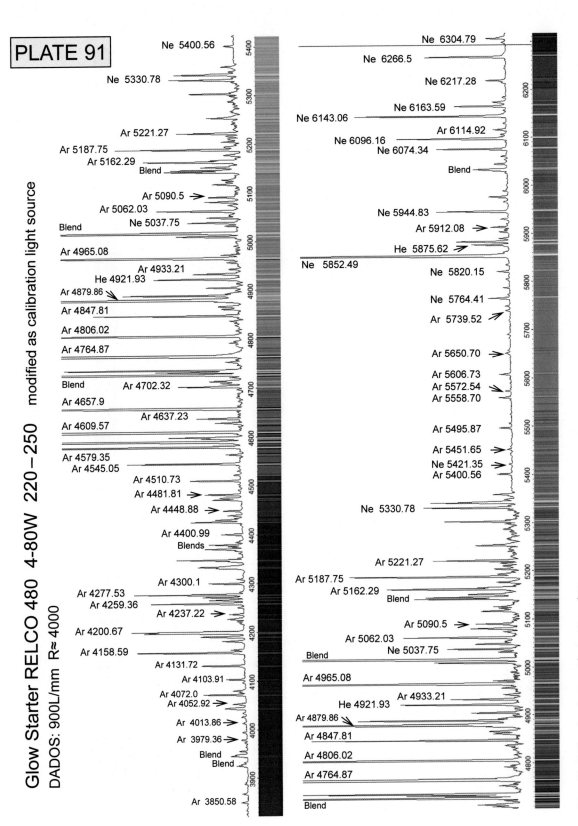

PLATE 91

Glow Starter RELCO 480 4-80W 220–250   modified as calibration light source
DADOS: 900L/mm R≈4000

Ne 5400.56
Ne 5330.78
Ar 5221.27
Ar 5187.75
Ar 5162.29
Blend
Ar 5090.5
Ar 5062.03
Ne 5037.75
Blend
Ar 4965.08
Ar 4933.21
He 4921.93
Ar 4879.86
Ar 4847.81
Ar 4806.02
Ar 4764.87
Blend   Ar 4702.32
Ar 4657.9
Ar 4609.57
Ar 4637.23
Ar 4579.35
Ar 4545.05
Ar 4510.73
Ar 4481.81
Ar 4448.88
Ar 4400.99
Blends
Ar 4300.1
Ar 4277.53
Ar 4259.36
Ar 4237.22
Ar 4200.67
Ar 4158.59
Ar 4131.72
Ar 4103.91
Ar 4072.0
Ar 4052.92
Ar 4013.86
Ar 3979.36
Blend
Blend
Ar 3850.58

Ne 6304.79
Ne 6266.5
Ne 6217.28
Ne 6163.59
Ne 6143.06
Ar 6114.92
Ne 6096.16
Ne 6074.34
Blend
Ne 5944.83
Ar 5912.08
He 5875.62
Ne 5852.49
Ne 5820.15
Ne 5764.41
Ar 5739.52
Ar 5650.70
Ar 5606.73
Ar 5572.54
Ar 5558.70
Ar 5495.87
Ar 5451.65
Ne 5421.35
Ar 5400.56
Ne 5330.78
Ar 5221.27
Ar 5187.75
Ar 5162.29
Blend
Ar 5090.5
Ar 5062.03
Ne 5037.75
Blend
Ar 4965.08
Ar 4933.21
He 4921.93
Ar 4879.86
Ar 4847.81
Ar 4806.02
Ar 4764.87
Blend

Plate 91 Calibration Lamp with Ar, Ne and He

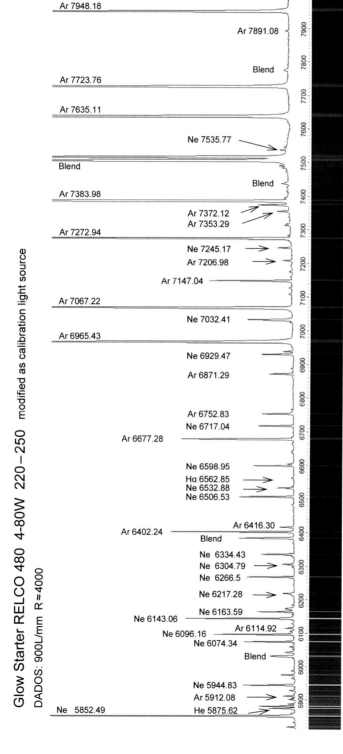

PLATE 92

Blend
Ar 7948.18
Ar 7891.08
Blend
Ar 7723.76
Ar 7635.11
Ne 7535.77
Blend
Blend
Ar 7383.98
Ar 7372.12
Ar 7353.29
Ar 7272.94
Ne 7245.17
Ar 7206.98
Ar 7147.04
Ar 7067.22
Ne 7032.41
Ar 6965.43
Ne 6929.47
Ar 6871.29
Ar 6752.83
Ne 6717.04
Ar 6677.28
Ne 6598.95
Hα 6562.85
Ne 6532.88
Ne 6506.53
Ar 6416.30
Ar 6402.24
Blend
Ne 6334.43
Ne 6304.79
Ne 6266.5
Ne 6217.28
Ne 6163.59
Ne 6143.06
Ar 6114.92
Ne 6096.16
Ne 6074.34
Blend
Ne 5944.83
Ar 5912.08
Ne 5852.49
He 5875.62

Glow Starter RELCO 480 4-80W 220 – 250 modified as calibration light source

DADOS: 900L/mm R≈4000

Plate 92 Calibration Lamp with Ar, Ne and He

RELCO 480 SQUES Echelle Orders 28–32

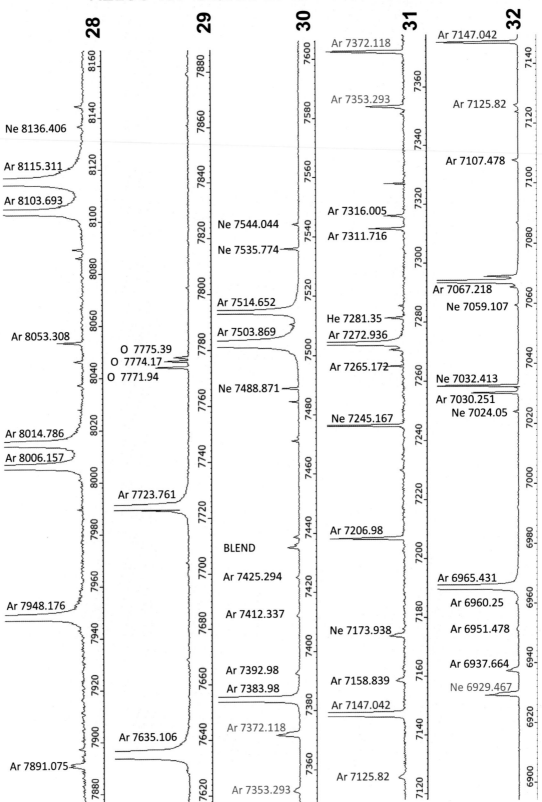

28

Ne 8136.406

Ar 8115.311

Ar 8103.693

Ar 8053.308

Ar 8014.786

Ar 8006.157

Ar 7948.176

Ar 7891.075

29

O 7775.39
O 7774.17
O 7771.94

Ar 7723.761

Ar 7635.106

30

Ne 7544.044

Ne 7535.774

Ar 7514.652

Ar 7503.869

Ne 7488.871

BLEND

Ar 7425.294

Ar 7412.337

Ar 7392.98
Ar 7383.98

Ar 7372.118

Ar 7353.293

31

Ar 7372.118

Ar 7353.293

Ar 7316.005

Ar 7311.716

He 7281.35

Ar 7272.936

Ar 7265.172

Ne 7245.167

Ar 7206.98

Ne 7173.938

Ar 7158.839

Ar 7147.042

Ar 7125.82

32

Ar 7147.042

Ar 7125.82

Ar 7107.478

Ar 7067.218

Ne 7059.107

Ne 7032.413

Ar 7030.251

Ne 7024.05

Ar 6965.431

Ar 6960.25

Ar 6951.478

Ar 6937.664

Ne 6929.467

Plates 93–98 Calibration Lamp with Ar, Ne and He

PLATE 94 RELCO 480 SQUES Echelle Orders 33–37

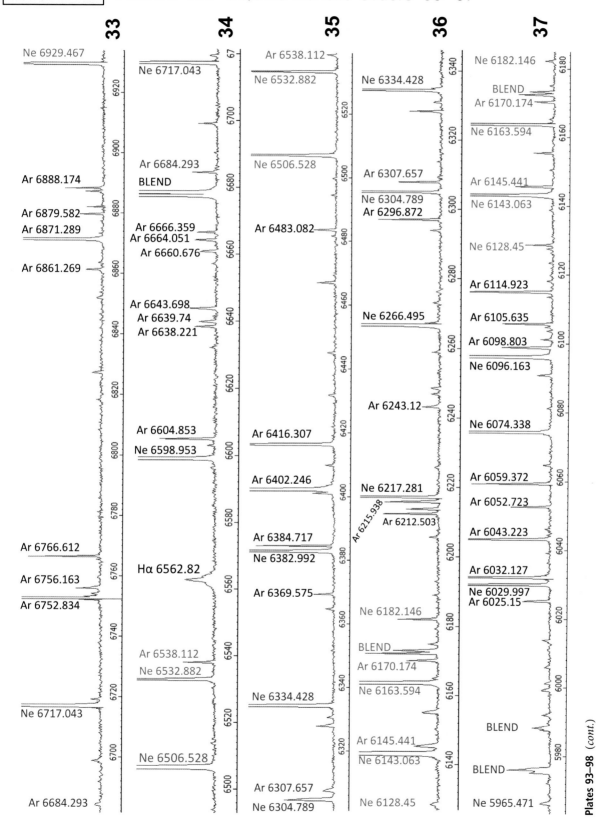

33 34 35 36 37

Ne 6929.467

Ne 6717.043

Ar 6538.112
Ne 6532.882

Ne 6334.428

Ne 6182.146

BLEND
Ar 6170.174

Ne 6163.594

Ar 6888.174

Ar 6684.293
BLEND

Ne 6506.528

Ar 6307.657

Ne 6304.789
Ar 6296.872

Ar 6145.441
Ne 6143.063

Ar 6879.582
Ar 6871.289

Ar 6666.359
Ar 6664.051
Ar 6660.676

Ar 6483.082

Ne 6128.45

Ar 6861.269

Ar 6114.923

Ar 6643.698
Ar 6639.74
Ar 6638.221

Ne 6266.495

Ar 6105.635

Ar 6098.803

Ne 6096.163

Ar 6243.12

Ne 6074.338

Ar 6604.853
Ne 6598.953

Ar 6416.307

Ar 6059.372

Ar 6402.246

Ne 6217.281

Ar 6215.938

Ar 6212.503

Ar 6052.723

Ar 6766.612

Hα 6562.82

Ar 6384.717
Ne 6382.992

Ar 6043.223

Ar 6756.163

Ar 6032.127

Ar 6752.834

Ar 6369.575

Ne 6182.146

Ne 6029.997
Ar 6025.15

BLEND
Ar 6170.174

Ar 6538.112
Ne 6532.882

Ne 6163.594

Ne 6717.043

Ne 6334.428

Ar 6145.441
Ne 6143.063

BLEND

Ne 6506.528

Ar 6307.657
Ne 6304.789

Ne 6128.45

BLEND

Ar 6684.293

Ne 5965.471

6920 6900 6880 6860 6840 6820 6800 6780 6760 6740 6720 6700

67 6700 6680 6660 6640 6620 6600 6580 6560 6540 6520 6500

6520 6500 6480 6460 6440 6420 6400 6380 6360 6340 6320

6340 6320 6300 6280 6260 6240 6220 6200 6180 6160 6140

6180 6160 6140 6120 6100 6080 6060 6040 6020 6000 5980

RELCO 480 SQUES Echelle Orders 38–42

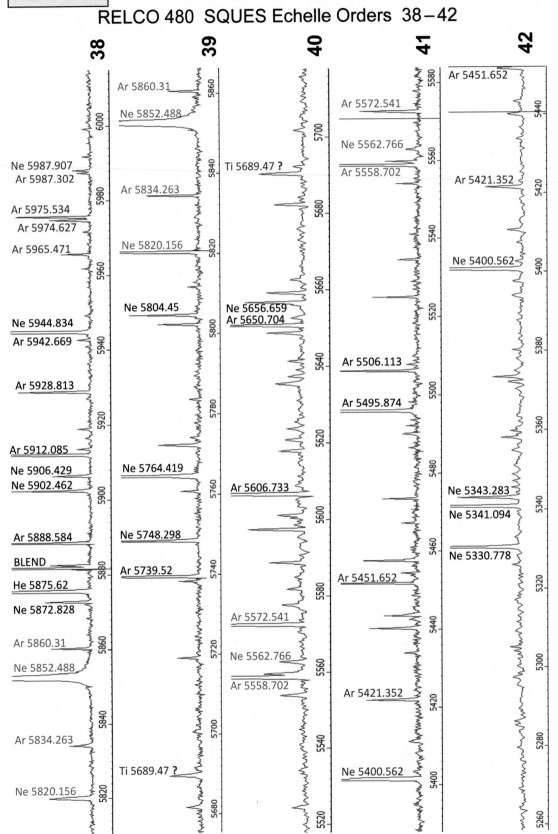

38

Ne 5987.907
Ar 5987.302

Ar 5975.534
Ar 5974.627

Ar 5965.471

Ne 5944.834
Ar 5942.669

Ar 5928.813

Ar 5912.085
Ne 5906.429
Ne 5902.462

Ar 5888.584

BLEND

He 5875.62

Ne 5872.828

Ar 5860.31

Ne 5852.488

Ar 5834.263

Ne 5820.156

39

Ar 5860.31

Ne 5852.488

Ar 5834.263

Ne 5820.156

Ne 5804.45

Ne 5764.419

Ne 5748.298

Ar 5739.52

Ti 5689.47 ?

40

Ti 5689.47 ?

Ne 5656.659
Ar 5650.704

Ar 5606.733

Ar 5572.541

Ne 5562.766

Ar 5558.702

41

Ar 5572.541

Ne 5562.766

Ar 5558.702

Ar 5506.113

Ar 5495.874

Ar 5451.652

Ar 5421.352

Ne 5400.562

42

Ar 5451.652

Ar 5421.352

Ne 5400.562

Ne 5343.283

Ne 5341.094

Ne 5330.778

Plates 93–98 *(cont.)*

PLATE 96 RELCO 480 SQUES Echelle Orders 43–47

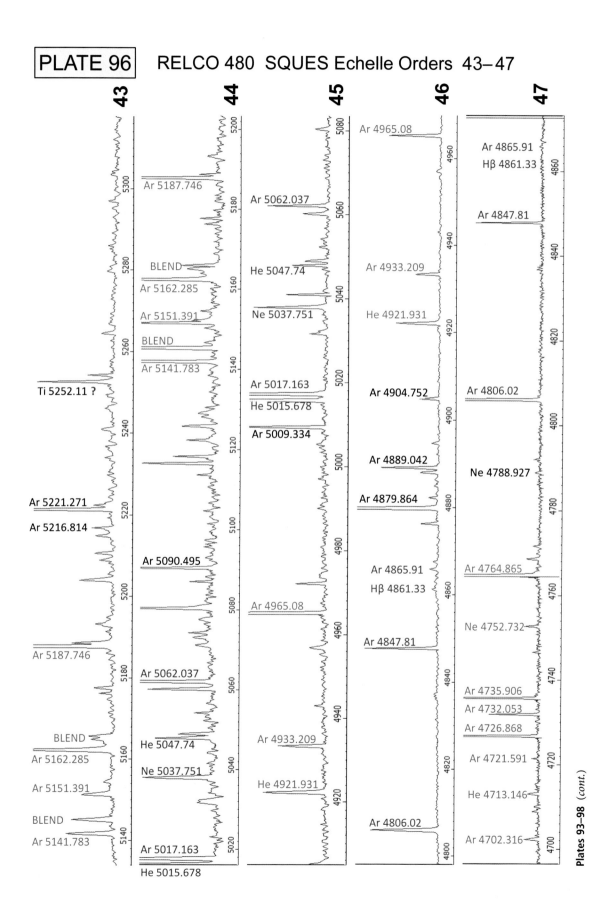

43

Ar 5187.746

BLEND
Ar 5162.285

Ar 5151.391

BLEND

Ar 5141.783

Ti 5252.11 ?

Ar 5221.271

Ar 5216.814

Ar 5187.746

BLEND
Ar 5162.285

Ar 5151.391

BLEND
Ar 5141.783

44

Ar 5187.746

BLEND

Ar 5162.285

Ar 5151.391

BLEND

Ar 5141.783

Ar 5090.495

Ar 5062.037

He 5047.74

Ne 5037.751

Ar 5017.163

He 5015.678

45

Ar 5062.037

He 5047.74

Ne 5037.751

Ar 5017.163
He 5015.678

Ar 5009.334

Ar 4965.08

Ar 4933.209

He 4921.931

Ar 4806.02

46

Ar 4965.08

Ar 4933.209

He 4921.931

Ar 4904.752

Ar 4889.042

Ar 4879.864

Ar 4865.91

Hβ 4861.33

Ar 4847.81

Ar 4806.02

47

Ar 4865.91
Hβ 4861.33

Ar 4847.81

Ar 4806.02

Ne 4788.927

Ar 4764.865

Ne 4752.732

Ar 4735.906

Ar 4732.053

Ar 4726.868

Ar 4721.591

He 4713.146

Ar 4702.316

Plates 93–98 (cont.)

PLATE 97

RELCO 480 SQUES Echelle Orders 48–52

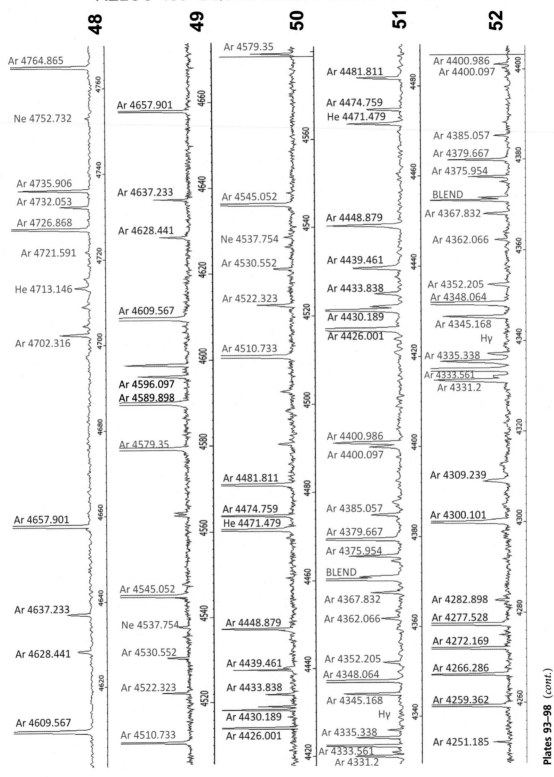

48

Ar 4764.865
Ne 4752.732
Ar 4735.906
Ar 4732.053
Ar 4726.868
Ar 4721.591
He 4713.146
Ar 4702.316
Ar 4657.901
Ar 4637.233
Ar 4628.441
Ar 4609.567

4760
4740
4720
4700
4680
4660
4640
4620

49

Ar 4657.901
Ar 4637.233
Ar 4628.441
Ar 4609.567
Ar 4596.097
Ar 4589.898
Ar 4579.35
Ar 4545.052
Ne 4537.754
Ar 4530.552
Ar 4522.323
Ar 4510.733

4660
4640
4620
4600
4580
4560
4540
4520

50

Ar 4579.35
Ar 4545.052
Ne 4537.754
Ar 4530.552
Ar 4522.323
Ar 4510.733
Ar 4481.811
Ar 4474.759
He 4471.479
Ar 4448.879
Ar 4439.461
Ar 4433.838
Ar 4430.189
Ar 4426.001

4560
4540
4520
4500
4480
4460
4440
4420

51

Ar 4481.811
Ar 4474.759
He 4471.479
Ar 4448.879
Ar 4439.461
Ar 4433.838
Ar 4430.189
Ar 4426.001
Ar 4400.986
Ar 4400.097
Ar 4385.057
Ar 4379.667
Ar 4375.954
BLEND
Ar 4367.832
Ar 4362.066
Ar 4352.205
Ar 4348.064
Ar 4345.168
Hγ
Ar 4335.338
Ar 4333.561
Ar 4331.2

4480
4460
4440
4420
4400
4380
4360
4340
4420

52

Ar 4400.986
Ar 4400.097
Ar 4385.057
Ar 4379.667
Ar 4375.954
BLEND
Ar 4367.832
Ar 4362.066
Ar 4352.205
Ar 4348.064
Ar 4345.168
Hγ
Ar 4335.338
Ar 4333.561
Ar 4331.2
Ar 4309.239
Ar 4300.101
Ar 4282.898
Ar 4277.528
Ar 4272.169
Ar 4266.286
Ar 4259.362
Ar 4251.185

4400
4380
4360
4340
4320
4300
4280
4260

Plates 93–98 (*cont.*)

PLATE 98

RELCO 480 SQUES Echelle Orders 53–57

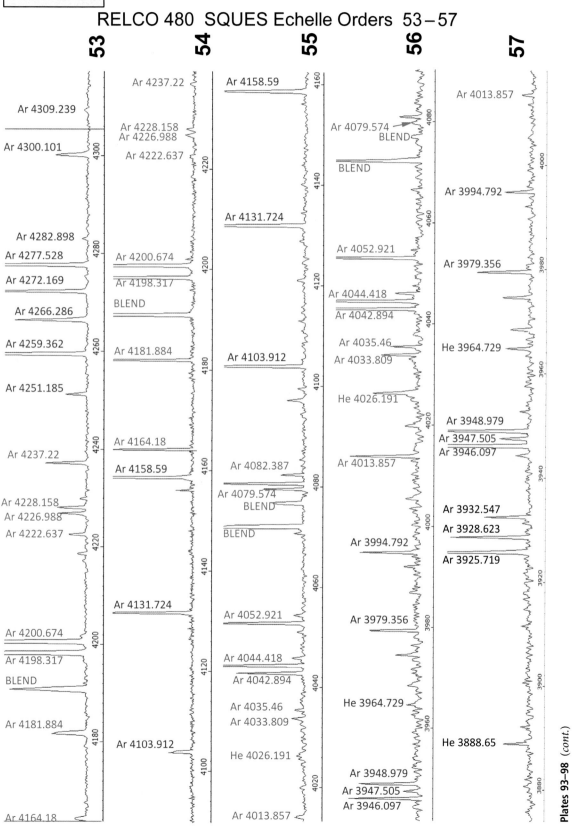

PLATE 99

RELCO 480 SQUES Echelle Orders 28–37

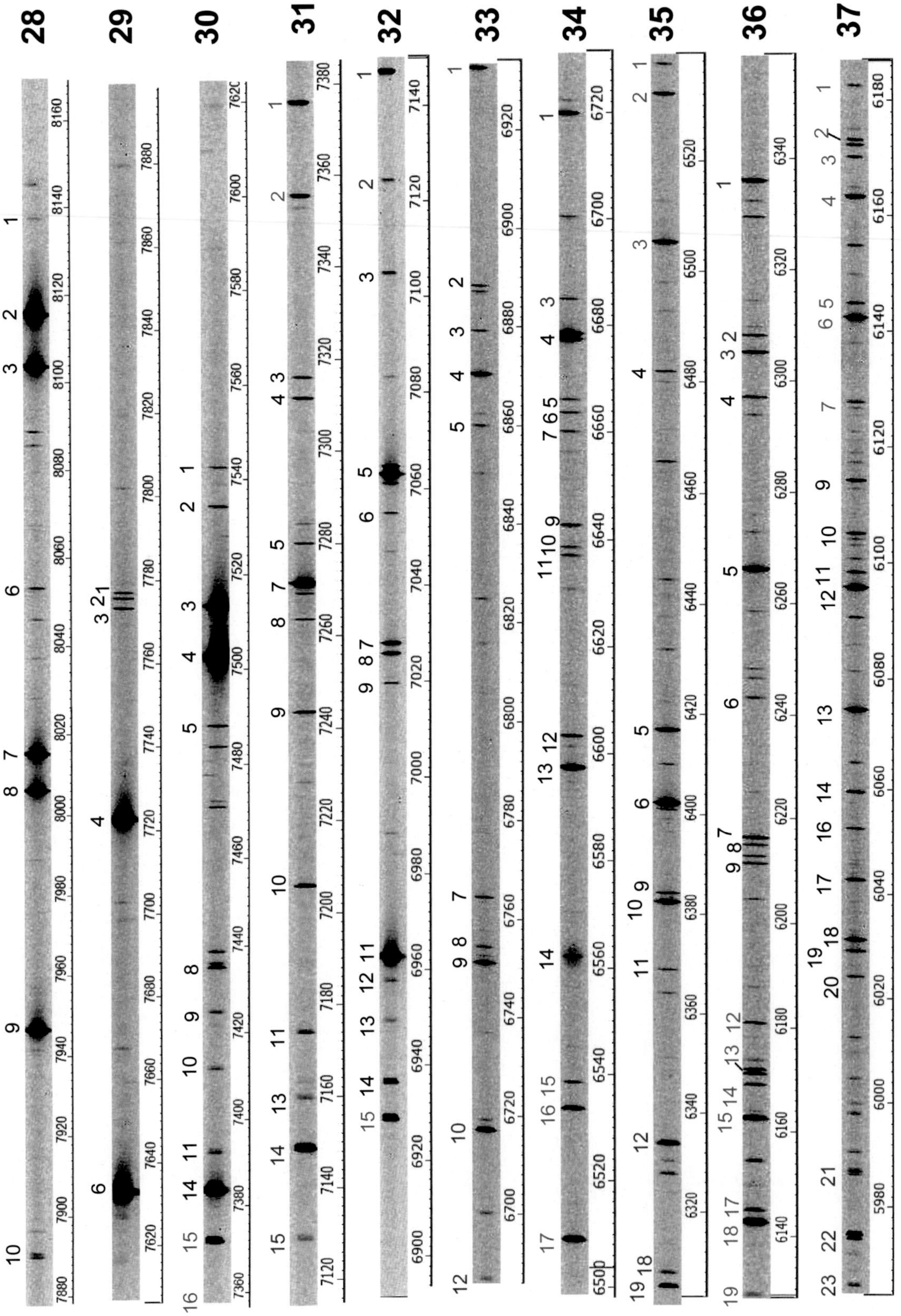

Plates 99–101 Calibration Lamp with Ar, Ne and He

PLATE 100 RELCO 480 SQUES Echelle Orders 38–47

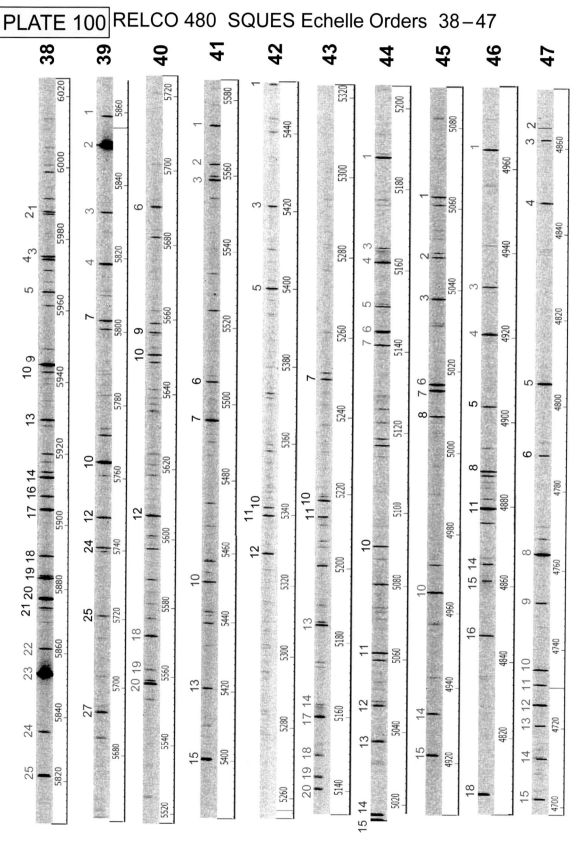

PLATE 101

RELCO 480 SQUES Echelle Orders 48–57

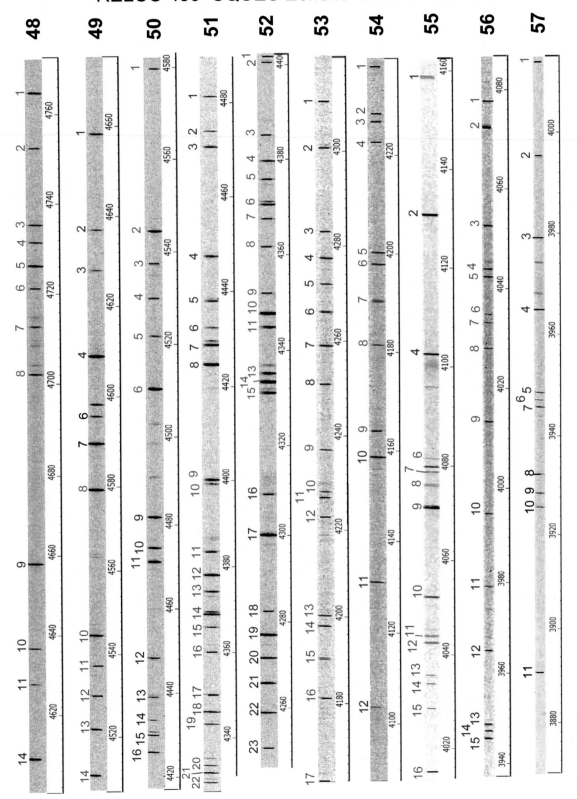

Table 33.2 Wavelengths RELCO calibration light source, Plates 99–101

Order	Line No.	λ	Element	Order	Line No.	λ	Element
28	1	8136.406	Ne	31	8	7265.172	Ar
28	2	8115.311	Ar	31	9	7245.167	Ne
28	3	8103.693	Ar	31	10	7206.98	Ar
28	6	8053.308	Ar	31	11	7173.938	Ne
28	7	8014.786	Ar	31	13	7158.839	Ar
28	8	8006.157	Ar	31	14	7147.042	Ar
28	9	7948.176	Ar	31	15	7125.82	Ar
28	10	7891.075	Ar				
				32	1	7147.042	Ar
29	1	7775.39	O	32	2	7125.82	Ar
29	2	7774.17	O	32	3	7107.478	Ar
29	3	7771.94	O	32	5	7067.218	Ar
29	4	7723.761	Ar	32	6	7059.107	Ne
29	6	7635.106	Ar	32	7	7032.413	Ne
				32	8	7030.251	Ar
30	1	7544.044	Ne	32	9	7024.05	Ne
30	2	7535.774	Ne	32	11	6965.431	Ar
30	3	7514.652	Ar	32	12	6960.25	Ar
30	4	7503.869	Ar	32	13	6951.478	Ar
30	5	7488.871	Ne	32	14	6937.664	Ar
30	8	BLEND	Ar	32	15	6929.467	Ne
30	9	7425.294	Ar				
30	10	7412.337	Ar	33	1	6929.467	Ne
30	11	7392.98	Ar	33	2	6888.174	Ar
30	14	7383.98	Ar	33	3	6879.582	Ar
30	15	7372.118	Ar	33	4	6871.289	Ar
30	16	7353.293	Ar	33	5	6861.269	Ar
				33	7	6766.612	Ar
31	1	7372.118	Ar	33	8	6756.163	Ar
31	2	7353.293	Ar	33	9	6752.834	Ar
31	3	7316.005	Ar	33	10	6717.043	Ne
31	4	7311.716	Ar	33	12	6684.293	Ar
31	5	7281.35	He				
31	7	7272.936	Ar	34	1	6717.043	Ne

Table 33.2 (*cont.*)

Order	Line No.	λ	Element	Order	Line No.	λ	Element
34	3	6684.293	Ar	36	9	6212.503	Ar
34	4	BLEND	Ar / Ne	36	12	6182.146	Ne
34	5	6666.359	Ar	36	13	BLEND	Ar
34	6	6664.051	Ar	36	14	6170.174	Ar
34	7	6660.676	Ar	36	15	6163.594	Ne
34	9	6643.698	Ar	36	17	6145.441	Ar
34	10	6639.74	Ar	36	18	6143.063	Ne
34	11	6638.221	Ar	36	19	6128.45	Ne
34	12	6604.853	Ar				
34	13	6598.953	Ne	37	1	6182.146	Ne
34	14	6562.82	Hα	37	2	BLEND	Ar
34	15	6538.112	Ar	37	3	6170.174	Ar
34	16	6532.882	Ne	37	4	6163.594	Ne
34	17	6506.528	Ne	37	5	6145.441	Ar
				37	6	6143.063	Ne
35	1	6538.112	Ar	37	7	6128.45	Ne
35	2	6532.882	Ne	37	9	6114.923	Ar
35	3	6506.528	Ne	37	10	6105.635	Ar
35	4	6483.082	Ar	37	11	6098.803	Ar
35	5	6416.307	Ar	37	12	6096.163	Ne
35	6	6402.246	Ar	37	13	6074.338	Ne
35	9	6384.717	Ar	37	14	6059.372	Ar
35	10	6382.992	Ne	37	16	6052.723	Ar
35	11	6369.575	Ar	37	17	6043.223	Ar
35	12	6334.428	Ne	37	18	6032.127	Ar
35	18	6307.657	Ar	37	19	6029.997	Ne
35	19	6304.789	Ne	37	20	6025.15	Ar
				37	21	BLEND	Ne/Ar
36	1	6334.428	Ne	37	22	BLEND	Ne
36	2	6307.657	Ar	37	23	5965.471	Ne
36	3	6304.789	Ne				
36	4	6296.872	Ar	38	1	5987.907	Ne
36	5	6266.495	Ne	38	2	5987.302	Ar
36	6	6243.12	Ar	38	3	5975.534	Ne
36	7	6217.281	Ne	38	4	5974.627	Ne
36	8	6215.938	Ar	38	5	5965.471	Ne

Table 33.2 (*cont.*)

Order	Line No.	λ	Element	Order	Line No.	λ	Element
38	9	5944.834	Ne	41	3	5558.702	Ar
38	10	5942.669	Ar	41	6	5506.113	Ar
38	13	5928.813	Ar	41	7	5495.874	Ar
38	14	5912.085	Ar	41	10	5451.652	Ar
38	16	5906.429	Ne	41	13	5421.352	Ar
38	17	5902.462	Ne	41	15	5400.562	Ne
38	18	5888.584	Ar				
38	19	BLEND	Ne/Ar	42	1	5451.652	Ar
38	20	5875.62	He	42	3	5421.352	Ar
38	21	5872.828	Ne	42	5	5400.562	Ne
38	22	5860.31	Ar	42	10	5343.283	Ne
38	23	5852.488	Ne	42	11	5341.094	Ne
38	24	5834.263	Ar	42	12	5330.778	Ne
38	25	5820.156	Ne				
				43	7	5252.11	Ti ?
39	1	5860.31	Ar	43	10	5221.271	Ar
39	2	5852.488	Ne	43	11	5216.814	Ar
39	3	5834.263	Ar	43	13	5187.746	Ar
39	4	5820.156	Ne	43	14	BLEND	Ar
39	7	5804.45	Ne	43	17	5162.285	Ar
39	10	5764.419	Ne	43	18	5151.391	Ar
39	12	5748.298	Ne	43	19	BLEND	Ne/Ar
39	24	5739.52	Ar	43	20	5141.783	Ar
39	25	5719.225	Ne				
39	27	5689.47	Ti ?	44	1	5187.746	Ar
				44	3	5165.773	Ar
40	6	5689.47	Ti ?	44	4	5162.285	Ar
40	9	5656.659	Ne	44	5	5151.391	Ar
40	10	5650.704	Ar	44	6	BLEND	Ne/Ar
40	12	5606.733	Ar	44	7	5141.783	Ar
40	18	5572.541	Ar	44	10	5090.495	Ar
40	19	5562.766	Ne	44	11	5062.037	Ar
40	20	5558.702	Ar	44	12	5047.74	He
				44	13	5037.751	Ne
41	1	5572.541	Ar	44	14	5017.163	Ar
41	2	5562.766	Ne	44	15	5015.678	He

Table 33.2 (*cont.*)

Order	Line No.	λ	Element	Order	Line No.	λ	Element
45	1	5062.037	Ar	48	2	4752.732	Ne
45	2	5047.74	He	48	3	4735.906	Ar
45	3	5037.751	Ne	48	4	4732.053	Ar
45	6	5017.163	Ar	48	5	4726.868	Ar
45	7	5015.678	He	48	6	4721.591	Ar
45	8	5009.334	Ar	48	7	4713.146	He
45	10	4965.08	Ar	48	8	4702.316	Ar
45	14	4933.209	Ar	48	9	4657.901	Ar
45	15	4921.931	He	48	10	4637.233	Ar
				48	11	4628.441	Ar
46	1	4965.08	Ar	48	14	4609.567	Ar
46	3	4933.209	Ar				
46	4	4921.931	He	49	1	4657.901	Ar
46	5	4904.752	Ar	49	2	4637.233	Ar
46	8	4889.042	Ar	49	3	4628.441	Ar
46	11	4879.864	Ar	49	4	4609.567	Ar
46	14	4865.91	Ar	49	6	4596.097	Ar
46	15	4861.33	Hβ	49	7	4589.898	Ar
46	16	4847.81	Ar	50	5	4522.323	Ar
46	18	4806.02	Ar	50	6	4510.733	Ar
				50	9	4481.811	Ar
47	2	4865.91	Ar	50	10	4474.759	Ar
47	3	4861.33	Hβ	50	11	4471.479	He
47	4	4847.81	Ar	50	12	4448.879	Ar
47	5	4806.02	Ar	50	13	4439.461	Ar
47	6	4788.927	Ne	50	14	4433.838	Ar
47	8	4764.865	Ar	50	15	4430.189	Ar
47	9	4752.732	Ne	50	16	4426.001	Ar
47	10	4735.906	Ar				
47	11	4732.053	Ar	51	1	4481.811	Ar
47	12	4726.868	Ar	51	2	4474.759	Ar
47	13	4721.591	Ar	51	3	4471.479	He
47	14	4713.146	He	51	4	4448.879	Ar
47	15	4702.316	Ar	51	5	4439.461	Ar
				51	6	4433.838	Ar
48	1	4764.865	Ar	51	7	4430.189	Ar

Table 33.2 (*cont.*)

Order	Line No.	λ	Element	Order	Line No.	λ	Element
51	8	4426.001	Ar	52	22	4259.362	Ar
51	9	4400.986	Ar	52	23	4251.185	Ar
51	10	4400.097	Ar				
51	11	4385.057	Ar	53	1	4309.239	Ar
51	12	4379.667	Ar	53	2	4300.101	Ar
51	13	4375.954	Ar	53	3	4282.898	Ar
51	14	BLEND	Ar	53	4	4277.528	Ar
51	15	4367.832	Ar	53	5	4272.169	Ar
51	16	4362.066	Ar	53	6	4266.286	Ar
51	17	4352.205	Ar	53	7	4259.362	Ar
51	18	4348.064	Ar	53	8	4251.185	Ar
51	19	4345.168	Ar	53	9	4237.22	Ar
51	20	4335.338	Ar	53	10	4228.158	Ar
51	21	4333.561	Ar	53	11	4226.988	Ar
51	22	4331.2	Ar	53	12	4222.637	Ar
				53	13	4200.674	Ar
52	1	4400.986	Ar	53	14	4198.317	Ar
52	2	4400.097	Ar	53	15	BLEND	Ar/Hf
52	3	4385.057	Ar	53	16	4181.884	Ar
52	4	4379.667	Ar	53	17	4164.18	Ar
52	5	4375.954	Ar				
52	6	BLEND	Ar	54	1	4237.22	Ar
52	7	4367.832	Ar	54	2	4228.158	Ar
52	8	4362.066	Ar	54	3	4226.988	Ar
52	9	4352.205	Ar	54	4	4222.637	Ar
52	10	4348.064	Ar	54	5	4200.674	Ar
52	11	4345.168	Ar	54	6	4198.317	Ar
52	13	4335.338	Ar	54	7	BLEND	Ar/Hf
52	14	4333.561	Ar	54	8	4181.884	Ar
52	15	4331.2	Ar	54	9	4164.18	Ar
52	16	4309.239	Ar	54	10	4158.59	Ar
52	17	4300.101	Ar	54	11	4131.724	Ar
52	18	4282.898	Ar	54	12	4103.912	Ar
52	19	4277.528	Ar				
52	20	4272.169	Ar	55	1	4158.59	Ar
52	21	4266.286	Ar	55	2	4131.724	Ar

Table 33.2 (*cont.*)

Order	Line No.	λ	Element	Order	Line No.	λ	Element
55	4	4103.912	Ar	57	8	3932.547	Ar
55	6	4082.387	Ar	57	9	3928.623	Ar
55	7	4079.574	Ar	57	10	3925.719	Ar
55	8	BLEND	Ar/?	57	11	3888.65	He
55	9	BLEND	Ar/Ar				
55	10	4052.921	Ar				
55	11	4044.418	Ar				
55	12	4042.894	Ar				
55	13	4035.46	Ar				
55	14	4033.809	Ar				
55	15	4026.36	He				
55	16	4013.857	Ar				
56	1	BLEND	Ar/?				
56	2	BLEND	Ar/Ar				
56	3	4052.921	Ar				
56	4	4044.418	Ar				
56	5	4042.894	Ar				
56	6	4035.46	Ar				
56	7	4033.809	Ar				
56	8	4026.36	He				
56	9	4013.857	Ar				
56	10	3994.792	Ar				
56	11	3979.356	Ar				
56	12	3964.729	He				
56	13	3948.979	Ar				
56	14	3947.505	Ar				
56	15	3946.097	Ar				
57	1	4013.857	Ar				
57	2	3994.792	Ar				
57	3	3979.356	Ar				
57	4	3964.729	He				
57	5	3948.979	Ar				
57	6	3947.505	Ar				
57	7	3946.097	Ar				

PLATE 102

Butane gas torch, comet Hyakutake and carbon star WZ cassiopeiae

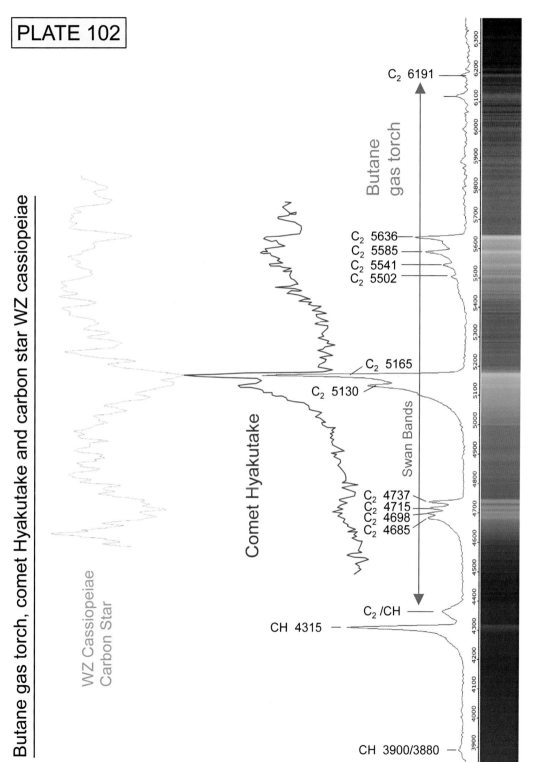

Plate 102 Swan Bands/Hydrocarbon gas flames in Comparison with a Montage of Spectra

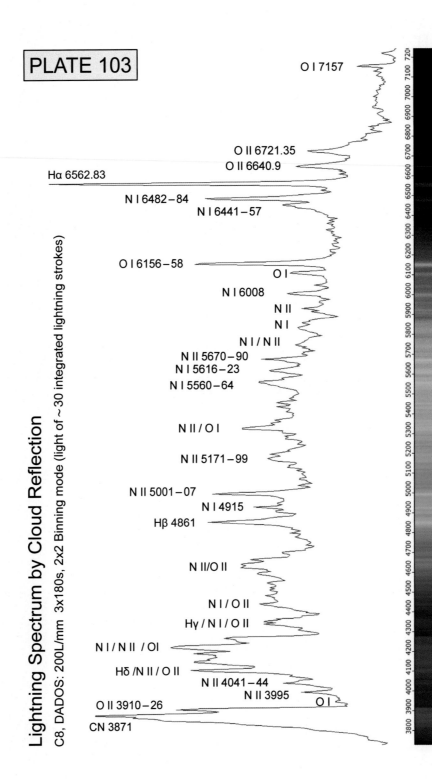

PLATE 103

Lightning Spectrum by Cloud Reflection

C8, DADOS: 200L/mm 3x180s, 2x2 Binning mode (light of ~30 integrated lightning strokes)

O I 7157

O II 6721.35
O II 6640.9

Hα 6562.83

N I 6482–84
N I 6441–57

O I 6156–58
O I
N I 6008
N II
N I
N I / N II
N II 5670–90
N I 5616–23
N I 5560–64

N II / O I

N II 5171–99

N II 5001–07
N I 4915
Hβ 4861

N II/O II

N I / O II
Hγ / N I / O II

N I / N II / OI

Hδ /N II / O II
N II 4041–44
N II 3995
O I

O II 3910–26
CN 3871

Plate 103 Spectra of Terrestrial Lightning Discharges

APPENDIX A
Spectral Classes and *v* sin *i* Values of Bright Stars

Stars of the spectral classes O, B, A, F, G, K, M, mostly according to CDS/SIMBAD [11]

Spectral class	Luminosity class	m_v	$v \sin i$ [km s^{-1}]	Bayer designation	Proper name
O4	If(n)p	2.3		ζ Pup	Naos
O8	III ((f))	3.5	75	λ Ori A	Meissa
O9	III	2.8	122	ι Ori	Nair al Saif
O9	V	2.2	133	δ Ori A	Mintaka
O9.5	V	3.7	94	σ Ori A	
O9.7	Ib	2.1		ζ Ori A	Alnitak
B0	Ia	1.7	65	ε Ori	Alnilam
	V	2.8	3	τ Sco	
B0.5	Ia	2.1	65	κ Ori	Saiph
	IV	1.25	35	β Cru	Becrux, Mimosa
	IVpe	2.5	295	γ Cas	Tsih
	IVe	2.3	165	δ Sco	Dschubba
	V	2.9	130	ε Per	Adid
B1	Ib	2.9	40	ζ Per	
	II–III	2.0	17	β CMa	Mirzam
	III	3.2	20	β Cep	Alfirk
	III	0.6	161	β Cen	Agena
	III–IV	1.0	130	α Vir	Spica
	V	3.4	47	η Ori	Algiebbah
B2	V	1.6	55	γ Ori	Bellatrix
	IV	2.8	0	γ Peg	Algenib
	IV	1.9	24	α Pav	Peacock
	V	2.0	165	σ Sgr	Nunki
B3	III	3.4	30	ε Cas	Segin

(cont.)

Spectral class	Luminosity class	m_v	$v \sin i$ [km s^{-1}]	Bayer designation	Proper name
B5	Ia	2.5	50	η CMa	Aludra
	III	3.0	190	δ Per	
B6	IIIe	3.7	90	17 Tau	Electra
	IVe	4.2	195	23 Tau	Merope
	Vpe	0.5		α Eri	Achernar
B7	III	3.2	40	ζ Dra	Aldhibah
	III	1.7	60	β Tau	Alnath
	IIIe	2.9	145	η Tau	Alcyone
B8	Iae	0.1	40	β Ori	Rigel
	II–IIIep	3.4	0	β Lyr	Sheliak
	IIIp Hg Mn	2.6	30	γ Crv	Gienah Corvi
	III	3.9	30	20 Tau	Maia
	III	3.6	170	27 Tau	Atlas
	III	3.2	35	φ Sgr	
	IVp Hg Mn	2.1	50	α And	Alpheratz
	IVn	1.4	300	α Leo	Regulus
	Ve	2.9	210	β CMi	Gomeisa
	Ve	5.1	215	β Cyg	Albireo B
	V	2.1	50	β Per	Algol
B9	III	3.3	60	γ Lyr	Sulafat
A0	V	0.0	24	α Lyr	Vega
A1	III–IVp	1.8	33	ε UMa	Alioth
	IV	1.9	15	γ Gem	Alhena
	IVps	2.4	46	β UMa	Merak
	IV–Vp	1.9	40	β Aur	Menkalinan
	V	1.9	18	α Gem A	Castor A
	Vm	−1.5	17	α CMa	Sirius
A2	Vm	3.0	33	α Gem B	Castor B
	Ia	1.3	39	α Cyg	Deneb
	Vn	3.4	222	ζ Vir	Heze
A4	V	1.2	92	α PsA	Formalhaut
A5	III	2.1	228	α Oph	Ras Alhague
	III–IV	2.7	123	δ Cas	Ruchbah
	IVn	2.5	180	δ Leo	Zosma
	V	2.6	73	β Ari	Sharatan

(cont.)

Spectral class	Luminosity class	m_v	$v \sin i$ [km s^{-1}]	Bayer designation	Proper name
A7	III	3.0	128	γ Boo	Seginus
	Vn	0.8	203	α Aql	Altair
A8	Vn	2.5	196	α Cep	Alderamin
A9	II	−0.7		α Car	Canopus
	IIIbn	3.8	140	γ Her	
F0	III	3.4	72	ζ Leo	Adhafera
	IV	2.8	31	γ Vir	Porrima
F2	III	2.3	71	β Cas	Caph
F5	IV–V	0.4	5	α CMi	Procyon
	Ib	1.8	18	α Per	Mirfak
F6	IV	3.4	82	α Tri	Mothallah
F8	Ia	1.8	13	δ CMa	Wezen
	Iab	2.2	20	γ Cyg	Sadr
	Ib	2.0	17	α UMi	Polaris
	V	3.6	3	β Vir	Zavijah
G0	Ib	2.9	18	β Aqr	Sadalsuud
	IV	2.7	15	η Boo	Muphrid
G1	II	3.0	8	ε Leo	Raselased
G2	Iab	2.8	13	β Dra A	Rastaban
	Ib	3.0	<17	α Aqr	Rigil Kentaurus
	V	0.01	2.6	α Cen A	Sadalmelik
	V	−26.75	1.9	Sol	Sun
G5	II	2.6	<17	β Crv	Kraz
G8	Ib	3.0	<17	ε Gem	Mebsuta
	III	2.8	0	ε Vir	Vindemiatrix
	III	2.8	3	β Her	Kornephoros
	IIIa	3.5	0	β Boo	Nekkar
	IV	3.5	4	δ Boo	
K0	III	1.1	2.8	β Gem	Pollux
	IIIa	2.2	8.5	α Cas	Shedar
K1	IIIb	2.0	4.2	α Ari	Hamal
K1.5	IIIpe	-0.0	4.2	α Boo	Arcturus

(cont.)

Spectral class	Luminosity class	m_v	$v \sin i$ [km s^{-1}]	Bayer designation	Proper name
K2	II–III	1.9		α Tra	
	III	2.8	5.4	β Oph	Cebalrai
	IIIb	2.7	4.3	α Ser	Unukalhai
K3	II	2.7	<17	γ Aql	Tarazed
	II	3.1		β Cyg A	Albireo A
	II–III	2.0	5.6	α Hya	Alphard
	III	2.8		α Tuc	
	IIIa	2.7		δ Sgr	Kaus Media
K4	III	2.1	<17	β UMi	Kochab
	III	3.5	6.9	β Cnc	Altarf
K4.5	III	4.3	<17	λ Leo	Alterf
K5	III	0.9	4.3	α Tau	Aldebaran
	III	2.2	6.0	γ Dra	Etamin
	V	5.2	4.7	61 Cyg A	
K7	III	3.1	6.4	α Lyn	Alsciaukat
	V	6.0	3.2	61 Cyg B	
M0	IIIa	2.1	7.2	β And	Mirach
	III	3.1	7.5	μ UMa	Tania Australis
M0.5	Ia–Ib	1.0	<20	α Sco	Antares
	III	2.8	7.0	δ Oph	Yed Prior
M1–2	Ia–Iab	0.4		α Ori	Betelgeuse
M1.5	IIIa	2.5	6.9	α Cet	Menkar
M2.5	II–IIIe	2.4	9.7	β Peg	Scheat
M3	III	3.3		η Gem	Propus
	III	3.4	6.0	δ Vir	Auva
	IIIab	2.9		μ Gem	Tejat Posterior
M3.5	III	1.6		γ Cru	Gacrux
M5	Ib–II	3.1		α Her	Ras Algethi

Lower case letters of the Greek alphabet

α	alpha	η	eta	ν	nu	τ	tau
β	beta	θ	theta	ξ	xi	υ	upsilon
γ	gamma	ι	iota	ο	omicron	φ	phi
δ	delta	κ	kappa	π	pi	χ	chi
ε	epsilon	λ	lambda	ρ	rho	ψ	psi
ζ	zeta	μ	mu	σ	sigma	ω	omega

APPENDIX B
The 88 IAU Constellations

Constellation Latin name	Abbreviation	English name	Constellation Latin name	Abbreviation	English name
Andromeda	And	Chained Maiden	Columba	Col	Dove
Antlia	Ant	Air Pump	Coma Berenices	Com	Bernice's Hair
Apus	Aps	Bird of Paradise	Corona Australis	CrA	Southern Crown
Aquarius	Aqr	Water Bearer	Corona Borealis	CrB	Northern Crown
Aquila	Aql	Eagle	Corvus	Crv	Crow
Ara	Ara	Altar	Crater	Crt	Cup
Aries	Ari	Ram	Crux	Cru	Southern Cross
Auriga	Aur	Charioteer	Cygnus	Cyg	Swan
Bootes	Boo	Herdsman	Delphinus	Del	Dolphin
Caelum	Cae	Engraving Tool	Dorado	Dor	Swordfish
Camelopardalis	Cam	Giraffe	Draco	Dra	Dragon
Cancer	Cnc	Crab	Equuleus	Equ	Little Horse
Canes Venatici	CVn	Hunting Dogs	Eridanus	Eri	River
Canis Major	CMa	Great Dog	Fornax	For	Furnace
Canis Minor	CMi	Lesser Dog	Gemini	Gem	Twins
Capricornus	Cap	Sea Goat	Grus	Gru	Crane
Carina	Car	Keel	Hercules	Her	Hercules
Cassiopeia	Cas	Seated Queen	Horologium	Hor	Clock
Centaurus	Cen	Centaur	Hydra	Hya	Female Water Snake
Cepheus	Cep	King	Hydrus	Hyi	Male Water Snake
Cetus	Cet	Sea Monster	Indus	Ind	Indian
Chamaeleon	Cha	Chameleon	Lacerta	Lac	Lizard
Circinus	Cir	Compass	Leo	Leo	Lion

(*cont.*)

Constellation Latin name	Abbreviation	English name	Constellation Latin name	Abbreviation	English name
Leo Minor	LMi	Lesser Lion	Puppis	Pup	Stern
Lepus	Lep	Hare	Pyxis	Pyx	Compass
Libra	Lib	Scales	Reticulum	Ret	Reticle
Lupus	Lup	Wolf	Sagitta	Sge	Arrow
Lynx	Lyn	Lynx	Sagittarius	Sgr	Archer
Lyra	Lyr	Lyre	Scorpius	Sco	Scorpion
Mensa	Men	Table Mountain	Sculptor	Scl	Sculptor
Microscopium	Mic	Microscope	Scutum	Sct	Shield
Monoceros	Mon	Unicorn	Serpens	Ser	Serpent
Musca	Mus	Fly	Sextans	Sex	Sextant
Norma	Nor	Carpenter's Square	Taurus	Tau	Bull
Octans	Oct	Octant	Telescopium	Tel	Telescope
Ophiuchus	Oph	Serpent Bearer	Triangulum	Tri	Triangle
Orion	Ori	Hunter	Triangulum Australe	TrA	Southern Triangle
Pavo	Pav	Peacock	Tucana	Tuc	Toucan
Pegasus	Peg	Winged Horse	Ursa Major	UMa	Great Bear
Perseus	Per	Hero	Ursa Minor	UMi	Little Bear
Phoenix	Phe	Phoenix	Vela	Vel	Sails
Pictor	Pic	Painter's Easel	Virgo	Vir	Maiden
Pisces	Psc	Fishes	Volans	Vol	Flying Fish
Piscis Austrinus	PsA	Southern Fish	Vulpecula	Vul	Fox

APPENDIX C
Spectral Classes and B–V Color Index

According to Fitzgerald [276], the following table shows the allocation of unreddened star colors, given in the photometric Johnson B–V color index, to the corresponding spectral and luminosity classes.

	V	IV	III	II	Ia		V	IV	III	II	Ia
O5	−0.32					F1	0.34	0.33	0.32		0.15
O6	−0.32					F2	0.35	0.37	0.36		0.18
O7	−0.32					F3	0.41	0.39	0.39		
O8	−0.31					F4	0.42	0.42	0.42		
O9	−0.31	−0.31	−0.31	−0.31	−0.28	F5	0.45	0.44	0.43	0.38	0.26
B0	−0.30	−0.30	−0.30	−0.29	−0.24	F6	0.48	0.46	0.46		
B1	−0.26	−0.26	−0.26	−0.24	−0.19	F7	0.50	0.50	0.48		
B2	−0.24	−0.24	−0.24	−0.21	−0.17	F8	0.53	0.53	0.52		0.55
B3	−0.20	−0.20	−0.20	−0.17	−0.13	F9	0.56	0.57			
B4	−0.18	−0.18	−0.18		−0.11	G0	0.60	0.63	0.64	0.73	0.82
B5	−0.16	−0.16	−0.16	−0.14	−0.09	G1	0.62	0.63	0.69	0.80	0.85
B6	−0.14	−0.14	−0.14	−0.12	−0.07	G2	0.63	0.64	0.77	0.87	0.88
B7	−0.13	−0.13	−0.12	−0.12	−0.04	G3	0.65	0.66	0.85	0.87	0.92
B8	−0.11	−0.10	−0.10	−0.10	−0.01	G4	0.66	0.68	0.88	0.87	
B9	−0.07	−0.07	−0.08	−0.06	0.00	G5	0.68	0.70	0.90	0.87	1.00
A0	−0.01	−0.02	−0.03	0.00	0.02	G6	0.72		0.92	0.91	1.04
A1	0.02	0.00	0.01		0.03	G7	0.73		0.94	0.95	1.10
A2	0.05	0.06	0.05		0.05	G8	0.74	0.82	0.95	0.99	1.14
A3	0.08	0.09	0.09		0.06	G9	0.76	0.90	0.98	1.02	1.16
A4	0.12	0.12	0.12		0.08	K0	0.81	0.91	1.01	1.06	1.18
A5	0.15	0.15	0.15	0.10	0.10	K1	0.86	0.99	1.09	1.14	1.20
A7	0.20	0.22	0.24	0.14	0.13	K2	0.92		1.16	1.29	1.23
A8	0.27	0.26	0.26		0.14	K3	0.95		1.26	1.40	1.42
A9	0.30	0.29	0.28	0.18	0.14	K4	1.0		1.43	1.42	1.50
F0	0.32	0.30	0.32		0.15	K5	1.15		1.51	1.45	1.60

(cont.)

	V	IV	III	II	Ia		V	IV	III	II	Ia
K7	1.33		1.53		1.62	M7	1.68		1.50		
K9	1.37		1.55	1.58	1.64	M8	1.77		1.50		
M0	1.37		1.57	1.58	1.65						
M1	1.47		1.60	1.59	1.65						
M2	1.47		1.60	1.59	1.65						
M3	1.47		1.60	1.60	1.67						
M4	1.52		1.63		1.75						
M5	1.61		1.65								
M6	1.64		1.49								

APPENDIX D
Spectral Classes and Effective Temperatures

T_{eff} values in [K], based on A Stellar Spectral Flux Library by A. J. Pickles [180]

Spectral Type	V	IV	III	II	I
O5	39,810				
O8			31,623		
O9	35,481				
B0	28,184				26,002
B1	22,387				20,701
B2	21,000	19,953		15,996	
B3	19,055		16,982		15,596
B5	14,125		14,791	12,589	13,397
B6		12,589			
B8	11,749				11,194
B9	10,715		11,092		
A0	9,550	9,727	9,572		9,727
A2	8,913				9,078
A3	8,790		8,974		
A5	8,492		8,453		
A7	8,054		8,054		
F0	7,211		7,586	7,943	7,691
F2	6,776		6,839	7,328	
F5	6,531	6,561	6,531		6,637
F6	6,281				
F8	6,039	6,152			6,095
G0	5,808	5,929	5,610		5,508

(cont.)

Spectral Type	V	IV	III	II	I
G2	5,636	5,688			5,297
G5	5,585	5,598	5,164	5,248	5,047
G8	5,333	5,309	5,012		4,592
K0	5,188	5,012	4,853		
K1		4,786	4,656	5,012	
K2	4,887		4,457		4,256
K3	4,498	4,571	4,365		4,130
K4	4,345		4,227		3,990
K5	4,188		4,009		
K7	3,999				
M0	3,802		3,819		
M1	3,681		3,776		
M2	3,548		3,707		3,451
M3	3,304		3,631	3,412	
M4	3,112		3,556		
M5	2,951		3,420		
M6	2,698		3,251		

APPENDIX E
Excitation Classes of Bright Planetary Nebulae

The following table contains planetary nebulae up to an apparent magnitude $m_v \approx 11$, sorted by excitation class E according to Gurzadyan *et al.* [157]. The temperatures of the central stars are from J. Kaler [14], the spectral classes from CDS/SIMBAD [11], apparent magnitudes from different sources, for example by CDS [11] and E. Karkoschka [181].

E	Catalog	Popular Name	Constellation	App. Mag. m_v	Central Star Spectral Class	Central Star Temperature T_{eff} [K]
1	IC418	Spirograph Nebula	Lepus	9.3	O7fp	35,000
2	IC2149		Auriga	10.6	O7.5	
2	IC4593		Hercules	10.9		35,000
2	IC4776		Sagittarius	11.5		
2	IC4997		Sagittarius	11.5		49,000
3	IC3568		Camelopardalis	11.5		57,000
4	NGC 6210	Turtle	Hercules	8.8	O7f	58,000
4	NGC 6790		Aquila	10.5		76,000
4	NGC 6891		Delphinus	10.5		
5	NGC 6543	Cat's Eye	Draco	8.1	O7+WR	80,000
5	NGC 6803		Aquila	11.4		
7	NGC 6884		Cygnus	10.9		87,000
7	NGC 7009	Saturn Nebula	Aquarius	8.3		90,000
7	NGC 6572	Blue Racquetball	Ophiuchus	9.0	Of+WR	100,000
7	NGC 7293	Helix Nebula	Aquarius	7.5	DA0	110,000
8	NGC 1514		Taurus	10.9	DB8	
8	NGC 1535		Eridanus	10.6		
8	NGC 3587	Owl Nebula	Ursa Major	9.9		
9	NGC 3132	South Ring Nebula	Vela	8.2		
9	NGC 6886		Sagitta	11.4		168,000
9	NGC 6741	Phantom Streak	Aquila	11.0		180,000
9	NGC 3242	Ghost of Jupiter	Hydra	7.7		90,000
9	NGC 3918		Centaurus	8.5		
10	NGC 7662	Blue Snowball	Andromeda	8.3		110,000

(*cont.*)

E	Catalog	Popular Name	Constellation	App. Mag. m_v	Central Star Spectral Class	Central Star Temperature T_{eff} [K]
10	NGC 7027		Cygnus	8.5	invisible	
10	NGC 6853	Dumbbell Nebula	Vulpecula	7.5	DO7	160,000
10	NGC 2438		Puppis	10.1		
10	NGC 650	Small Dumbbell Nebula	Perseus	10.1		
10	NGC 6818		Sagittarius	9.3		155,000
10	NGC 6302	Bug Nebula	Scorpius	9.6		
10	NGC 6720	Ring Nebula	Lyra	8.8		150,000
10	NGC 2392	Eskimo Nebula	Gemini	9.1		65,000
11	NGC 6826	Blinking Planetary	Cygnus	8.8		100,000
11	NGC 2818		Pyxis	8.2		
11	NGC 7008		Cygnus	10.7		
12	NGC 1360		Fornax	9.4		
12+	NGC 246		Cetus	10.9		
12+	NGC 4361		Corvus	10.9		

APPENDIX F
Ionization Energies of Important Elements

This table shows for a certain element the required energies in [eV], to achieve a certain stage of ionization, starting from the ground state.

Z	Element	Ionization stages and required excitation energies [eV], starting from the ground state							
		II	III	IV	V	VI	VII	VIII	IX
1	H	13.6							
2	He	24.6	54.4						
3	Li	5.4	75.6	122.5					
4	Be	9.3	18.2	153.9	217.7				
5	B	8.3	25.2	37.9	259.4	340.2			
6	C	11.3	24.4	47.9	64.5	392.1	490.0		
7	N	14.5	29.6	47.5	77.5	97.9	552.0	667.0	
8	O	13.6	35.1	54.9	77.4	113.9	138.1	739.3	871.4
9	F	17.4	35.0	62.7	87.1	114.2	157.2	185.2	953.9
10	Ne	21.6	41.0	63.5	97.1	126.2	157.9	207.3	239.1
11	Na	5.1	47.3	71.6	98.9	138.4	172.2	208.5	264.2
12	Mg	7.6	15.0	80.1	109.2	141.3	186.5	224.9	265.9
13	Al	6.0	18.8	28.4	120.0	153.7	190.5	241.4	284.6
14	Si	8.2	16.3	33.5	45.1	166.8	205.1	246.5	303.2
15	P	10.5	19.7	30.2	51.4	65.0	230.4	263.2	309.4
16	S	10.4	23.3	34.8	47.3	72.7	88.0	280.9	328.2
17	Cl	13.0	23.8	39.6	53.5	67.8	98.0	114.2	348.3
18	Ar	15.8	27.6	40.7	59.8	75.0	91.0	124.3	143.5
19	K	4.3	31.6	45.7	60.9	82.7	100.0	117.6	154.9
20	Ca	6.1	11.9	50.9	67.1	84.4	108.8	127.7	147.2
21	Sc	6.5	12.8	24.8	73.5	91.7	111.1	138.0	158.7
22	Ti	6.8	13.6	27.5	43.3	99.2	119.4	140.8	168.5
23	V	6.7	14.7	29.3	46.7	65.2	128.1	150.2	173.7
24	Cr	6.8	16.5	31.0	49.1	69.3	90.6	161.1	184.7

(*cont.*)

Z	Element	Ionization stages and required excitation energies [eV], starting from the ground state							
		II	III	IV	V	VI	VII	VIII	IX
25	Mn	7.4	15.6	33.7	51.2	72.4	95.0	119.3	196.5
26	Fe	7.9	16.2	30.7	54.8	75.0	99.0	125.0	151.1
27	Co	7.9	17.1	33.5	51.3	79.5	102	129	157
28	Ni	7.6	18.2	35.2	54.9	75.5	108	133	162
29	Cu	7.7	20.3	36.8	55.2	79.9	103	139	166
30	Zn	9.4	18.0	39.7	59.4	82.6	108	134	174
31	Ga	6.0	20.5	30.7	64				

APPENDIX G
Spectroscopic Measures and Units

Preliminary Remarks

According to recommendations of IAU the SI system (meter, kilogram, second) should be preferred. The cgs system (centimeter, gram, second) with the energy unit erg, ångström [Å] and gauss [G] should no longer be applied. However, and not only in the amateur sector, mainly ångström is in use for the wavelength. Further in astrophysics, for example, the surface gravity g and many other applications are frequently expressed in cgs units as well as the magnetic flux density in gauss [G].

Units for Energy and Wavelength

For some applications, such as the spectral flux [erg s^{-1} cm^{-2}], the cgs system is still very common:

$1\ \text{erg} = 10^{-7}\ \text{J}$

For the extremely low energies of electron transitions, instead of joule (J) almost always the electronvolt is applied:

$\text{eV} = 1.602 \times 10^{-19}\ \text{J}$
$\text{eV} = 1.602 \times 10^{-12}\ \text{erg}$

Within the relevant spectral domain the wavelengths are extremely small and therefore measured in ångströms [Å] or nanometers [nm], in the infrared range in micrometers [μm]:

$1\ \text{Å} = 10^{-10}\ \text{m}$
$1\ \text{nm} = 10\ \text{Å}$
$1\ \mu\text{m (micrometer)} = 1000\ \text{nm} = 10{,}000\ \text{Å}$

Converting Wavelength λ to Photon Energy E

Energy and frequency of a photon are related by Planck's energy equation $E = h\nu$ {28.1}. Wavelength and frequency are linked by the equation $\nu\lambda = c$ (where c is the speed of light) and so:

$$E[\text{eV}] = \frac{12403}{\lambda[\text{Å}]} \qquad \{\text{G.1}\}$$

Absolute Measures and Units Applied at Spectral Lines

The following measures and relationships are demonstrated by an absorption line with the rest wavelength λ_0 (Figure G1). With few exceptions the same applies analogously to emission lines as shown more detailed in [3].

The absorption process causes a drop of the local continuum level I_C between λ_1 and λ_2, expressed by the variable spectral flux density $I = f(\lambda)$, here also called "intensity I." The common unit for I is [erg s^{-1} cm^{-2} Å^{-1}]. The peak intensity I_P corresponds here to the difference between the minimum intensity $I(\lambda)$min at the lowest point of the absorption line, and the local continuum level $I_C(\lambda)$.

The total energy flux F, with the common unit [erg s^{-1} cm^{-2}], which is removed by the absorption process from the local continuum radiation, corresponds to the total area A of the line, measured below the continuum level:

$$F = \int_{\lambda_1}^{\lambda_2} [I_C(\lambda) - I(\lambda)]d\lambda \qquad \{\text{G.2}\}$$

Further the following measures are applied, indicated in Figure G1:

FWHM: Full Width at Half Maximum height [Å]
FWZI: Full Width at Zero Intensity [Å]

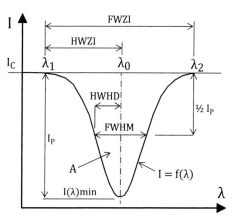

Figure G1 Spectroscopic measures of an absorption line

HWHD/HWHM: Half Width at Half Depth/at Half
        Maximum height = ½ FWHM [Å]

HWZI: Half Width at Zero Intensity = ½
        FWZI [Å]

Relative Measures Related to the Continuum Level I_c

The determination of the total energy flux F and the peak intensity I_P, requires a very demanding and time consuming calibration in absolute intensity units. Therefore in most cases, even in the professional field, these measures are relatively related to the local continuum level $I_C(\lambda)$.

The Continuum Related Peak Intensity P

Even in a non-intensity-calibrated profile the relative peak intensity P allows to compare the peak intensities of different spectral lines if related to the local continuum level $I_C(\lambda)$.

$$P = I_P/I_C \qquad \{G.3\}$$

Equivalent Width: The Continuum Related Energy Flux

Normally this measure is obtained in a rectified profile, normalized to unity $I_C = 1$ (Figure G2). The profile area between

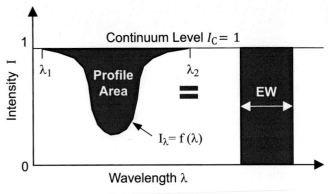

Figure G2 Definition of the equivalent width EW

the rectified continuum level, and the profile of the spectral line, has the same size as the rectangular area with the depth $I_C = 1$ and the equivalent width EW [Å].

$$EW = \frac{\text{Profile Area}}{I_C} \qquad \{G.4\}$$

Expressed in a general form:

$$EW = \int_{\lambda_1}^{\lambda_2} \frac{I_C(\lambda) - I(\lambda)}{I_C(\lambda)}\, d\lambda \qquad \{G.5\}$$

APPENDIX H
Distant AGN and Quasars Brighter than $m_v \approx 16$

This table, sorted by increasing z-values, is a compilation from different sources, supplemented with data from NED [12]. The covered distance range here is $z \approx 0.030$–3.912. Typical for such objects the huge diversity of designations may be somewhat confusing. This is mainly due to their discovery by various sky surveys at different wavelength domains. Some of these objects may be identified by more than 30 different designations!

To distinguish a quasar (QSO) or blazar (BL) from the fainter AGN of Seyfert galaxies (Sy), sometimes, as a very rough criterion, an absolute magnitude of at least $M_V = -23$ is applied. Anyway for such objects in the NED, this limit seems to be higher. It should further be noted that all of them show at least a significantly variable brightness. In some cases this range may reach up to several magnitudes! Therefore particularly the values for the apparent magnitudes m_v may

strongly differ between the individual databases, for example CDS [11] or NED [12].

Further, for many of these listed objects, NED also provides spectra in the optical domain. If available, for the planning of a spectroscopic observation it is advisable to consult NED. So it becomes clear, for example, that most of the BL objects show "flat spectra" which are not suited to determine any redshift. Just a few blazars may display some weak emissions.

Highly interesting are the two top listed, ultra-luminous quasars from the early times of the Universe with $z > 3$. They belong to the most luminous objects hosting supermassive black holes ($>20 \times 10^9 \ M_{\odot}$). APM 08279+5255 appears strongly deformed due to gravitational lensing effect. With larger telescopes ($\gtrsim 12$ inch) and excellent seeing conditions these impressive quasars should be achievable also by amateurs.

z	Object	Type	m_v	M_V	Const.	RA	DEC
0.030	Mrk421	BL	12.8	−25.96	UMa	11 04 27.3	+38 12 32
0.034	Mrk501	BL	13.3	−27.12	Her	16 53 52.2	+39 45 37
0.034	Mrk509	Sy	14.2	−23	Aqr	20 44 09.7	−10 43 25
0.044	MCG 11-19-006	Sy	15.0	−22.98	UMi	15 19 33.7	+65 35 59
0.045	Mrk180	BL	15.5	−26.78	Dra	11 36 26.4	+70 09 27
0.047	Mrk926	Sy	14.6	−22	Aqr	23 04 43.5	−08 41 09
0.047	1ES 1959+650	BL	14.7	−24.73	Dra	19 59 59.8	+65 08 55
0.049	PKS J1517-2422	BL	15.08	−26.25	Lib	15 17 41.8	−24 22 19
0.050	Mrk734	Sy	14.7	−23.16	Leo	11 21 47.1	+11 44 18
0.051	3C371	BL	14.2	−25.46	Dra	18 06 50.7	+69 49 28
0.053	IRAS 17596+4221	Sy	15.6	−21.15	Her	18 01 09.0	+42 21 43
0.055	IRAS 01072-0348	Sy	14.5	−19.73	Cet	01 09 45.1	−03 32 33
0.056	PKS 0521-36	BL	14.5	−22.55	Col	05 22 58.0	−36 27 31
0.057	1E 2124-149	Sy	14.7		Cap	21 27 32.4	−14 46 48
0.059	Mrk1502	Sy	14.4	−23.57	Psc	00 53 34.9	+12 41 36
0.062	Mrk1298	Sy	14.4	−23.65	Leo	11 29 16.6	−04 24 08

(cont.)

z	Object	Type	m_V	M_V	Const.	RA	DEC
0.063	2ZW 136	Sy	14.3	−24.92	Peg	21 32 27.8	+10 08 19
0.064	MR 2251−178	Sy	14.36		Aqr	22 54 05.8	−17 34 55
0.064	PG 0844+349	Sy1	13.5	−23.78	Lyn	08 47 42.4	+34 45 04
0.066	Mrk304	QSO	15.0	−23.24	Peg	22 17 12.2	+14 14 21
0.069	BL Lac	BL	14.7		Lac	22 02 43.3	+42 16 40
0.071	Mrk205	Sy1	14.5	−23.49	Dra	12 21 44.2	+75 18 39
0.079	Mrk478	Sy1	15.0	−23.79	Boo	14 42 07.4	+35 26 23
0.080	7ZW 118	Sy1	14.5	−23.55	Cam	07 07 13.1	+64 35 59
0.081	PG 1211+143	Sy1	14.3	−24.35	Com	12 14 17.7	+14 03 13
0.086	HE 1029−1401	QSO	13.6	−24.61	Hya	10 31 54.3	−14 16 51
0.087	Mrk1383	Sy	14.2	−23.86	Vir	14 29 06.6	+01 17 06
0.088	PG 1351+640	Sy	14.5	−24.30	Dra	13 53 15.8	+63 45 46
0.096	B3 0754+394	Sy	14.36	−24.24	Lyn	07 58 00.0	+39 20 29
0.100	PG 0804+762	QSO	14.3	−24.94	Cam	08 10 58.6	+76 02 43
0.112	IRAS 21219−1757	Sy	14.8	−23.56	Cap	21 24 41.6	−17 44 46
0.116	PKS 2155−304	BL	14.0		PsA	21 58 52.0	−30 13 32
0.142	PG 0026+129	Sy	15.4		Psc	00 29 13.7	+13 16 04
0.158	3C273	QSO	12.8	−27.70	Vir	12 29 06.7	+02 03 09
0.177	TON 1388	QSO	14.4	−25.34	Leo	11 19 08.7	+21 19 18
0.266	PKS J0757+0956	BL	14.5		Cnc	07 57 06.6	+09 56 35
0.297	KUV 18217+6419	QSO	14.1	−27.17	Dra	18 21 57.3	+64 20 36
0.306	PKS J0854+2006	BL	14.0	−28.14	Cnc	08 54 48.9	+20 06 31
0.370	HS 0624+6907	QSO	14.2	−27.61	Cam	06 30 02.5	+69 05 04
0.424	PKS 0735+17	BL	14.8		Gem	07 38 07.4	+17 42 19
0.573	PKS 0405−123	QSO	14.9	−27.91	Eri	04 07 48.4	−12 11 37
1.084	PG 1718+481	QSO	14.6	−30.02	Her	17 19 38.2	+48 04 12
1.334	PG 1634+706	QSO	14.7	−31.14	Dra	16 34 29.0	+70 31 32
3.366	S5 0014+81	BL	16.5	−31.40	Cep	00 17 08.5	+81 35 08
3.912	APM 08279+5255	QSO	15.2	−33.80	Lyn	08 31 41.7	+52 45 18

Catalogs:

1E: Einstein Satellite X-ray Survey, 1979–81

APM: Automatic Plate Measuring Facility, Cambridge

B3: Third Bologna Survey (B3) (Ficarra+ 1985)

HE: Hamburg/ESO Survey

HS: Hamburg Quasar Survey

IRAS: Cataloged Galaxies + QSO observed in IRAS Survey

KUV: Kiso Ultraviolet Catalog (Noguchi+ 1980–1984)

MCG: Vorontsov–Velyaminov, 1974

Mrk: Catalog of Markarian Galaxies (Markarian 1967–1981)

PG: Palomar–Green Catalog UV-excess stellar objects (Green+ 1986)

PKS: Parkes Radio Sources Catalog (PKSCAT90) (Wright+ 1990)

S5:   Fifth Survey of Strong Radio Sources

3C: Third Cambridge Revised Catalog (Bennet A. S.: 1961)

TON: Tonantzintla Catalog of Blue Stellar Objects, 1957–1959

ZW: Zwicky Catalog of Galaxies and of Clusters of Galaxies

APPENDIX J
Excerpts from Historical Spectral Atlases

The very first spectral atlas was written by Father Angelo Secchi at the Vatican Observatory, illustrating his classification system. Many sources call him therefore the "father of the modern astrophysics." The plates with the hand drawn spectra are taken from the German translation of his book *Die Sterne, Grundzüge der Astronomie der Fixsterne* (1878).

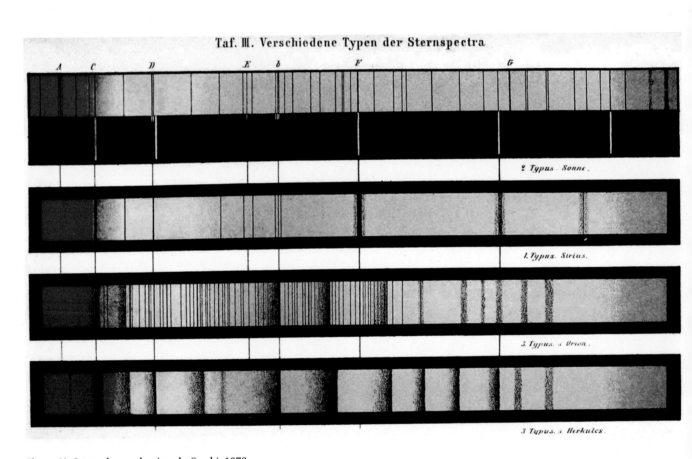

Figure J1 Spectral types by Angelo Secchi, 1878

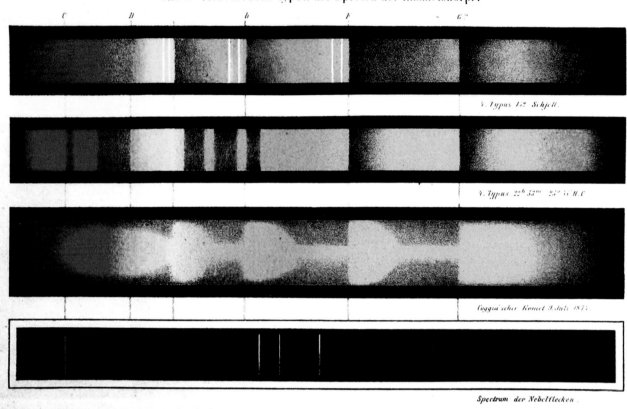

Figure J1 (*cont.*)

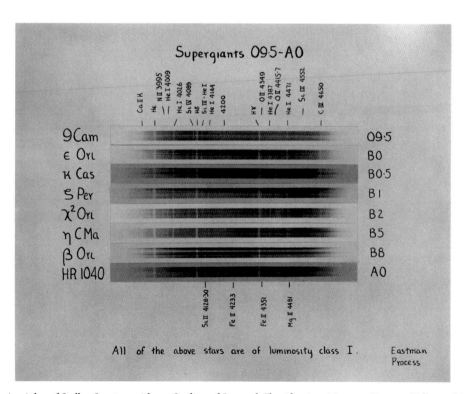

Figure J2 Plate from *An Atlas of Stellar Spectra, with an Outline of Spectral Classification*, Morgan, Keenan, Kellman (University of Chicago Press, 1943) [36]. The plates are all labeled by hand!

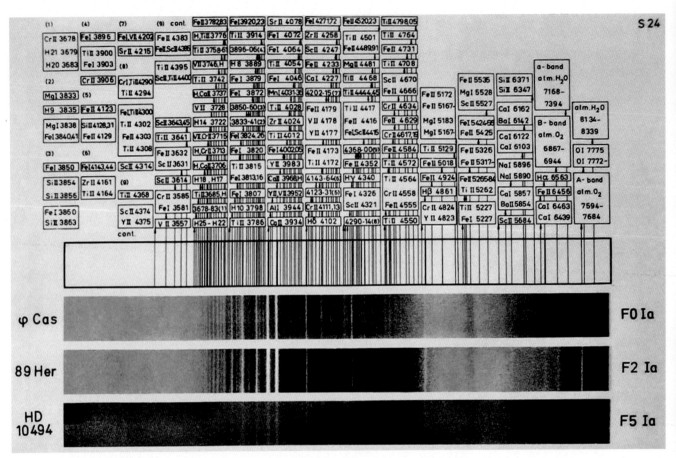

Figure J3 Plate from the *Bonner Spektralatlas* (out of print) (1975) by Waltraut Carola Seilter [10]. The presentation of the spectra is strictly ordered by luminosity classes (here Ia)

APPENDIX K
Instruments

Most of the spectra have been recorded by the following instruments:

Telescope: Celestron C8 Schmidt Cassegrain, aperture 8 inch, focal length ~200 cm, equatorial mount Vixen Sphinx SXD

Spectrographs: DADOS, Baader Planetarium [4], reflection gratings 200 and 900 L mm^{-1}, slit width: 25, 35, 50 µm $R_{200L} \approx 900$ $R_{900L} \approx 4000$ (at λ6160), dispersion [Å/pixel] with Atik 314L+: 200 L mm^{-1}: 2.55, 900 L mm^{-1}: 0.65

SQUES echelle, Eagleowloptics Switzerland [5], R ≈ 20,000 adjustable slit width: ~15–85 µm, dispersion with Atik 314L+: 0.18 [Å/pixel].

Cameras: Atik 314L+: Recording of the spectra, Meade DSI II: Slit camera

Location: Rifferswil, Switzerland

Elevation: 610 m above sea level

A few spectra of faint deep sky objects have been recorded in the Nasmith focus of the CEDES Cassegrain telescope of the Mirasteilas Observatory in Falera, Switzerland: aperture 36 inch, focal length ~900 cm.

Figure K1 C8 with DADOS spectrograph

Figure K3 CEDES Cassegrain telescope, Mirasteilas. With Martin Huwiler (left), technical head of the observatory.

Figure K2 SQUES echelle spectrograph with Ar, Ne, He glow starter calibration unit (12 V DC)

BIBLIOGRAPHY

[1] J. B. Kaler, *Stars and Their Spectra: An Introduction to the Spectral Sequence* (Cambridge: Cambridge University Press, 2011).

[2] R. O. Gray, Ch. J. Corbally, *Stellar Spectral Classification* (Princeton, NJ: Princeton University Press, 2009).

[3] M. Trypsteen, R. Walker, *Spectroscopy for Amateur Astronomers: Recording, Processing, Analysis and Interpretation* (Cambridge: Cambridge University Press, 2017).

[4] DADOS and Baches Spectrographs, Baader Planetarium: available online at: http://www.baader-planetarium.de/dados/download/dados_manual_english.pdf

[5] SQUES Echelle Spectrograph, Eagleowloptics Switzerland http://www.eagleowloptics.com/

[6] IRIS and ISIS: Christian Buil's home page http://www.astrosurf.com/buil/

[7] Visual Spec, Vspec: Valerie Désnoux's processing freeware http://astrosurf.com/vdesnoux/

[8] NIST Atomic Spectra Database, http://physics.nist.gov/PhysRefData/ASD/lines_form.html

[9] A. Lobel SpectroWeb, The Interactive Database of Spectral Standard Star Atlases, Royal Observatory of Belgium, available online at: http://spectra.freeshell.org/SpectroWeb_news.html

[10] W. C. Seitter, *Atlas für Objektiv Prismen Spektren: Bonner Spektralatlas* (Bonn: Astronomical Institute Bonn, 1975), available online at: http://www.archive.org/search.php?query=bonner%20atlas [accessed October, 2016].

[11] CDS Strasbourg: *SIMBAD Astronomical Database*, http://simbad.u-strasbg.fr/simbad/

[12] NASA Extragalactic Database (NED), http://nedwww.ipac.caltech.edu/

[13] The Bright Star Catalogue, 5th Revised Edn., data access by Alcyone Software: http://www.alcyone.de/search_in_bsc.html

[14] J. Kaler, STARS: a comprehensive suite of pages that tell the stories of stars and their constellations, http://stars.astro.illinois.edu/sow/sowlist.html

[15] A. Sota *et al.*, The Galactic O Star Spectroscopic Catalog (GOSC) and Survey (GOSSS), (2013), available online at: http://arxiv.org/abs/1306.6417

[16] A. Sota *et al.*, The Galactic O-Star Spectroscopic Survey (GOSSS). II. Bright Southern Stars, (2013), available online at: http://arxiv.org/pdf/1312.6222v1.pdf

[17] Versuchsanleitung zum Astrophysikalischen Praktikum, Grobe Klassifikation von Sternspektren, Kiepenheuer-Institut für Sonnenphysik, available online at: http://www.kis.uni-freiburg.de/fileadmin/user_upload/lehre/praktikum/sternspektren.pdf

[18] H. -G. Reimann, (Hrgb.), Astronomisches Praktikum, Astrophysikalisches Institut der Friedrich-Schiller-Universität Jena, (2000), available online at: http://www.wissenschaft-schulen.de/sixcms/media.php/1308/spektroskopie.pdf

[19] G. Ledrew, The Real Starry Sky, *Journal of the Royal Astronomical Society of Canada*, 95, (2001), 32, available online at: http://adsabs.harvard.edu/abs/2001JRASC..95...32L

[20] I. Grenier, *The Gould Belt, Star Formation and the Interstellar Medium*, (2004) available online at: http://arxiv.org/PS_cache/astro-ph/pdf/0409/0409096v1.pdf

[21] I. Grenier, C. Perrot, 3D Dynamical Evolution of the Interstellar Gas in the Gould Belt, *Astronomy & Astrophysics*, 404, (2003), 519–31, available online at: http://arxiv.org/pdf/astro-ph/0303516

[22] W. W. Morgan *et al.*, *Revised MK Spectral Atlas for Stars Earlier than the Sun*, (Yerkes Observatory, University of Chicago, Kitt Peak National Observatory, 1978), available online at: http://nedwww.ipac.caltech.edu/level5/March02/Morgan/frames.html

[23] N. R. Walborn, E. L. Fitzpatrick, Contemporary Optical Spectral Classification of the OB Stars: A Digital Atlas, *Astronomical Society of the Pacific*, 102, (1990), 379–411, available online at: http://adsabs.harvard.edu/full/1990PASP..102..379W

[24] V. Y. Alduseva, A. A. Aslanov *et al.*, Spectrum Variability of 68 Cygni, an O(f) Star at the Center of a Ring Nebula, *Pisma v Astronomicheskii Zhurnal*, 8, 717–21; *Soviet Astronomy Letters*, 8, (1982) 386–8, Translation, available online at: http://adsabs.harvard.edu/full/1982SvAL....8..386A

[25] S. Kraus, Y. Y. Balega, *et al.*, Visual/Infrared Interferometry of Orion Trapezium Stars: Preliminary Dynamical Orbit and Aperture Synthesis Imaging of the θ^1 Orionis C system, *Astronomy & Astrophysics*, 466, (2007), 649–59, available online at: www.skraus.eu/papers/kraus.T1OriC.pdf

[26] Harvard University Graduate Course 201B, 2011. *ISM and Star Formation: The Orion Bar*, http://ay201b.wordpress.com/2011/05/01/the-orion-bar/

[27] N. R. Walborn, An Atlas of Yellow-Red OB Spectra, *Astrophysical Journal Supplement Series*, 44, (1980), 535–8, available online at: http://articles.adsabs.harvard.edu/full/1980ApJS...44..535W

[28] D. F. Gray, *The Observation and Analysis of Stellar Photospheres*, Third Edition (Cambridge: Cambridge University Press, 2008).

[29] R. O. Gray, *A Digital Spectral Classification Atlas*, Appalachian State University available online at: http://nedwww.ipac.caltech.edu/level5/Gray/frames.html also available online at: http://aesesas.com/mediapool/142/1423849/data/DOCUMENTOS/digitalspectralatlasGray.pdf

[30] J. D. Monnier *et al.*, Imaging the Surface of Altair, *Science*, 317, 5836, (2007), 342–5, available online at: https://arxiv.org/pdf/0706.0867v2.pdf

[31] Bass2000: A download of the solar spectrum available online at: http://bass2000.obspm.fr/download/solar_spect.pdf

[32] NSO Digital Library, *Solar Spectral Atlases*, available online at: http://diglib.nso.edu/ftp.html

[33] J. Ryon *et al.*, Comparing the Ca II H and K Emission Lines in Red Giant Stars, *Publications of the Astronomical Society of Pacific*, 121, 882, (2009), 842–56, available online at: http://arxiv.org/pdf/0907.3346v2.pdf

[34] L. E. Allen, K. M. Strom, Moderate-Resolution Spectral Standards from λ5600 to λ9000, *Astronomical Journal* 109, 3, (1995), 1379–90, available online at: http://adsabs.harvard.edu/full/1995AJ....109.1379A

[35] J. D. Kirkpatrick *et al.*, A Standard Stellar Spectral Sequence in the Red/Near Infrared Classes: K5 to M9, *Astrophysical Journal Supplement Series*, 77, (1991), 417–40, available online at: http://articles.adsabs.harvard.edu/full/1991ApJS...77..417K

[36] W. W. Morgan, P. C. Keenan, E. Kellman, *An Atlas of Stellar Spectra, with an Outline of Spectral Classification*, Astrophysics Monographs, (Chicago: University of Chicago Press, 1943), available online at: http://nedwww.ipac.caltech.edu/level5/ASS_Atlas/frames.html

[37] A. V. Torres-Dodgen, W. B. Weaver, An Atlas of Low-Resolution Near-Infrared Spectra of Normal Stars, *Astronomical Society of the Pacific*, 105, 689, (1993), 693–720, available online at: http://adsabs.harvard.edu/abs/1993PASP..105..693T

[38] AAVSO, American Association of Variable Star Observers http://www.aavso.org/

[39] A. Valenti *et al.*, Spectral Synthesis of TiO Lines, *The Astrophysical Journal*, 498, 2, (1998), 851–62, available online at: http://adsabs.harvard.edu/abs/1998ApJ...498..851V

[40] J. M. Scalo, J. E. Ross, Synthetic Spectra of Red Giants. I – Representative Band Head Profiles of Diatomic Molecules, *Astronomy and Astrophysics*, 48, 2, (1976), 219–34, available online at: http://adsabs.harvard.edu/abs/1976A%26A....48..219S

[41] P. C. Keenan, P. C. Boeshaar, Spectral Types of S and SC Stars on the Revised MK system, *Astrophysical Journal Supplement Series*, 43, (1980), 379–91, available online at: http://articles.adsabs.harvard.edu/full/1980ApJS...43..379K

[42] S. J. Adelman, BVRI Photometry of the Extrinsic S Star HR 1105, *Astronomy and Astrophysics*, 333, (1998), 952–5, available online at: http://articles.adsabs.harvard.edu/full/1998A%26A...333..952A

[43] S. Wyckoff, R. E. S. Clegg, Molecular Spectra of Pure S-Stars, *Monthly Notices of the Royal Astronomical Society*, 184, (1978), 127–43, available online at: http://adsabs.harvard.edu/abs/1978MNRAS.184..127W

[44] A. G. Shcherbakov, I. Tuominen, Activity Modulation of the Red Giant HR 1105 as Observed in the He I Lambda 10830 A, *Astronomy and Astrophysics*, 255, 1–2, (1992), 215–20, available online at: http://adsabs.harvard.edu/abs/1992A&A...255..215S

[45] C. Albia *et al.*, Understanding Carbon Star Nucleosynthesis from Observations, *Publications of the Astronomical Society of Australia*, 20, 4, (2003), 314–23, available online at: http://www.publish.csiro.au/?act=view_file&file_id=AS03021.pdf

[46] P. Keenan, W. Morgan, Classification of the Red Carbon Stars, *Astrophysical Journal*, 94, (1941), 501, available online at: http://adsabs.harvard.edu/abs/1941ApJ....94..501K

[47] P. Keenan, Revised MK Classification of the Red Carbon Stars, *Astronomical Society of the Pacific*, 105, 691, (1993), 905–10, available online at: http://adsabs.harvard.edu/full/1993PASP..105..905K

[48] N. Mahmud *et al.*, Understanding Stellar Birth through the Photometric and Spectroscopic Variability of T Tauri Stars (2011), http://www.as.utexas.edu/ugrad_symposium/?a=3

[49] B. Kleman, Laboratory Excitation of the Blue-Green Bands Observed in the Spectra of N-Type Stars, *Astrophysical Journal*, 123, (1955), 162, available online at: http://adsabs.harvard.edu/doi/10.1086/146142

[50] P. J. Sarre *et al.*, SiC$_2$ in Carbon Stars: Merrill–Sanford Absorption Bands between 4100 and 5500 A, *Monthly Notices of the Royal Astronomical Society*, 319, 1, (2000), 103–10, available online at: http://articles.adsabs.harvard.edu/full/2000MNRAS.319..103S

[51] M. Hirai, Spectral Analysis of a Peculiar Star WZ Cassiopeiae, University of Tokyo, *Publications of the Astronomical Society of Japan*, 21, (1969), 91, available online at: http://adsabs.harvard.edu/full/1969PASJ...21...91H

[52] A. McKellar, The 6708 Resonance Line of Li I in the Spectrum of the N-Type Variable Star WZ Cassiopeiae, *The Observatory*, 64, (1941), 4–11, available online at: http://articles.adsabs.harvard.edu//full/1941Obs....64....4M/0000004.000.html

[53] C. Barnbaum *et al.*, A Moderate Resolution Spectral Atlas of Carbon Stars, *Astrophysical Journal Supplement*, 105, (1996), 419, available online at: http://articles.adsabs.harvard.edu/full/1996ApJS..105..419B

[54] D. Tyson *et al.*, Radial Velocity Distribution and Line Strengths of 33 Carbon Stars in the Galactic Bulge, *Astrophysical Journal*, 367, (1991), 547–60, available online at: http://articles.adsabs.harvard.edu/full/1991ApJ...367..547T

[55] L. Wallace, Band-Head Wavelengths of C$_2$, CH, CN, CO, NH, NO, O$_2$, OH and Their Ions, *Astrophysical Journal Supplement*, 7, (1962), 165, available online at: http://adsabs.harvard.edu/full/1962ApJS....7..165W

[56] S. Schneiderbauer, Carbon Stars, Hydrostatic Models and Optical/Near Infrared Interferometry, (2008), available online at: http://othes.univie.ac.at/1627/1/2008-10-13_9940129.pdf

[57] M. Vetesnik, Photometric and Spectroscopic Investigation of a New Carbon Star in the Auriga Region, *Astronomical Institutes of Czechoslovakia, Bulletin*, 30, 1, (1979), 1–12, available online at: http://articles.adsabs.harvard.edu/full/1979BAICz..30....1V

[58] J. B. Holberg *et al.*, A Determination of the Local Density of White Dwarf Stars, *The Astrophysical Journal*, 571, 1, (2002), 512–18, available online at: http://arxiv.org/pdf/astro-ph/0102120.pdf

[59] E. Falcon *et al.*, A Gravitational Redshift Determination of the Mean Mass of White Dwarfs DA Stars, *The Astrophysical Journal*, 712, 1, (2010), 585–95, available online at: http://arxiv.org/pdf/1002.2009v1.pdf

[60] G. P. McCook, E. M. Sion, A Catalog of Spectroscopically Identified White Dwarfs, *The Astrophysical Journal Supplement Series*, 121, 1, (1999), 1–130, available online at: http://heasarc.gsfc.nasa.gov/W3Browse/star-catalog/mcksion.html

[61] J. Liebert, E. M. Sion, The Spectroscopic Classification of White Dwarfs: Unique Requirements and Challenges, *Astronomical Society of the Pacific Conference Series*, 60, (1994), 64–72, available online at: http://adsabs.harvard.edu/full/1994ASPC...60...64L

[62] J. B. Holberg *et al.*, A New Look at the Local White Dwarf Population, *The Astronomical Journal*, 135, 4, (2008), 1225–38, available online at: http://iopscience.iop.org/1538-3881/135/4/1225/pdf/1538-3881_135_4_1225.pdf

[63] B. Wolff *et al.*, Element Abundances in Cool White Dwarfs II. Ultraviolet Observations of DZ White Dwarfs, *Astronomy and Astrophysics*, 385, (2002), 995–1007, available online at: http://arxiv.org/pdf/astro-ph/0204408v1.pdf

[64] Bonn University: *Pre-supernova Evolution of Massive Stars*, https://astro.uni-bonn.de/~nlanger/siu_web/ssescript/new/chapter11.pdf

[65] K. A. Van der Hucht, The VIIth Catalogue of Galactic Wolf–Rayet Stars, *New Astronomy Reviews*, 45, (2001), 135–232, available online at: www.astrosurf.com/luxorion/Documents/wrcat.pdf

[66] P. A. Crowther, Physical Properties of Wolf-Rayet Stars, *Annual Review of Astronomy and Astrophysics*, 45, (2007), 177–219, available online at: http://www.stsci.de/wr140/pdf/crowther2006.pdf

[67] A. B. Underhill, C IV λ5806 in Wolf Rayet Stars, *Astronomical Society of the Pacific*, 100, (1988), 1269–72, available online at: http://articles.adsabs.harvard.edu//full/1988PASP..100.1269U/0001269.000.html

[68] A. Underhill *et al.*, A Study of the Moderately Wide Wolf Rayet Spectroscopic Binary HD 190918, *Astrophysical Journal*, 432, 2, (1994), 770–84, available online at: http://adsabs.harvard.edu/full/1994ApJ...432..770U

[69] CFA, Harvard-Smithsonian Center for Astrophysics: *Atlas for Wolf–Rayet Stars*, https://www.cfa.harvard.edu/~pberlind/atlas/htmls/wrstars.html

[70] W. A. Hiltner, R. E. Schild, Spectral Classification of Wolf–Rayet Stars, *Astrophysical Journal*, 143, (1966), 770, available online at: http://articles.adsabs.harvard.edu/full/1966ApJ...143..770H

[71] S. V. Marchenko *et al.*, The Unusual 2001 Periastron Passage in the "Clockwork" Colliding Wind Binary WR 140, *Astrophysical Journal*, 596, (2003), 1295–304, available online at: http://www.stsci.de/wr140/pdf/marchenko2003.pdf

[72] P. M. Williams *et al.*, Multi Frequency Variations of Wolf-Rayet System HD 193793, *Monthly Notices of the Royal Astronomical Society*, 243, (1990), 662–84, available online at: http://esoads.eso.org/abs/1990MNRAS.243..662W

[73] W. R. Hamann *et al.*, Non-LTE Spectral Analyses of Wolf-Rayet Stars: The Nitrogen Spectrum of the WN6 Prototype HD 192163 (WR136), *Astronomy and Astrophysics* 281, 1, (1994), 184–98, available online at: http://adsabs.harvard.edu/abs/1994A&A...281..184H

[74] C. E. Cappa *et al.*, The Interaction of NGC 6888 and HD 192163 with the Surrounding Interstellar Medium, *Astronomical Journal* 112, (1996), 1104, available online at: http://articles.adsabs.harvard.edu/full/1996AJ....112.1104C

[75] S. A. Zhekov *et al.*, Suzaku Observations of the Prototype Wind-Blown Bubble NGC 6888, *The Astrophysical Journal*, 728, (2011), 135, available online at: http://arxiv.org/pdf/1012.3917

[76] N. Sanduleak, On Stars Having Strong O VI Emission, *Astrophysical Journal*, 164, (1971), L71, available online at: http://adsabs.harvard.edu/abs/1971ApJ...164L..71S

[77] C. B. Stephenson, Search for New Northern Wolf–Rayet Stars *Astronomical Journal*, 71, (1966), 477–81, available online at: http://adsabs.harvard.edu/abs/1966AJ....71..477S

[78] K. R. Sokal *et al.*, Chandra Detects the Rare Oxygen Type Wolf–Rayet Star WR 142 and OB Stars in Berkeley 87, *The Astrophysical Journal*, 715, 2, (2010), 1327–37, available online at: http://arxiv.org/pdf/1004.0462v1.pdf

[79] J. E. Drew *et al.*, Discovery of a WO Star in the Scutum-Crux Arm of the Inner Galaxy, *Monthly Notices of the Royal Astronomical Society*, 351, 1, (2004), 206–14, available online at: http://arxiv.org/pdf/astro-ph/0403482.pdf

[80] A. M. Cherepashchuk, D. N. Rustamov, WR stars with the O VI λλ3811, 3834 Emission Doublet, *Astrophysics and Space Science* 167, 2, (1990), 281–96, http://articles.adsabs.harvard.edu/full/1990Ap%26SS.167..281C

[81] V. F. Polcaro *et al.*, The WO Stars IV. Sand 5: A Variable WO Star?, *Astronomy and Astrophysics*, 325, (1996), 178–88, available online at: http://adsabs.harvard.edu/full/1997A%26A...325..178P

[82] O. Stahl, *Spectral Atlas of P Cygni*, Landessternwarte Heidelberg, http://maps.seds.org/Stars_en/Spectra/pcyg.html

[83] L. S. Luud, The Spectrum of P Cygni 1964, *Soviet Astronomy*, 11, (1964), 211, available online at: http://adsabs.harvard.edu/full/1967SvA....11..211L

[84] Maart de Groot, On the Spectrum and Nature of P Cygni, *Bulletin of the Astronomical Institutes of the Netherlands*, 20, (1968), 225, available online at: http://adsabs.harvard.edu/full/1969BAN....20..225D

[85] A. Miroshnichenko, Presentation: *Spectra of the Brightest Be Stars and Objects Description*, available as a download from: www.astrospectroscopy.de/Heidelbergtagung/Miroshnichenko2.ppt

[86] A. Miroshnichenko, Presentation: *Summary of Experiences from Observations of the Be-binary δ Sco*, available as a download from: www.astrospectroscopy.de/Heidelbergtagung/Miroshnichenko1.ppt

[87] J. Chauville *et al.*, High and Intermediate-Resolution Spectroscopy of Be Stars, *Astronomy & Astrophysics*, 378, (2001),

861–82, available online at: http://www.aanda.org/articles/aa/pdf/2001/42/aa1599.pdf

[88] J. Canto *et al.*, Carbon Monoxide Observations of R Monocerotis, NGC 2261 and Herbig Haro 39: The Interstellar Nozzle, *Astrophysical Journal*, 244, (1981), 102–14, available online at: http://articles.adsabs.harvard.edu/full/1981ApJ...244..102C

[89] K. H. Böhm *et al.*, Spectrophotometry of R Monocerotis, *Astronomy and Astrophysics*, 50, 3, (1976), 361–6, available online at: http://adsabs.harvard.edu/full/1976A%26A....50..361B

[90] N. J. J. Turner *et al.*, Models of the Spectral Energy Distributions of FU Orionis Stars, *The Astrophysical Journal*, 480, 2, (1996), 754–66, available online at: http://iopscience.iop.org/0004-637X/480/2/754/

[91] G. H. Herbig *et al.*, Line Structure in the Spectrum of FU Orionis, *The Astronomical Journal*, 136, (2008), 676–83, available online at: http://arxiv.org/abs/0806.4053

[92] S. J. Kenyon *et al.*, The FU Orionis Phenomenon, *Annual Review of Astronomy Astrophysics* 34, (1996), 207–40, available online at: www.ifa.hawaii.edu/~reipurth/reviews/hartmann.ps

[93] S. L. Powell *et al.*, The Periodic Spectroscopic Variability of FU Orionis, *Monthly Notices of the Royal Astronomical Society*, 426, 4, (2012), 3315–33, available online at: http://arxiv.org/abs/1209.0981

[94] P. Wellmann, Veränderungen im Spektrum von FU Orionis, *Zeitschrift für Astrophysik*, 29, (1951), 154, available online at: http://adsabs.harvard.edu/full/1951ZA....29..154W

[95] C. A. Grady *et al.*, Iron Emission Lines in the Spectra of Herbig Ae/Be Stars Viewed through Their Proto-Planetary Disks. *Bulletin of the American Astronomical Society*, 25, (1993), 1353, available online at: http://articles.adsabs.harvard.edu/full/1993AAS...183.4109G

[96] N. N. Samus *et al.*, *General Catalogue of Variable Stars (Samus+ 2007–2011)*, Online catalog available at: http://www.sai.msu.su/gcvs/cgi-bin/search.htm

[97] K. C. Smith, Chemically Peculiar Hot Stars, *Astrophysics and Space Science*, 237, 1–2, (1996), 77–105, available online at: http://articles.adsabs.harvard.edu/full/1996Ap%26SS.237...77S

[98] G. W. Preston, The Chemically Peculiar Stars of the Upper Main Sequence, *Annual Review of Astronomy and Astrophysics*, 12, (1974), 257–7, available online at: http://adsabs.harvard.edu/abs/1974ARA%26A..12..257P

[99] S. L. S. Shorlin *et al.*, A Highly Sensitive Search for Magnetic Fields in B, A and F Stars, *Astronomy and Astrophysics*, 392, (2002), 637–52, available online at: http://www.aanda.org/articles/aa/pdf/2002/35/aah3570.pdf

[100] C. W. Allen, *Astrophysical Quantities*, 3rd edn., (London: The Athlone Press, 1973), available online at: https://archive.org/details/AstrophysicalQuantities

[101] U. K. Gehlich, Differential Fine Analysis Sirius versus Vega, *Astronomy and Astrophysics*, 3, (1969), 169–78, available online at: http://articles.adsabs.harvard.edu/full/1969A%26A....3..169G

[102] H. M. Qiu *et al.*, The Abundance Patterns of Sirius and Vega, *The Astrophysical Journal*, 548, 2, (2001), 953–65, available online at: http://iopscience.iop.org/0004-637X/548/2/953/pdf/0004-637X_548_2_953.pdf

[103] H. W. Babcock, S. Burd, The Variable Magnetic Field of α^2 Canum Venaticorum, *The Astrophysical Journal*, 116, (1952), 8–17, available online at: http://adsabs.harvard.edu/abs/1952ApJ...116....8B

[104] H. W. Babcock, The 34-kilogauss Magnetic Field of HD 215441, *The Astrophysical Journal*, 132, (1960), 521, available online at: http://adsabs.harvard.edu/full/1960ApJ...132..521B

[105] F. Borra, J. D. Landstreet, The Magnetic Field Geometry of HD 215441, *The Astrophysical Journal*, 1, 222, (1978), 226–33, available online at: http://adsabs.harvard.edu/full/1978ApJ...222..226B

[106] V. G. Elkin *et al.*, A Rival for Babcock's Star: the Extreme 30-kG Variable Magnetic Field in the Ap Star HD 75049, *Monthly Notices of the Royal Astronomical Society*, 402, 3, (2010), 1883–91, available online at: http://arxiv.org/pdf/0908.0849v2.pdf

[107] D. M. Peterson *et al.*, The Spectroscopic Orbit of β Scorpii A, *Astronomical Society of the Pacific*, 91, (1979), 87–94, available online at: http://adsabs.harvard.edu/abs/1979PASP...91...87P

[108] M. Palate *et al.*, Spectral Modelling of the Alpha Virginis (Spica) Binary System, *Astronomy & Astrophysics*, 556, (2013), 1–9, available online at: http://arxiv.org/pdf/1307.1970.pdf

[109] D. Pourbaix, A. Tokovinin *et al.*, SB9: The Ninth Catalogue of Spectroscopic Binary Orbits, (2003), database available online at: http://sb9.astro.ulb.ac.be/

[110] S. K. Gray, R. Ignace, Hα-Study of β Lyrae, *Journal of the Southeastern Association for Research in Astronomy*, 2, (2008), 71–8, available online at: http://jsara.org/volume02-080501/ms0214-Gray.pdf

[111] J. L. Hoffman *et al.*, Spectropolarimetric Evidence for a Bipolar Flow in β Lyrae, *The Astronomical Journal*, 115, 4, (1998), 1576–91 available online at: http://adsabs.harvard.edu/abs/1998AJ....115.1576H

[112] R. E. Mennickent, G. Djurašević, On the accretion disc and Evolutionary Stage of β Lyrae, *Monthly Notices of the Royal Astronomical Society*, 432, 1, (2013), 799–809, available online at: http://arxiv.org/abs/1303.5812

[113] M. Skulsky *et al.*, On the Dynamics of Circumstellar Gaseous Structures and Magnetic Field of β Lyrae, *Magnetic Stars*, (2011), 259–63, available online at: https://www.sao.ru/Docen/Science/Public/Conf/magstars-2010/p259.pdf

[114] R. E. Williams, The Formation of Novae Spectra, *Astronomical Journal* 104, 2, (1992), 725–33, available online at: http://adsabs.harvard.edu/full/1992AJ....104..725W

[115] A. Skopal *et al.*, The Early Evolution of the Extraordinary Nova Del 2013 (V339 Del), *Astronomy and Astrophysics*, 569, (2014), 14 pp., available online at: http://arxiv.org/abs/1407.8212

[116] V. Buat-Ménard *et al.*, The Nature of Dwarf Nova Outbursts, *Astronomy and Astrophysics*, 366, (2000), 612–22, available online at: http://arxiv.org/abs/astro-ph/0011312

[117] L. Morales-Rueda, T. R. Marsh, Spectral Atlas of Dwarf Novae in Outburst, *Monthly Notices of the Royal Astronomical*

Society, 332, 4, (2002), 814–26, available online at: http://arxiv.org/pdf/astro-ph/0201367v1.pdf

[118] J. Roche, *EG Andromedae, A Symbiotic System as an Insight into Red Giant Chromospheres*, Ph.D. thesis, Trinity College Dublin, (2012), available online at: http://arxiv.org/abs/1210.7699

[119] K. Kolb *et al.*, Synthetic Spectral Analysis of the Hot Component in the S-Type Symbiotic Variable EG Andromeda, *The Astronomical Journal*, 128, 4, (2004), 1790–4, available online at: http://arxiv.org/abs/astro-ph/0407536

[120] R. E. Williams *et al.*, The Evolution and Classification of Post-outburst Novae Spectra, *The Astrophysical Journal*, 376, (1991), 721–37, available online at: http://adsabs.harvard.edu/full/1991ApJ...376..721W

[121] R. E. Williams *et al.*, The Tololo Nova Survey: Spectra of Recent Novae, *Astrophysical Journal Supplement Series*, 90, 1, (1993), 297–316, available online at: http://adsabs.harvard.edu/full/1994ApJS...90..297W

[122] G. C. Anupama, T. P. Prabhu, Hα Variability in the Quiescence Spectrum of the Recurrent Nova T Coronae Borealis, *Monthly Notices of the Royal Astronomical Society*, 253, (1991), 605–9, available online at: http://adsabs.harvard.edu/full/1991MNRAS.253..605A

[123] D. R. Van Rossum, The Nature of the Si II 6150Å, Ca II HK, Ca II IR-Triplet, and Other Spectral Features in Supernova Type Ia Spectra, (2012), eprint arXiv:1208.3781 available online at: http://arxiv.org/pdf/1208.3781.pdf

[124] A. Filippenko, Optical Spectra of Supernovae, *Annual Review of Astronomy and Astrophysics*, 35, (1997), 309–55, available online at: http://www.astrouw.edu.pl/~nalezyty/semistud/Artykuly/annurev.astro.35.1.309_ss.pdf

[125] S. Woolsley, H. T. Janka, The Physics of Core Collapse Supernovae, *Nature Physics*, 1, 3, (2006), 147–54, available online at: http://arxiv.org/abs/astro-ph/0601261

[126] V. Grinberg, Presentation *Supernovae-Explosionsmechanismen*, La Villa (2006), available online at: http://pulsar.sternwarte.uni-erlangen.de/wilms/teach/lavilla06/07_grinberg.pdf

[127] O. Eberhardt, Entfernungsmessung mit Supernovae, 2008, available online at: http://www.physik.uni-regensburg.de/forschung/gebhardt/gebhardt_files/skripten/Entfernungsbest.pdf

[128] D. Richardson *et al.*, A Comparative Study of the Absolute Magnitude Distributions of Supernovae, *The Astronomical Journal*, 123, 2, (2001), 745–52, available online at: http://arxiv.org/pdf/astro-ph/0112051v1.pdf

[129] A. Filippenko *et al.*, Estimating The First-Light Time of the Type IA Supernova 2014J in M82, *The Astrophysical Journal Letters*, 783, 1, (2014), L24, 5 pp., available online at: http://arxiv.org/pdf/1401.7968.pdf

[130] K. E. H. Kjær, *Integral Field Spectroscopy Observations of SN 1987A*, Ph.D. thesis, Ludwig-Maximilian University of Munich (2007), available online at: https://www.imprs-astro.mpg.de/sites/default/files/2004_kjaer_karina.pdf

[131] ESO Messenger Nr. 47, *The Spectrum of Supernova 1987A*, available online at: https://www.eso.org/sci/publications/messenger/archive/no.47-mar87/messenger-no47-32-33.pdf

[132] E. M. Schlegel *et al.*, The Type Ib Supernova 1984L in NGC 991, *The Astronomical Journal* 98, (1989), 577–89, 740, 741, available online at: http://adsabs.harvard.edu/abs/1989AJ.....98..577S

[133] A. Filippenko *et al.*, The Type Ic (Helium-Poor Ib) Supernova 1987M - Transition to the Supernebular Phase, *The Astronomical Journal*, 100, (1990), 1575–87, 1757, available online at: http://adsabs.harvard.edu/full/1990AJ.....100.1575F

[134] R. C. Kennicutt, Spectrophotometric Atlas of Galaxies, *Astrophysical Journal Supplement Series*, 79, 2, (1992), 255–84, available online at: http://articles.adsabs.harvard.edu/full/1992ApJS...79..255K

[135] N. W. Evans *et al.*, Dynamical Mass Estimates for the Halo of M31 from Keck Spectroscopy, *The Astrophysical Journal*, 540, 1, (2000), L9–L12, available online at: https://arxiv.org/pdf/astro-ph/0008155v1.pdf

[136] I. Trujillo *et al.*, Unveiling the Nature of M94's (NGC 4736) Outer Region: A Panachromatic Perspective, *The Astrophysical Journal*, 704, 1, (2009), 618–28, available online at: http://arxiv.org/pdf/0907.4884v1.pdf

[137] I. Appenzeller, *Introduction to Astronomical Spectroscopy* (Cambridge: Cambridge University Press, 2012).

[138] S. Veilleux, D. Osterbrock, *Spectral Classification of Emission Line Galaxies*, 1986, available online at: http://adsabs.harvard.edu/abs/1987ApJS...63..295V

[139] B. Garcia-Lorenzo *et al.*, Spectroscopic Atlas of the Central 24"x 20" of the Seyfert 2 Galaxy NGC 1068, *The Astrophysical Journal* 518, 1, (1999), 190, available online at: http://iopscience.iop.org/0004-637X/518/1/190/

[140] P. M. Ogle *et al.*, Testing the Seyfert Unification Theory: Chandra HETGS Observations of NGC 1068, *Astronomy and Astrophysics*, 402, (2003), 849–64, available online at: http://arxiv.org/abs/astro-ph/0211406

[141] J. M. Shuder, D. E. Osterbrock, Empirical Results from a Study of Active Galactic Nuclei, *The Astrophysical Journal*, 250, (1981), 55–65, available online at: http://adsabs.harvard.edu/abs/1981ApJ...250...55S

[142] S. Paltani, M. Türler, The Mass of the Black Hole in 3C273, *Astronomy and Astrophysics*, 435, 3, (2005), 811–20, available online at: http://arxiv.org/abs/astro-ph/0502296

[143] A. Boksenberg *et al.*, New Spectrometric Results on the Quasar 3C273, *Monthly Notices of the Royal Astronomical Society*, 172, (1974), 289–303, available online at: http://adsabs.harvard.edu/full/1975MNRAS.172..289B

[144] B. M. Peterson *et al.*, Are Forbidden Lines Present in the Optical Spectrum of the QSO 3C 273?, *The Astrophysical Journal*, 283, (1984), 529–31, available online at: http://articles.adsabs.harvard.edu/full/1984ApJ...283..529P

[145] J. B. Oke, The Optical Spectrum of 3C273, *The Astrophysical Journal*, 141, (1964), 6, available online at: http://adsabs.harvard.edu/abs/1965ApJ...141....6O

[146] M. J. Avara, *Precision X-Ray Spectroscopy of 3C273 Jet Knots*, B.Sc. dissertation, Massachusetts Institute of Technology, (2008), available online at: http://dspace.mit.edu/bitstream/handle/1721.1/44464/297176629.pdf?sequence=1

[147] J. L. Schmitt, BL Lac Identified as a Radio Source, *Nature*, 218, (1968), 663.

[148] P. W. Gorham *et al.*, Markarian 421's Unusual Satellite Galaxy, *The Astronomical Journal*, 119, (2000), 1677–86, available online at: http://arxiv.org/pdf/astro-ph/9908077v3.pdf

[149] C. Foster *et al.*, Deriving Metallicities from the Integrated Spectra of Extragalactic Globular Clusters Using the Near-Infrared Calcium Triplet, *The Astronomical Journal*, 139, 4, (2009), 1566–78, available online at: http://arxiv.org/pdf/1002.1107v2.pdf

[150] T. Idiart *et al.*, The Infrared Ca II Triplet as a Metallicity Indicator of Stellar Populations, *The Astronomical Journal*, 113, (1997), 1066, available online at: http://adsabs.harvard.edu/abs/1997AJ....113.1066I

[151] F. Mauro *et al.*, Deriving Metallicities from Calcium Triplet Spectroscopy in Combination with Near Infrared Photometry, *Astronomy and Astrophysics*, 563, (2014), A76, 22 pp., available online at: http://arxiv.org/pdf/1401.0014v1.pdf

[152] H. A. Abt, H. Levato, Spectral Types in the Pleiades, *Publications of the Astronomical Society of the Pacific*, 90, (1978), 201–3, available online at: http://adsabs.harvard.edu/full/1978PASP...90..201A

[153] W. Morgan, The Integrated Spectral Types of Globular Clusters, *Publications of the Astronomical Society of the Pacific*, 68, 405, (1956), 509, available online at: http://adsabs.harvard.edu/full/1956PASP...68..509M

[154] R. P. Schiavon *et al.*, A Library of Integrated Spectra of Galactic Globular Clusters, *The Astrophysical Journal Supplement Series*, 160, (2005), 163–75, available online at: https://www.noao.edu/ggclib/ms.pdf

[155] S. Katsuda *et al.*, Discovery of a Pulsar Wind Nebula Candidate in the Cygnus Loop, *The Astrophysical Journal Letters*, 754, 1, (2012), L7, 5 pp.

[156] G. A. Gurzadyan, *The Physics and Dynamics of Planetary Nebulae* (Berlin, Heidelberg: Springer-Verlag, 2010).

[157] G. A. Gurzadyan, Excitation Class of Nebulae: An Evolution Criterion? *Astrophysics and Space Science*, 181, 1, (1990), 73-88, available online at: http://articles.adsabs.harvard.edu/full/1991Ap%26SS.181...73G

[158] W. A. Reid *et al.*, An Evaluation of the Excitation Parameter for the Central Stars of Planetary Nebulae, *Publications of the Astronomical Society of Australia*, 27, 2, (2010), 187–98, available online at: http://arxiv.org/PS_cache/arxiv/pdf/0911/0911.3689v2.pdf

[159] J. Schmoll, AIP: *3D Spektrofotometrie Extragalaktischer Emissionslinienobjekte*, Ph.D. thesis, University of Potsdam (2001), available online at: http://www.aip.de/groups/publications/schmoll.pdf

[160] F. Gieseking, *Planetarische Nebel*, six-part article series, in Sterne und Weltraum, 1983.

[161] J. P. Rozelot, C. Neiner, *Astronomical Spectrography for Amateurs*, EAS Publication Series, 47 (2011).

[162] Y. P. Varshni *et al.*, Emission Lines Identified in Planetary Nebulae, (2006), table available online at: http://laserstars.org/data/nebula/identification.html

[163] J. Kaler, *The Planetary Nebulae*, part of Jim Kaler's STARS website, available online at: http://stars.astro.illinois.edu/sow/pn.html

[164] M. Cohen *et al.*, The Peculiar Object HD 44179 "The Red Rectangle," *Astrophysical Journal*, 196, (1975), 179–89, available online at: http://adsabs.harvard.edu/abs/1975ApJ...196..179C

[165] J. D. Thomas, *Spectroscopic Analysis and Modeling of the Red Rectangle*, Ph.D. thesis, University of Toledo, (2012), available online at: https://etd.ohiolink.edu/ap:10:0::NO:10:P10_ACCESSION_NUM:toledo1341345222

[166] R. J. Glinski *et al.*, On the Red Rectangle Optical Emission Bands, *Monthly Notices of the Royal Astronomical Society*, 332, 2, (2002), L17–L22, available online at: http://adsabs.harvard.edu/full/2002MNRAS.332L..17G

[167] N. Wehres *et al.*, C_2 Emission Features in the Red Rectangle, *Astronomy and Astrophysics*, 518, (2010), A36, 10 pp., available online at: http://adsabs.harvard.edu/abs/2010A%26A...518A..36W.php

[168] K. Davidson, Emission-line Spectra of Condensations in the Crab Nebula, *The Astrophysical Journal*, 228, (1979), 179–90, available online at: http://adsabs.harvard.edu/abs/1979ApJ...228..179D

[169] J. R. Walsh *et al.*, Complex Ionized Structure in the Theta-2 Orionis Region, *Monthly Notices of the Royal Astronomical Society*, 201, (1981), 561-7, available online at: http://articles.adsabs.harvard.edu/full/1982MNRAS.201..561W

[170] NASA Website: *NASA Education/The Planet Venus* http://www.nasa.gov/audience/forstudents/5-8/features/F_The_Planet_Venus_5-8.html

[171] M. E. Brown *et al.*, A High Resolution Catalogue of Cometary Emission Lines, *The Astronomical Journal*, 112, (1996), 1197–202, available online at: http://www.gps.caltech.edu/~mbrown/comet/echelle.html

[172] G. Catanzaro, High Resolution Spectral Atlas of Telluric Lines, *Astrophysics and Space Science*, 257, 1, (1997), 161–70, available online at: http://webusers.ct.astro.it/gca/papers/telluric.pdf

[173] W. D. Vacca *et al.*, A Method of Correcting Near-Infrared Spectra for Telluric Absorption. *The Publications of the Astronomical Society of the Pacific*, 115, 805, (2003), 389–409, available online at: http://www.journals.uchicago.edu/doi/pdf/10.1086/346193

[174] P. Schlatter, "Trocknen" von Sternspektren, *Spektrum 43*, (2012), available online at: http://spektroskopie.fg-vds.de/pdf/Spektrum43.pdf

[175] H. D. Babcock, L. Herzberg, Fine Structure of the Red System of Atmospheric Oxygen Bands, *The Astrophysical Journal*, 108, (1948), 167, available online at: http://adsabs.harvard.edu/full/1948ApJ...108..167B

[176] C. R. Benn, Measuring Light Pollution on La Palma, (2007), available online at: http://www.starlight2007.net/pdf/proceedings/C_Benn.pdf

[177] P. Schlatter, A. Ulrich, Der Natrium Flash, *Mitteilungsblatt Spektrum*, 47, (2014), 8–13, available online at: http://www.astrospectroscopy.de/Spektrum%2047.pdf

[178] V. M. Slipher, The Spectrum of Lightning, *Lowell Observatory Bulletin No. 79*, 3, (1917), 55–8, available online at: http://adsabs.harvard.edu/full/1917LowOB...3...55S

[179] R. E. Orville, R. W. Henderson, Absolute Spectral Irradiance Measurements of Lightning from 375 to 880 nm, *Journal of Atmospheric Sciences*, 41, (1984), 21, 3180, available online at: http://adsabs.harvard.edu/abs/1984JAtS...41.3180O also available online at: http://arxiv.org/pdf/1206.4367.pdf

[180] A. J. Pickles, A Stellar Spectral Flux Library, 1150–25000 Å. *Publications of the Astronomical Society of the Pacific*, 110, 749, (1998), 863–78, available online at: http://adsabs.harvard.edu/abs/1998PASP··110··863P

[181] E. Karkoschka, *The Observer's Sky Atlas*, (New York: Springer-Verlag, 2007).

Further Reading

C. Abia *et al.*, The Chemical Composition of the Rare J-Type Carbon Stars, *Memorie della Societa Astronomica Italiana*, 71, (1999), 631–8, available online at: http://arxiv.org/PS_cache/astro-ph/pdf/9912/9912025v1.pdf

R. Ansgar, Observations of Cool-Star Magnetic Fields, *Living Reviews in Solar Physics*, 8, (2012), 1, available online at: http://solarphysics.livingreviews.org/Articles/lrsp-2012-1/download/lrsp-2012-1Color.pdf

J. P. Aufdenberg *et al.*, The Nature of the Na I D-lines in the Red Rectangle, *Monthly Notices of the Royal Astronomical Society*, 417, 4, (2011), 2860–73, available online at: http://arxiv.org/pdf/1107.4961.pdf

C. Barnbaum, A High Resolution Spectral Atlas of Carbon Stars, *Astrophysical Journal Supplement Series*, 90, 1, (1993), 317–432, available online at: http://articles.adsabs.harvard.edu/full/1994ApJS...90··317B

Caltech, Images, Catalogs and Atlases: A Collection of Spectral Atlases of Extragalactic Objects, available online at: http://nedwww.ipac.caltech.edu/level5/catalogs.html

D. Chochol *et al.*, Photometry and Spectroscopy of the Classical Nova V339 Del in the First Month After Outburst, *Contributions of the Astronomical Observatory Skalnaté Pleso*, 43, 3, (2014), 330–7, available online at: http://adsabs.harvard.edu/abs/2014CoSka··43··330C

I. J. Danziger *et al.*, Optical Spectra of Supernova Remnants, *Astronomical Society of the Pacific*, 88, (1976), 44–9, available online at: http://articles.adsabs.harvard.edu/full/1976PASP...88...44D

I. J. Danziger *et al.*, Optical and Radio Studies of SNR in the Local Group Galaxy M33, *The Messenger*, 21, (1980), 7, available online at: http://www.eso.org/sci/publications/messenger/archive/no.21-sep80/messenger-no21-7-11.pdf

J. R. Ducsati *et al.*, Intrinsic Colors of Stars in the Near Infrared, *The Astrophysical Journal*, 558, 1, (2001), 309, available online at: http://iopscience.iop.org/0004–637X/558/1/309/fulltext/

DVAA Quasar List, Sorted by magnitude, available online at: http://dvaa.org/AData/DVAA_Quasars_Appendix_B_Magnitude.pdf

T. Eversberg, K. Vollmann, *Spectroscopic Instrumentation, Fundamentals and Guidelines for Astronomers* (Berlin, Heidelberg: Springer-Verlag, 2015).

A. Filippenko, Optical Spectra and Light Curves of Supernovae, *Proceedings of the ESO/MPA/MPE Workshop, ESO Astrophysics Symposia*, (2003), p. 171, available online at: http://arxiv.org/pdf/astro-ph/0307138v1.pdf

M. Fitzgerald, The Intrinsic Colours of Stars and Two-Colour Reddening Lines, *Astronomy and Astrophysics*, 4, (1970), 234, available online at: http://articles.adsabs.harvard.edu/full/1970A%26A....4··234F

A. Gianninas *et al.*, A Spectroscopic Survey and Analysis of Bright, Hydrogen-Rich White Dwarfs, *The Astrophysical Journal*, 743, 2, (2011), 138, available online at: http://arxiv.org/abs/1109.3171

B. J. L. Greenstein, The Gravitational Redshift of 40 Eridani B, *Astrophysical Journal*, 175, (1972), L1, available online at: http://adsabs.harvard.edu/abs/1972ApJ...175L...1G

C. S. Hansen-Ruiz, F. van Leeuwen, Definition of the Pleiades Main Sequence in the HR Diagram, *Proceedings From The Hipparcos Venice '97 Symposium* (1997), available online at: http://www.cosmos.esa.int/documents/532822/546798/poster03_05.pdf

R. W. Hanuschik, High Resolution Emission line Spectroscopy of Be Stars, I. Evidence for a Two-Component Structure of the Hα Emitting Envelope, *Astronomy and Astrophysics*, 166, 1–2, (1986), 185–94, available online at: http://articles.adsabs.harvard.edu/full/1986A%26A...166··185H

K. M. Harrison, *Astronomical Spectroscopy for Amateurs* (New York: Springer, 2011).

S. Harrold, J. Kajubi, Active Nuclei and Their Host Galaxies: Observations of Seyfert Galaxies, 2008, available online at: www.pas.rochester.edu/~advlab/reports/harrold_kajubi_astro2.pdf

J. B. Hearnshaw, *The Analysis of Starlight: One Hundred and Fifty Years of Astronomical Spectroscopy* (Cambridge: Cambridge University Press, 1990).

J. B. Hearnshaw, *Astronomical Spectrographs and Their History* (Cambridge: Cambridge University Press, 2009).

F. V. Hessmann, The Spectrum of SS Cygni During a Dwarf Nova Eruption, *The Astrophysical Journal*, 300, (1986), 794–803, available online at: http://adsabs.harvard.edu/full/1986ApJ...300··794H

M. Hogerheijde, Leiden Observatory, *Radiative Processes, Appendix E, Extra Problem Set*, available online at: http://home.strw.leidenuniv.nl/~michiel/ismclass_files/radproc07/extra_problem_set_3.pdf

IRAF, NOAO, Image Reduction and Analysis Facility, http://iraf.noao.edu

G. H. Jacobi *et al.*, *A Library of Stellar Spectra*, http://cdsarc.u-strasbg.fr/viz-bin/Cat?III/92

M. A. Jolin *et al.*, Toward Understanding the Environment of R Monocerotis from High Resolution Near-Infrared Polarimetric Observations, *The Astrophysical Journal*, 721, (2010), 1748–54, available online at: http://iopscience.iop.org/0004–637X/721/2/1748/fulltext/apj_721_2_1748.text.html

F. Joos, H. M. Schmid, Limb Polarization of Uranus and Neptune II: Spectropolarimetric Observations, *Astronomy and Astrophysics*, 463, 3, (2007), 1201–10, available online at: http://adsabs.harvard.edu/abs/2007A%26A...463.1201J

C. R. Kitchin, *Optical Astronomical Spectroscopy* (New York: Taylor and Francis Group, 1995).

F. Lamareille *et al.*, Spectral Classification of Emission-Line Galaxies from the Sloan Digital Sky Survey I, *Astronomy and Astrophysics*, 509, (2010), A53, 5 pp. and Spectral Classification of Emission-Line Galaxies from the Sloan Digital Sky Survey II, *Astronomy and Astrophysics*, 531, (2011), A71, 10 pp., available online at: Part I: 2010 http://adsabs.harvard.edu/abs/2009arXiv0910.4814L, Part II: 2011 http://arxiv.org/abs/1105.0488

E. Landi Degl'Innocenti, On the Effective Landé Factor of Magnetic Lines, *Solar Physics*, 77, 1–2, (1980), 285–9, available online at: http://adsabs.harvard.edu/full/1982SoPh...77..285L

J. D. Landstreet, The Measurement of Magnetic Fields in Stars, *The Astronomical Journal*, 85, (1980), 611–20, available online at: http://adsabs.harvard.edu/full/1980AJ.....85..611L

R. Lorenz *et al.*, Backyard Spectroscopy and Photometry of Titan, Uranus and Neptune, *Planetary and Space Science*, 51, (2003), 113–25, available online at: http://www.lpl.arizona.edu/~rlorenz/amateur.pdf

B. T. Lynds, The Spectra of White Dwarfs, *The Astrophysical Journal*, 125, (1957), 719, available online at: http://adsabs.harvard.edu/full/1957ApJ...125..719L

G. Mathys, Ap Stars with Resolved Zeeman Split Lines, *Astronomy and Astrophysics*, 232, 1, (1990), 151–72, available online at: http://articles.adsabs.harvard.edu/full/1990A%26A...232..151M

G. Mathys, T. Lanz, Ap Stars with Resolved Magnetically Split Lines, *Astronomy and Astrophysics*, 256, 1, (1992), 169–84, available online at: http://adsabs.harvard.edu/full/1992A%26A...256..169M

G. Mathys *et al.*, The Mean Magnetic Field Modulus of Ap Stars, Astronomy *and Astrophysics Supplement Series*, 123, (1996) 353–402, available online at: http://aas.aanda.org/articles/aas/abs/1997/08/ds1257/ds1257.html

A. McKay, *Spectra of T Tauri Stars*, http://astronomy.nmsu.edu/

MIDAS, ESO, http://www.eso.org/sci/software/esomidas//

MILES Spectral Library, containing ~1000 spectra of reference stars, http://miles.iac.es/pages/stellar-libraries/miles-library.php

A. Miroshnichenko *et al.*, Properties of the δ Scorpii Circumstellar Disk from Continuum Modeling, *The Astrophysical Journal*, 652, 2, (2006), 1617, available online at: http://libres.uncg.edu/ir/uncg/f/A_Miroshnichenko_Properties_2006.pdf

B. D. Moore *et al.*, HST Observations of the Wolf-Rayet Nebula NGC 6888, *The Astronomical Journal*, 119, (2000), 2991–3002, available online at: http://arxiv.org/abs/astro-ph/0003053

L. A. Morgan, The Emission Line Spectrum of the Orion Nebula, *Monthly Notices of the Royal Astronomical Society*, 153, (1971), 393–9, available online at: http://articles.adsabs.harvard.edu/cgi-bin/nph-iarticle_query?1971MNRAS.153..393M&defaultprint=YES&filetype=.pdf

S. Noll, *Spektroskopische Variationen des Herbig Ae/Be Sterns HD 163296*. Diploma thesis, University of Heidelberg (1999), http://www.lsw.uni-heidelberg.de/projects/hot-stars/Diplom_Noll.pdf

D. K. Ojha, S. C. Joshi *et al.*, On the Shell Star Pleione (BU Tauri), *Journal of Astrophysics and Astronomy*, 12, 3, (1991), 213–23, available online at: http://www.ias.ac.in/jarch/jaa/12/213–223.pdf

B. M. Peterson *et al.*, Central Masses and Broad-Line Region Sizes of Active Galactic Nuclei. II. A Homogeneous Analysis of a Large Reverberation-Mapping Database, *The Astrophysical Journal*, 613, 2, (2004), 682–99, available online at: http://arxiv.org/abs/astro-ph/0407299

E. Pollmann, *Spektroskopische Beobachtungen der Hα- und der He I 6678-Emission am Doppelsternsystem δ Scorpii*, http://www.bav-astro.de/rb/rb2009-3/151.pdf

A. K. Pradhan, S. N. Nahar, *Atomic Astrophysics and Spectroscopy* (Cambridge: Cambridge University Press, 2011).

A. Reiners, Magnetic Fields in Low-Mass Stars: An Overview of Observational Biases, *Proceedings of the International Astronomical Union, IAU Symposium*, 302, (2014), 156–63, available online at: http://arxiv.org/pdf/1310.5820v1.pdf

K. Robinson, *Spectroscopy: The Key to the Stars* (New York: Springer-Verlag, 2007).

J. P. Rozelot, C. Neiner, The Rotation of Sun and Stars, *Lecture Notes in Physics 765*, (Berlin, Heidelberg: Springer, 2009).

RSpec: Tom Field's Real-time Spectroscopy, http://www.rspec-astro.com/

Shelyak Instruments: Alpy, Lhires, LISA and eShel, http://www.shelyak.com/

H. L. Shipman *et al.*, The Mass and Radius of 40 Eridani B from HIPPARCOS: An Accurate Test of Stellar Interior Theory, *The Astrophysical Journal*, 488, (1997), L43–6, available online at: http://iopscience.iop.org/1538–4357/488/1/L43

T. G. Slanger, P. C. Cosby *et al.*, The High-Resolution Light-polluted Night-Sky Spectrum at Mount Hamilton, California, *The Publications of the Astronomical Society of the Pacific*, 115, 809, (2003), 869–78, available online at: http://adsabs.harvard.edu/abs/2003PASP..115..869S

J. L. Sokoloski, Symbiotic Stars as Laboratories for the Study of Accretion and Jets: A Call for Optical Monitoring, *Journal of the American Association of Variable Star Observers*, 31, 2, (2003), 89–102, available online at: http://arxiv.org/ftp/astro-ph/papers/0403/0403004.pdf

Spectra L200, Users Manual for the Spectra L200 by Ken Harrison, http://www.jtwastronomy.com/products/manuals/L200%20User%20Manual.pdf

SpectroTools: freeware program by Peter Schlatter, http://www.peterschlatter.ch/SpectroTools/

S. Stefl *et al.*, *V/R Variations of Binary Be Stars*, ESO 2007, http://www.arc.hokkai-s-u.ac.jp/~okazaki/Meetings/sapporo/361-0274.pdf

A. Stockton *et al.*, Spectroscopy of R Monocerotis and NGC 2261, *The Astrophysical Journal*, 199, (1974), 406–10, available online at: http://adsabs.harvard.edu/full/1975ApJ...199..406S

M. Tanaka *et al.*, Near-Infrared Spectra of 29 Carbon Stars: Simple Estimates of Effective Temperature, *Publications of* the *Astronomical Society of Japan*, 59, (2007), 939–53, available online at: http://pasj.asj.or.jp/v59/n5/590508/590508.pdf

The SAO/NASA Astrophysics Data System, http://adsabs.harvard.edu/index.html

G. B. Thompson, *Time-series Analysis of Line Profile Variability in Optical Spectra of ε Orionis*, Ph.D. Thesis, University of Toledo, (2009), available online at: https://etd.ohiolink.edu/!etd.send_file?accession=toledo1249511358&disposition=inline

S. F. Tonkin, *Practical Amateur Spectroscopy* (London: Springer-Verlag, 2004).

UCM, University of Madrid: Libraries of Stellar Spectra, available online at: http://www.ucm.es/info/Astrof/invest/actividad/spectra.html

B. D. Warner, *A Practical Guide to Lightcurve Photometry and Analysis* (New York: Springer-Verlag, 2006).

Williams College, USA, *Gallery of Planetary Nebula Spectra*, available online at: http://www.williams.edu/astronomy/research/PN/nebulae/ http://www.williams.edu/astronomy/research/PN/nebulae/legend

O. C. Wilson, The Wolf–Rayet Spectroscopic Binary HD 190918, *The Astrophysical Journal*, 109, (1948), 76, available online at: http://articles.adsabs.harvard.edu/full/1949ApJ...109...76W

SUBJECT INDEX

STELLAR INDEX

OBJECT INDEX: DEEP SKY AND SOLAR SYSTEM

Printed in the United States
By Bookmasters